Bayesian Methods
for Measures
of Agreement

Chapman & Hall/CRC Biostatistics Series

Editor-in-Chief

Shein-Chung Chow, Ph.D.
Professor
Department of Biostatistics and Bioinformatics
Duke University School of Medicine
Durham, North Carolina, U.S.A.

Series Editors

Byron Jones
Senior Director
Statistical Research and Consulting Centre
(IPC 193)
Pfizer Global Research and Development
Sandwich, Kent, UK

Jen-pei Liu
Professor
Division of Biometry
Department of Agronomy
National Taiwan University
Taipei, Taiwan

Karl E. Peace
Georgia Cancer Coalition
Distinguished Cancer Scholar
Senior Research Scientist and
Professor of Biostatistics
Jiann-Ping Hsu College of Public Health
Georgia Southern University
Statesboro, GA

Bruce W. Turnbull
Professor
School of Operations Research
and Industrial Engineering
Cornell University
Ithaca, NY

Chapman & Hall/CRC Biostatistics Series

Published Titles

Chapman & Hall/CRC Biostatistics Series

Bayesian Methods for Measures of Agreement

Lyle D. Broemeling

Director
Broemeling & Associates Inc.
Medical Lake
Washington, U.S.A.

CRC Press
Taylor & Francis Group
Boca Raton London New York

CRC Press is an imprint of the
Taylor & Francis Group, an **informa** business
A CHAPMAN & HALL BOOK

First published 2009 by Chapman & Hall

Published 2019 by CRC Press
Taylor & Francis Group
6000 Broken Sound Parkway NW, Suite 300
Boca Raton, FL 33487-2742

First issued in paperback 2020

© 2009 by Taylor & Francis Group, LLC
CRC Press is an imprint of Taylor & Francis Group, an Informa business

No claim to original U.S. Government works

ISBN-13: 978-0-367-57738-4 (pbk)
ISBN-13: 978-1-4200-8341-5 (hbk)

Library of Congress Cataloging-in-Publication Data

Broemeling, Lyle D., 1939-
    Bayesian methods for measures of agreement / Lyle D. Broemeling.
        p. cm. -- (Chapman & Hall/CRC biostatistics series ; 29)
    Includes bibliographical references and index.
    ISBN 978-1-4200-8341-5 (hardcover : alk. paper)
    1. Bayesian statistical decision theory. 2. Clinical medicine--Decision making.
    I. Title. II. Series.

    QA279.5.B76 2009
    519.5'42--dc22   2008049374

Visit the Taylor & Francis Web site at
http://www.taylorandfrancis.com

and the CRC Press Web site at
http://www.crcpress.com

Contents

Preface

Bayesian methods are being used more frequently in medicine and biology. At the University of Texas MD Anderson Cancer Center, for example, Bayesian sequential stopping rules are implemented somewhat routinely in the design of clinical trials. In addition, Bayesian techniques are used more frequently in diagnostic medicine, such as for estimating the accuracy of diagnostic tests and for screening large populations for various diseases. Bayesian methods are quite attractive in many areas of medicine because they are based on prior information that is usually available in the form of related previous studies. An example of this is in the planning of Phase II clinical trials during which a new therapy will be administered to patients who have an advanced disease. Such therapies are developed by pharmaceutical companies and their success depends on the success of previous Phase I or other relevant Phase II trials. Bayes theorem allows for a logical way to incorporate the previous information with the information that will accrue in the future. Agreement among the various players who give the diagnosis is an integral part of the diagnostic process and is the key issue of this book. Of course, medicine and biology are not the only areas where the concept of agreement plays the paramount role. In most sports, agreement or disagreement among judges and referees can dominate the outcome of the contest, for example, in ice skating, the judges determine the "winners" as they do in diving, gymnastics, and boxing.

Bayesian Methods and Measures of Agreement is intended as a textbook for graduate students of statistics and as a reference for consulting statisticians. The book is very practical and students will learn many useful methods for the design and analysis of agreement studies. Students should already have had a year of introductory probability and mathematical statistics and several introductory methods courses, such as regression and the analysis of variance. An introduction to Bayesian inference is included as an Appendix, and will provide the novice with the necessary theoretical foundation for Bayesian methods.

This book will provide consulting statisticians working in the areas of medicine, biology, sports, and sociology with an invaluable reference that will supplement the books *Measures of Interobserver Agreement* by Shoukri and *Analyzing Rater Agreement* by Alexander Von Eye and Eun Young Mun. These two references are not presented from a Bayesian viewpoint, thus, this volume is unique and will develop methods of agreement that should prove to be very useful to the consultant. Another unique feature of the book is that all computing and analysis are based on the WinBUGS package, which allows the user a platform that efficiently uses prior information. Many of the ideas in this volume are presented here for the first time and go far beyond the two standard references. For the novice, Appendix B introduces the fundamentals of programming and executing BUGS, and as a result the reader will have the tools and experience necessary to successfully analyze information from agreement studies.

Acknowledgments

The author gratefully acknowledges many departments and people at the University of Texas MD Anderson Cancer Center in Houston who assisted during the writing of this book. Many of the analyses are based on studies performed at the Division of Diagnostic Imaging and in particular the author would like to thank Drs. Gayed, Munden, Kundra, Marom, Ng, Tamm, and Gupta. Also, thanks to the Department of Biostatistics and Applied Mathematics, which provided the time off and the computing and secretarial facilities that were invaluable in order to successfully collaborate on the many agreement and diagnostic studies for the Department of Diagnostic Imaging.

Special thanks to Ana Broemeling for proofreading and organizing the early versions of the manuscript and for her encouragement during the writing of this book.

Author

Lyle D. Broemeling, PhD, is Director of Broemeling and Associates Inc. and is a consulting biostatistician. He has been involved with academic health science centers for about twenty years and has taught, and been a consultant, at the University of Texas Medical Branch in Galveston, the University of Texas MD Anderson Cancer Center and the University of Texas School of Public Health. His main interests are in developing Bayesian methods for use in medical and biological problems and in authoring textbooks in statistics. His previous books are *Bayesian Analysis of Linear Models, Econometrics and Structural Change*, written with Hiroki Tsurumi, and *Bayesian Biostatistics and Diagnostic Medicine*.

1

Introduction to Agreement

1.1 Introduction

This book is an introduction to measures of agreement using a Bayesian approach to inference. The subject of agreement began as a separate topic some 45–50 years ago and has developed into a sophisticated area of statistics, both theoretically and methodologically. It has proven to be quite useful in such areas as medicine; the sports of boxing, ice skating, diving, and gymnastics; to many areas of science, including sociology and psychology; and also to law and wine tasting. The major emphasis in agreement studies is to estimate the agreement between several raters (judges, radiologists, observers, pathologists, jury members, boxing judges, etc.). Large differences between raters can be problematic; thus, it is important to study why such differences exist and to take remedial action if necessary.

The book is unique in that the Bayesian approach will be employed to provide statistical inferences based on various models of intra and inter rater agreement. Bayesian inferential ideas are introduced and many examples illustrate the Bayesian mode of reasoning. The elements of a Bayesian application are explained and include prior information, experimental information, the likelihood function, the posterior distribution, and the predictive distribution. The software package WinBUGS is used to implement Bayesian inferences of estimation and testing hypotheses. The package is used to compute the posterior distribution of various measures of agreement, such as Kappa, and to analyze cross-classified data with models that involve complex experiments involving several raters, using logistic and log linear models. The WinBUGS code is provided for many examples, making the book very appealing to students and providers of statistical support. Generally, there are two cases: (1) when the data are categorical (nominal or ordinal); and (2) when the observations are continuous.

When the observations are continuous, agreement is analyzed with regression and correlation models and with the analysis of variance. Shoukri[1] and Fleiss[6] are two excellent books that introduce methods to analyze agreement when observations are continuous. Both explain how the intra class correlation coefficient measures both intra and inter rater agreement.

The ideas of reliability and agreement and their interface are explained by both authors.

The literature on agreement continues to grow, but there are very few books. This book will fill a gap in that there are no books from a Bayesian viewpoint and there are many topics covered in this book that are not covered in the others. There are two books which have appeared somewhat recently, namely Shoukri[1] and Von Eye and Mun[2]. The former describes the basic measures of association with many examples, and the latter focuses primarily on the analysis of data using the log linear model. An earlier book on agreement is Liebetrau[3], who emphasizes measures like Kappa, plus other measures of association that are not usually found in the recent literature. All three are informative and should be read by anyone interested in the topic. In addition, all three present many examples from various areas of science, and some will be analyzed in the present work from a Bayesian viewpoint. Some books on categorical data have chapters devoted to agreement. For example, Fleiss Levin, and Paik[4] and Agresti[5] introduce the foundation of various measures of agreement such as Kappa and intra class correlation.

The author was introduced to the subject when providing support to the Division of Diagnostic Imaging at the University of Texas MD Anderson Cancer Center in Houston, Texas. The diagnostic process in a major medical center always involves a team approach to making decisions. For example, in Phase II clinical trials, the effect of treatment is measured by imaging devices like computed tomography (CT) or magnetic resonance imaging (MRI), and the success of the treatment is usually decided by a team of radiologists, oncologists, and surgeons. Obviously, agreement among the radiologists plays a major role in cancer treatment. Consensus between members of the diagnostic team is necessary in order to judge a treatment as a failure, a success, or worthy of further consideration. DeGroot[6] gives one of the first Bayesian models for consensus.

In medicine, agreement between various raters can be a problem, and is well recognized among radiologists because such variation is always present and needs to be recognized and steps taken to study it and to adjust for it. It takes time to train a radiologist and differences between experienced and inexperienced radiologists are to be expected. Shoukri[1] references several instances where disagreement has been a problem in medicine. For example, Birkelo[7] found a large variation between five readers diagnosing pulmonary tuberculosis via X-ray. A more recent example is reported by Erasmus et al.[8] for five readers of CT images to diagnose lung cancer in 33 patients, where each radiologist reads the images and each reader also examines each of the 40 lesion images twice. It was found that the intra observer error (between duplicate readings) could be as much as 40%. The medical literature often reports such discrepancies in diagnostic decision making; however, generally speaking, the public and most medical patients are quite unaware of such problems. The general public is more aware of such variation in other areas such as boxing, where three judges grade each round on a 10-point must system, then at the end of the fight, at the last round, (if the fight is not stopped earlier due to a knock out, a technical knock out, or a disqualification) the scores are summed over the 12 rounds of the fight. The fight fans become quite agitated if there are large differences in the scores of the judges. The same large variation occurs frequently in ice skating, diving, and gymnastics. Several examples will highlight these inter observer differences.

1.2 Agreement and Statistics

The notation of agreement plays a central role in human existence. During the day, one constantly evaluates their agreement to many external stimuli. For example, do I agree with the latest news about global warming? Or, do I agree with Hilary Clintons' latest position on foreign affairs? Do I agree more with Fred Thompson's view of the Iraq war than the position of John Edwards'? I could go on and on with the many daily mundane encounters with agreement or disagreement on hundreds of topics or actions. I agreed with my boss's latest salary increase. Thus, to agree is to make a decision, and some agreements are good for one's welfare while others are not so beneficial. There are uses of the word agreement that will not be considered. For example, agreement in a legal sense will not be addressed, as for example, a service agreement attached to a purchase. Agreement in this sense is a contract, agreed to by all parties. Our use of agreement is in a more mathematical sense of the word.

Agreement had its beginnings as a formal statistical topic many years ago, and among the pioneers were Scott[9] and Cohen[10] who introduced what was to become known as the Kappa coefficient. This is a chance corrected measure of overall agreement and best explained in the context of two raters who are judging n subjects with a binary nominal rating. The number n of subjects is selected at random from some population. The Kappa coefficient has become the most popular way to measure overall agreement between two raters.

Consider the 2×2 in Table 1.1 giving a binary score to n subjects. Each subject is classified as either positive, denoted by $X=1$, or negative, denoted by $X=0$, by both readers.

TABLE 1.1

Classification Table

Rater 1	Rater 2	
	$X=0$	$X=1$
$X=0$	(n_{00}, θ_{00})	(n_{01}, θ_{01})
$X=1$	(n_{10}, θ_{10})	(n_{11}, θ_{11})

Let θ_{ij} be the probability that rater 1 gives a score of i and rater 2 a score of j, where $i,j = 0$ or 1, and let n_{ij} be the corresponding number of subjects. The experimental results have the structure of a multinomial distribution. Obviously, the probability of agreement is the sum of the diagonal probabilities $\theta_{00} + \theta_{11}$, however this measure is usually not used. Instead, the Kappa parameter is often employed as an overall measure, and is defined as

$$\kappa = [(\theta_{00} + \theta_{11}) - (\theta_{0.}\theta_{.0} + \theta_{1.}\theta_{.1})]/[1 - (\theta_{0.}\theta_{.0} + \theta_{1.}\theta_{.1})] \qquad (1.1)$$

where the dot indicates summation of the θ_{ij} over the missing subscript. Thus, the numerator of Kappa is the difference in two terms: the first the sum of the diagonal elements and the second assumes that the raters are independent in their assignment of rating to subjects. This second term gives the probability of agreement that will occur by chance, thus Kappa is a chance corrected measure of agreement that varies over $[-\infty,1]$. Actually, Scott and Cohen used different approaches to develop Kappa, and these developments and others will be described in Chapter 2, when the early history of the subject is presented. In the non Bayesian approach, the Kappa is estimated with the MLE estimates of the θ_{ij}, namely the relevant proportion from Table 1.1, and the standard deviation is known and reported in Shoukri[1], to make various inferences about agreement. On the other hand, the Bayesian approach is based on the posterior distribution of Kappa, which in turn is based on prior information and the multinomial likelihood function. More about the Bayesian is given in the next section.

When the outcomes are continuous variables, regression and analysis of variance methods are appropriate for measuring agreement. For example, suppose there are two readers measuring the tumor size (which is measured in centimeters) imaged by CT. Suppose each reader reads one image from each of n patients, selected at random from a population of readers. The observations from one can be regressed on the other. Assuming the assumptions are met for a normal simple linear regression, agreement can be tested by examining the slope and intercept of the fitted regression. If the agreement is 'good', one would expect the fitted line to go through the origin with a slope of unity. As will be described later, this approach is related to the Bland–Altman[11] method to test agreement between pairs of observers. The Pearson correlation coefficient is sometimes used to measure agreement and of course goes hand and hand with the regression analysis. A scatter plot of one readers' values on the other readers' values are an essential component of the analysis.

The usual statistical methods are used for the analysis of agreement. As mentioned above, regression and ANOVA techniques are the standard fare with continuous observations, on the other hand, for categorical data, cross-classification techniques for contingency tables are appropriate. For example, the log linear model is very popular for comparing raters in complex experimental situations, where subject covariates and other information can have an effect on how two raters agree, and thus the effects on the raters can be adjusted for the other factors in the study. Graphical techniques, such as

Box plots, bar graphs, error-bar graphs, and scatter plots should be used to examine agreement. Along with the graphical evidence, descriptive statistics for each rater, and those describing their joint distribution should be used. When modeling with log linear or logistic models, methods appropriate for diagnostic checks should be employed. Bayesian methods for diagnostic checks based on the predictive distribution will be described in Appendix A, and can be found in Box and Tiao[12]. The Kappa statistic continues to be very popular, especially in the social and medical sciences, and the log linear model greatly advanced the way one analyzes agreement with categorical data. Log and logistic linear models are used much more frequently than latent variable models, but the latter have some advantages. See Johnson and Albert[13] on the Bayesian use of latent variable models for comparing raters.

This book will use Bayesian methods throughout, including the posterior analysis for multiple linear regression models, the one-way and two-way analysis of variance models with fixed and/or random factors. These as such are recognized as "older" techniques and are presented in Zellner[13], Box and Tiao[14], and Broemeling[15]. Also, regression models for more complex scenarios, such as hierarchical models with many levels and encompassing random effects, will be appropriate for many situations in modeling both continuous and categorical observations. Such an approach is useful for agreement studies in a time series and/or repeated measures situation. The Congdon[16-18] books are references explaining the "newer" modeling based on Monte Carlo Markov Chain (MCMC) sampling of the posterior distribution. In order to implement Bayesian inferences, WinBUGS the Bayesian software package, will provide the necessary computations.

1.3 The Bayesian Approach

As mentioned in the introductory section, this book will use the Bayesian approach to statistics, where Bayes theorem is implemented to produce inferential procedures like estimation of parameters, testing hypotheses, and predictive inferences for forecasting and for diagnostic checks. Bayes theorem can be expressed as follows: Suppose X is a continuous observable random vector and $0 \in \Omega \subset R^m$ is an unknown parameter vector, and suppose the conditional density of X, given θ is denoted by $f(x/\theta)$, then the conditional density of θ given $X = x$ is

$$\xi(\theta/x) = cf(x/\theta)\xi(\theta), \quad \theta \in \Omega \quad \text{and} \quad x \in R^n \tag{1.2}$$

The normalizing constant is $c > 0$ is chosen so that the integral of $f(x/\theta)\xi(\theta)$ with respect to θ is unity. The above equation is referred to as Bayes theorem and is the foundation of all statistical inferences to be employed in the analysis of data. If X is discrete, $f(x/\theta)$ is the probability mass function of X.

The density $\xi(\theta)$ is the prior density of θ and represents the knowledge one possesses about the parameter before one observes X. Note that θ is considered a random variable and that Bayes theorem transforms one's prior knowledge of θ, represented by its prior density, to the posterior density, and that the transformation is the combining of the prior information about θ with the sample information represented by $f(x/\theta)$.

Note the various components of Bayes theorem: (1) The prior density $\xi(\theta)$ for the unknown parameter θ; (2) the conditional density of X given θ, $f(x/\theta)$, and as a result of Equation 1.1; (3) the posterior density of θ, or the conditional density of θ given $X=x$. Having observed $X=x$, $f(x/\theta)$ as a function of θ is referred to as the likelihood function. The parameter θ may be a vector, such as the vector comprising the multinomial parameters θ_{ij} of the classification Table 1.1, for agreement of two raters, in which case the objective would be to determine the posterior density of θ. The random variable X would be the vector of observations n_{ij} of Table 1.1. Note, however, the prior density must be known before the posterior density can be determined by Bayes theorem. Typically, the prior information is based on either an informative situation, where previous experimental results are available, or in a non informative case, where little prior information is available. Usually in the latter case a uniform density for θ is taken, or in the former informative situation, a Dirichlet type prior for the parameter is adopted. In either scenario the posterior density represents a Dirichlet distribution for θ.

Since the properties of the Dirichlet are well-known, posterior inferences are easily determined, either analytically, from the formulas for the moments and other characteristics of the distribution, or numerically by sampling directly form the posterior distribution. Since the posterior distribution of θ is known, the posterior density of Kappa can be computed by generating samples from the posterior distribution of θ, then transforming to Kappa from Equation 1.1. The number of samples generated will depend on how much accuracy one wants for the posterior distribution of Kappa. This will be described in detail in Chapter 4.

A Bayesian model, as represented by Equation 1.2, is called a parametric Bayesian approach, as opposed to non parametric Bayesian. The former will be employed in this book, but for additional information about some non parametric techniques with Dirichlet process priors, Congdon[17] is a good reference.

Of special interest for model checking and diagnostic checks, is the predictive distribution of a future observation Z, namely the conditional distribution of Z, given $X=x$ with density

$$g(z/X = x) = \int_{\Omega} f(z/\theta, x)\xi(\theta/x)d\theta, \qquad z \in R^n \qquad (1.3)$$

When log linear (for categorical data) or normal theory regression (for continuous outcomes) models are employed to compare two or more raters in complex experimental settings, the predictive distribution is an important component in choosing the proper model and for detecting outliers and spurious observations.

Using the Bayesian approach, the posterior analysis of some early measures of agreement (precursors to Kappa) are presented followed by the analysis for Kappa itself, which in turn is followed by a posterior analysis for the weighted Kappa, the conditional Kappa, the stratified Kappa, and the intra class coefficient, all for two raters and binary outcomes. The computations for the analysis is done with WinBUGS, which provides a detailed posterior analysis consisting of the posterior mean, median, standard deviation, the lower and upper 2½ percentiles of the posterior distribution, and indices that measure the convergence of the sequence (to the posterior distribution) of the observations generated from the desired distribution. Accompanying this are graphical displays of the posterior densities and plots that monitor the convergence of the generated sequence. The simulations done by WinBUGS are based on MCMC methods such as Gibbs sequences etc. Appendix A gives an introduction on the use of WinBUGS. Of course, the same Bayesian method is taken for the analysis of generalizations of Kappa to multiple raters and multiple outcomes, and for modeling agreement studies with logistic linear and normal theory regression models.

One of the few Bayesian investigations of Kappa is Basu, Banerjee and Sen[19] who determined the posterior density of Kappa by MCMC techniques with WinBUGS and developed a test for the homogeneity of Kappa across several different experiments. In addition, simulations compared the Bayesian to some frequentist techniques. An interesting aspect of the study was the choice of the prior distribution for the probabilities θ_{ij} of Table 1.1, where priors were specified on θ_{11}, $\theta_{1.}$, $\theta_{.1}$.

1.4 Some Examples of Agreement

There are many interesting examples of agreement and some will be presented here and will include the world of clinical trials for cancer, the sport of boxing with a description of the Lennox Lewis–Evander Holyfield fight in 1999, an example measuring lung tumor sizes with CT, and the Lindley example of wine tasting, that was presented as a contest between French and American wines. The last example to be presented is from the law and involves a famous case of forgery. This is just a sample of what is to come in the remainder of the book and shows why the subject of agreement is one of the most fascinating in statistics.

1.4.1 The Lennox Lewis–Evander Holyfield Bout

The Lennox Lewis–Evander Holyfield bout of March 13, 1999 is a good example where the disagreement between the three judges provoked a controversy that excited and agitated fans. The fight was scored a draw, whereas many

observers believed that Lewis had clearly won, and the crux of the controversy centered on Eugenia Williams, the American judge, who gave seven
rounds to Holyfield and the remaining five to Lewis. One judge scored in
favor of Lewis, with seven rounds, while the third scored the bout a draw.
The main question is: Was Williams biased in favor of Holyfield?

To better understand the problem a brief description of the rules of boxing
and the 10-point scoring system is in order.

Each judge for each round scores a 10 for the winner and a nine or less for
the loser and the scores are summed at the end of 12 rounds. The results of
the fight are illustrated in Table 1.2 from Lee, Cork, and Algranati[20]. L indicates Lewis won the round, H indicates Holyfield, and E is an even round,
thus for round 7, Lewis won the round according to two judges, while the
third scored it even. The judges hand in their results after each round and
are supposed to score each round independently of the earlier rounds. The
judges are advised to base their scores equally on (1) clean punches landed;
(2) defense; (3) effective aggressiveness; and (4) ring generalship. Obviously,
the judges will differ on the degree they take these four criteria into consideration and how the scores from previous rounds will affect their scores of
the present round. It is not surprising that the judge's scores will differ sometimes from the fight fan and other unofficial observers, such as reporters and
broadcasters. The judges are focusing on each round independently of earlier
rounds and of other judges, whereas others usually base their impression on
the totality of all the rounds taken together and are influenced by others, for
example by other reporters, broadcasters, and fans. Since one judge scored
the fight even, and the other two split, the fight was scored a draw. The boxing rules are quite specific about how the winner is determined from the
judges' scores.

A look at the table shows that the judges agreed on six rounds. The main
focus of Lee et al.[20] was to determine if judge #1(Williams) was biased, and to
investigate this they considered three approaches: (1) exact tests; (2) logistic
regression; and (3) a direct Bayesian test.

In the first alternative, let x_{ij} be binary, where $x_{ij}=1$ indicates judge j
scores round i for fighter 1 and 0 otherwise. Assuming that the chance

TABLE 1.2

Score Cards for the Lewis–Holyfield Bout

| | Round | | | | | | | | | | | | |
Judge	1	2	3	4	5	6	7	8	9	10	11	12	Total
1	L	L	H	H	H	L	L	H	H	H	H	L	115–113
2	L	L	H	L	L	H	E	H	H	E	H	L	115–115
3	L	L	H	L	L	L	L	H	H	H	E	L	113–116

Source: Lee, H. K. H., D. L. Cork, and D. J. Algranati. 2002. Did Lennox Lewis beat Evander
Holyfield?: Methods for analyzing small samples interrater agreement problems. *The
Statistician.* 51(2), 129.

a fighter wins a particular round depends only on the round and that no even rounds are permissible, the probability of observing each of the 2^{12} possible outcomes can be computed. For example, the probability that the first judge (Williams) scored as indicated in Table 1.2 is 0.0019. This is also the probability of the second judge treating an even round as one won by Holyfield. How extreme are the outcomes of these two score cards relative to all possible sequences? It can be shown that the first two judges have score cards that are highly unlikely to be observed by chance, with a p-value less than 0.05.

The second alternative, assumes that the probability that judge j scores round i for fighter 1 is

$$\log it(p_{ij}) = \mu + \alpha_i + \beta_j + \varepsilon_{ij}$$

where μ is the mean, α_i is the indicator variable for round i, β_j the indicator covariate for judge j, and ε_{ij} is the error term. The biases between the judges are compared by comparing the fitted values of β_j.

Lastly for the third alternative, Lee et al. considers direct Bayesian methods, based on a latent variable method. Let y_i be distributed $N(0,1)$ and indicate the difference between the performance of the two fighters for round i, where $y_i = 0$ indicates an even round , a value $y_i > 0$ indicates fighter 1 won the round, and $y_i < 0$ indicates fighter 2 won. Now y_i is unobserved, thus let y_{ij} be the perceived difference (of fighter 1 versus fighter 2) of judge j in round i of the fighters performance, and let

$$y_{ij} \sim N(y_i + b_j, \sigma_j^2)$$

where b_j is the bias for judge j. A normal prior is put on the b_j and an inverse gamma is placed on the σ_j^2. Note that the y_{ij} are observable and should indicate the degree to which fighter 1 is better than fighter 2, expressed as a difference on a normal scale.

The bias of Williams is clearly indicated with this model, as it was with the logistic model of the second alternative, and a more detailed analysis will be presented in a following chapter. Lee et al.[20] also examined, in a similar fashion, the Oscar de la Hoya–Felix Trinidad bout, which investigated the losing strategy of de la Hoya. De la Hoya took a defensive stance in the later rounds of the fight, because he thought he had scored enough points to stay away from Trinidad and still win. The judges thought otherwise.

Lee et al. did not measure the overall agreement with one measure such as Kappa, but instead took a more appropriate modeling approach to compare the three judges and in particular to investigate the bias of Williams relative to the other two, using as a standard the scores of a group of unofficial "judges" (among others, the HBO broadcasters and associated press). One overall measure of agreement would not answer the major question of interest: Was Ms. Williams biased in favor of Holyfield? Such disagreement between judges is common in other sports including gymnastic, diving, ice skating.

1.4.2 Phase II Clinical Trials

Our next example is taken from a Phase II clinical trial for cancer research. In such a trial the main purpose is to study the efficacy of a new chemotherapy treatment for patients with advanced stage cancer, where the treatment had earlier been tested in Phase I trials. A Phase I trial would have determined an optimal dose of the new treatment, and the treatment would have been also tested for safety. One is interested in the response to therapy, where response is clearly defined in the protocol of the trial. Such a protocol would usually state exactly how the response to the therapy is to be measured and would use the so-called response evaluation criteria in solid tumors (RECIST) criteria to guide the radiologists as how to determine the response. Briefly, these criteria inform the radiologist that the response to therapy is to be classified into the following groups of patients: (1) a CR or complete response, where the target lesions disappear over the course of treatment; (2) a PR or partial response, were the size of the tumors have decreased by at least 30%; (3) SD or stable disease, where the size of the target tumors have not increased sufficiently to qualify as PD or decreased sufficiently to be classified as a PR; and (4) PD or progressive disease, where the size of the target lesions increase by at least 20%. This is only a brief description of the RECIST criteria and the reader is referred to Padhani and Ollivier[21] or to Broemeling[22] for a more complete description.

Several radiologists use RECIST to classify each patient as a CR, PR, SD, or PD, however, the criteria do not state the number of radiologists nor how they are to arrive at a consensus for an overall judgement about treatment response of each patient. I should mention that others, including surgeons, pathologists, and oncologists, are also involved in the determination of the response to treatment. As one could well imagine, there is much uncertainty and disagreement between the radiologists in the classification of a patient as to their response to therapy. Consider Table 1.3 as reported by Thiesse et al.[23] which gives the degree of agreement between the original report of a clinical Phase II European clinical trial and a review committee for 126 renal cancer patients receiving cytokine therapy. A glance at the table reveals a low level of agreement between the two committees, in fact the Kappa coefficient is only 0.32. The implications for this are interesting. When the results of a Phase II trial are reported, one should be sure to know the number of CR, PR, SD, and PD patients. Of course, the disagreement between the various radiologists will not be reported, unlike a prize fight, where the judge's scores are public information. This example will be reexamined from a Bayesian viewpoint in a later chapter.

1.4.3 Agreement on Tumor Size

The discussion of uncertainty and disagreement between radiologists is continued with a recent study from MD Anderson Cancer Center in Houston, Texas, by Erasmus et al.[8] This study demonstrates the problem of agreement

TABLE 1.3

Agreement Between the Review Committee and the Original Report

	Response by Review Committee					
Original Report	CR	PR	MR	SD	PD	Total
CR	14	2	1	0	2	19
PR	4	38	4	6	10	62
MR*	0	7	9	3	5	20
SD	0	0	1	4	15	20
PD	0	1	0	3	1	5
Total	18	48	11	16	33	126

Source: Thiese, P., L. Ollivier, D. DiStefano-Louieheau, S. Negrier, J. Savary, K. Pignard, C. Lesset, and B. Escudier. 1997. Response rate accuracy in oncology trials: reasons for interobserver variability. *J. Clin. Oncol.* 15, 3507.

* MR is marginal response.

and how such disagreement affects the efficacy of a treatment in a Phase II clinical trial. Responses of solid tumors to therapy is usually determined by serial measurements of lesion size. The purpose of this study is to assess the consistency of measurements performed by readers measuring the size of lung tumors.

There were 33 patients with non-small-cell lung cancer and with 40 lung lesions whose size exceeded at least 1.5 cm. Using CT, unidimensional (UD) and bidimensional (BD) determinations were performed to estimate the size of the lesion, where the longest diameter of a lesion determined the UD size and where the longest diameter and the corresponding largest perpendicular diameter of a lesion determined the BD size. These measurements are defined by World Health Organization (WHO) for BD sizes and by the RECIST criteria for UD measurements. Using printed film, five radiologists independently performed the UD and BD measurements, which were repeated on the same image 5–7 days later. Thus each radiologist read each of 40 images, then repeated the process. It should be stressed, that the replication consisted of looking at the same image twice. The actual tumor size does not change!

There were 40 tumors with an average size from 1.8 to 8.0 cm and Table 1.4 summarizes the study results. These are taken from Table 1.3 of Erasmus et al.[8] where the BD sizes were also reported. The tumor means were averaged over the 40 lesions. For the moment the BD sizes will be ignored.

The means varied from 3.7 to 4.42 cm, or 19% for replication 1, and from 3.48 to 4.25 cm of replication 2 for a 22% change, however this does not tell us the misclassification errors that would be committed in say classifying a patient into the progressive disease category using the RECIST definition, where the target tumors increase in size by at least 20%.

Consider first the intraobserver error. Referring to Table 1.6 of Erasmus et al., or our Table 1.5 for UD sizes for progressive disease, the relative measurement change varied from 0 to 80%. The RD is the relative difference,

TABLE 1.4

UD Tumor Sizes for all Observers and Replications

Replication	Reader	Mean (cm)	Median	Range	SD (cm)
1	1	3.92	3.80	1.5–8.0	1.64
1	2	3.70	3.80	1.2–7.8	1.51
1	3	4.42	4.20	1.5–9.0	1.55
1	4	4.36	4.10	1.7–9.0	1.61
1	5	4.14	3.95	1.2–9.0	1.55
2	1	4.13	4.00	1.4–9.0	1.66
2	2	3.48	3.45	1.0–7.5	1.45
2	3	4.25	4.10	1.5–8.0	1.42
2	4	4.17	3.95	1.4–8.5	1.68
2	5	4.07	4.00	1.5–9.0	1.69

Source: Erasmus, J. J., G. W. Gladish, L. Broemeling, B. S. Sabloff, M. T. Truong, R. S. Herbst, and R. F. Munden. 2003. Interobserver variability in measurement of non-small cell carcinoma of the lung lesions: implications for assessment of tumor response. *J. Clin. Oncol.* 21, 2574.

TABLE 1.5

Intraobserver Errors of Progressive Disease

	Reader					
UD Errors	**1**	**2**	**3**	**4**	**5**	**Average**
Min. RD (%)	0.00	0.00	0.00	0.00	0.00	0.00
Max. RD (%)	50.00	80.00	68.18	25.00	75.00	59.64
Median RD (%)	5.20	4.08	7.32	6.11	4.60	5.46
# Misclassifications	7	5	2	1	4	3.80
% of lesions	17.5	12.5	5.00	2.50	10.00	9.50

i.e., the absolute value of the difference between the two sizes measured on the same image divided by the smaller of the two sizes. Remember there is actually no change in tumor size, but nevertheless, reader one has determined that 17.5% of the 40 tumors had changed in size by at least 20%. On the other hand reader four had a phenomenal record of committing only one error. These are the type of errors committed in a static experiment where there is actually no change in lesion size, thus, the errors committed can be considered on the conservative side. In reality, in an actual Phase II trial, the tumors are either increasing or decreasing in size and thus the actual errors of misclassification are probably greater than those represented in Table 1.5.

The story is much the same for interobserver error. Consider Table 1.6, where the minimum, maximum, and median relative differences (the

TABLE 1.6

Interobserver Error for Progressive Disease. Replication 1

UD	Reader Pair										
	1,2	1,3	1,4	1,5	2,3	2,4	2,5	3,4	3,5	4,5	Average
Min. (%)	0.00	0.00	0.00	0.00	0.00	0.00	0.00	0.00	0.00	0.00	0.00
Max. (%)	130.0	170.5	88.6	47.5	90.9	150	109	194	123	66.6	116.9
Median (%)	13.2	8.6	11.8	7.49	8.05	22	13.1	14.6	10.6	13.8	12.3
# Misclassifications	14	9	10	7	11	20	14	15	9	10	11.9
% of lesions	35	22.5	25.00	17.5	27.5	50	35	37.5	22.5	25	29.7

absolute value of the difference between the two sizes of the same tumor, divided by the smaller of the two) are listed, along with the number of mis-classifications and the percentage of incorrect classifications of progressive disease.

Thus, for readers 1 and 2, the maximum error was 130%, with 14 misclassi-fications for progressive disease at a rate of 35%, and compared to the intraob-server error, the interrater differences are much larger. This scenario is much more realistic, because there is no true replication in an actual Phase II trial. But again, there is actually no change in tumor size, however the perceptions (of lesion size) of the radiologists say there is, and in some cases as much as a 50% difference, as for example with readers 2 and 4. Given these results, how would a consensus be reached among the five on how to classify a patient into one of the RECIST categories, CR, PR, SD, or PD? Such variability is a fact of life and not a condemnation of medical practice. Obviously, priority is set on training radiologists, and this continues to be the positive side of fighting the inter and intraobserver variability.

Another way to estimate intra and interobserver variability is with the analysis of variance, which was also reported by Erasmus et al. Consider Table 1.7, which gives the results of the three-say ANOVA for the UD mea-surement of lesion size, where the three factors are readers (the five radiolo-gists), nodule (the 40 lesions), and time (the two replications).

The differences between readers and between nodules was significant with $p < 0.001$, while the differences between the two times (replications) was not significant. Also, using the three-way ANOVA, but assuming the three factors were random, the variance components for readers, nodules, replication, and error were estimated as 0.077, 0.165, 0.002, and 0.297, respec-tively, indicating most of the variation is taken up between nodules, and that there was more interobserver variability than intraobserver variability. To get a better idea of the between observer agreement, the scatter plot in Figure 1.1 compares the unidimensional lesion sizes of reader 1 with those of reader 2.

The fitted line has an R^2 of 0.851, a slope $= 0.890$ (0.088) with a 95% con-fidence interval of (0.712, 1.068), and the intercept is estimated as 0.643 (0.349), with a 95% confidence interval of $(-0.062, 1.340)$, which indicates

TABLE 1.7

Analysis of Variance of UD Measurements

Source	Type III SS	d.f.	Mean Square	F	p
Readers	27.29	4	6.82	23.01	0.000
Nodule	872.6	39	22.37	75.46	0.000
Time	0.714	1	0.714	2.408	0.122
Error	105.260	355	0.297		
Total		400			

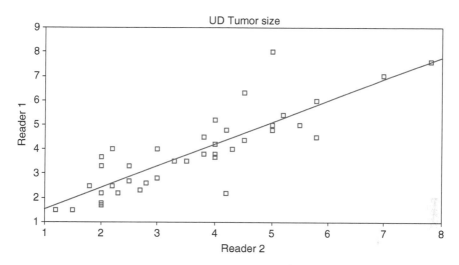

FIGURE 1.1
Reader 1 versus reader 2.

fairly good agreement. There are many ways to study agreement including graphical techniques like scatter plots with simple linear regression plots superimposed, ANOVA methods, using fixed and random models, and has been explained, estimating the number of misclassifications for a particular RECIST criterion for both intra and interobserver errors. The Erasmus et al.[8] study did not employ the so-called Bland–Altman[24] method of investigating agreement, and also that study did not employ the Bayesian approach. However, the Bayesian method will be done in a later chapter and I will use as prior information the results of earlier studies of agreement and will establish the posterior distribution of the interclass correlation coefficient in order to estimate intra and interrater disagreement.

1.4.4 Wine Tasting

Judging the best Californian and French wines provides an interesting diversion to the examples described in this section. It is taken from the internet with Lindley[25] as the author, who analyzed the data of a famous wine tasting event which was reported in the Underground Wine Journal in July 1976. The tasting was organized by Steven Spurrier in Paris, where some Californian wines were introduced, and along with some French wines were judged by 11 tasters, savoring ten Chardonnay and ten Cabernet wines. Two of the tasters were American and six of the Chardonnay and six of the Cabernet wines were Californian.

The results of the Chardonnay and Cabernet tasting are reported in Table 1.8 and Table 1.9, respectively.

The order the wines were tasted is not known and I do not know to what extent the tasters compared notes, but it looks as if tasters 7 and 11 collaborated, because their scores are identical. For additional information about the wines

TABLE 1.8

Chardonnay Scores

Wines	Tasters										
	1	2	3	4	5	6	7	8	9	10	11
A	10	18	3	14	16.5	18.5	17	14	14	16.5	17
B	15	15	12	16	16	15	14	11	13	16	14.5
C	16	10	16	15	12	11	13	14	16	14	13
D	10	13	10	16	10	15	12	10	13	9	12
E	13	12	4	13	13	10	12	15	9	15	12
F	8	13	10	11	9	9	12	15	14	7	12
G	11	10	5	15	8	8	12	14	13	9	12
H	5	15	4	7	16	14	14.5	7	12	6	14.5
I	6	12	7	12	5	16	10	10	14	8	10
J	0	8	2	11	1	4	7	10	8	5	7

Source: Taken from http://www.liquidasset.com/lindley.htm. March 2008.

TABLE 1.9

Cabernet Scores

Wines	Tasters										
	1	2	3	4	5	6	7	8	9	10	11
A	14	15	10	14	15	16	14	14	13	16.5	14
B	16	14	15	15	12	16	12	14	11	16	14
C	12	16	11	14	12	17	14	14	14	11	15
D	17	15	12	12	12	13.5	10	8	14	17	15
E	13	9	12	16	7	7	12	14	17	15.5	11
F	10	10	10	14	12	11	12	12	12	8	12
G	12	7	11.5	17	2	8	10	13	15	10	9
H	14	5	11	13	2	9	10	11	13	16.5	7
I	5	12	8	9	13	9.5	14	9	12	3	13
J	7	7	15	15	5	9	8	13	14	6	7

Source: Taken from http://www.liquidasset.com/lindley.htm. March 2008.

and tasters, refer to http://www.liquidasset.com/lindley.htm. Tasters 4 and 8 were American and the remaining nine were from France.

The mean score for the five Californian white wines is 11.2 and 11.58 for the French, while for the red wines, the mean Californian score is 11.04 and 13.03 for the five French wines. Apparently, Californian wines performed quite well in comparison to the French. Remember at this time in 1976,

Californian wines were just beginning to be well-received by the world wine tasting community.

Descriptive statistics for both white and red wines were computed by Lindley and are reported in Table 1.10. White wines A, C, D, E, I, and J are Californian, while among the red wines, A, E, G, H, I, and J are from California. Californian wine J received a poor score of 5.7 and 9.6 for red and whites, respectively. For example, compared with the average score of 11.36 (3.93) for Chardonnay, wine J is quite small.

When comparing scores by country of origin, the median score is the same for both countries. The medians are the same and the inter quartile range is almost the same for both, although there are more extreme scores, both large and small, for the six Californian Chardonnay wines (see Figure 1.2).

In order to compare tasters, Lindley[25] reported the descriptive statistics of Table 1.10. It is interesting to note that the American taster #4 gave the highest mean (averaged over ten wines) score for both Chardonnay and Cabernet wines, however, French judges #3 and #5 gave the lowest scores to the white and red wines, respectively (Table 1.11).

Lindley provides the basic analysis of the data with the descriptive statistics, an analysis of variance, computing the variance components, testing for outliers by examining the residuals generated by the analysis of

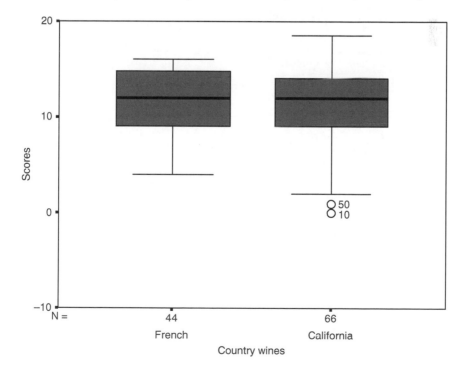

FIGURE 1.2
Boxplots of 11 taster scores versus origin of ten Chardonnay wines.

TABLE 1.10

Summaries

	Chardonnays				Cabernets		
	Means				Means		
Wines	Raw	Shrunk	Variances	Wines	Raw	Shrunk	Variances
A	14.4	13.7	10.64	A	14.1	13.6	5.25
B	14.3	13.7	4.69	B	14.1	13.6	4.07
C	13.6	13.2	10.39	C	13.6	13.2	4.77
D	11.8	11.8	3.04	D	13.2	12.9	9.18
E	11.6	11.7	7.51	E	12.1	12.1	6.34
F	10.9	11.1	4.71	F	11.2	11.3	2.90
G	10.6	10.9	4.32	G	10.4	10.8	8.46
H	10.5	10.8	15.06	H	10.1	10.6	11.50
I	10.0	10.5	6.41	I	9.8	10.3	15.87
J	5.7	–	6.50	J	9.6	10.2	9.67

Source: Taken from http://www.liquidasset.com/lindley.htm. March 2008.

TABLE 1.11

Tasters

	Chardonnays		Cabernets	
Tasters	Mean	Variance	Mean	Variance
1	9.4	8.84	12.0	7.73
2	12.6	5.25	11.0	7.07
3	7.3	15.06	11.6	6.76
4	13.0	6.66	13.9	7.08
5	10.7	10.97	9.2	15.08
6	12.1	10.48	11.6	5.77
7	12.4	2.05	11.6	3.98
8	12.0	8.89	12.2	5.52
9	12.6	4.34	13.5	6.83
10	10.6	6.94	12.0	16.82
11	12.4	1.93	11.7	4.00

Source: Taken from http://www.liquidasset.com/lindley.htm. March 2008.

variance. Some of the analysis is repeated by eliminating wine J, the one that received the lowest scores for both Chardonnay and Cabernet. There is very little of a Bayesian flavor, except for Bayes factor for each test of hypothesis of the ANOVA.

In order to compare tasters, what should one do for the Bayesian analysis? The usual approach is to perform the descriptive statistics, including an estimate of the correlation between all pairs of tasters, plot the scores of one taster versus the other, for all pairs of tasters, to determine if the regression line goes through the origin with a slope of one, perform a Bland–Altman analysis (for each pair, plot the differences in the paired scores versus the mean of the paired scores to determine if the fitted line is flat and if the observations are within two standard deviations of the mean), perform a one-way AOV to estimate the variance components and the intra class correlation coefficient, and always perform diagnostic checks to see if a normal theory linear regression is appropriate. Graphical displays such as the boxplots are always required to compare the medians of appropriate subsets.

The pairwise correlations between all pairs of tasters reveal that tasters 6 and 8 have a correlation of -0.211, indicating poor agreement, which is also demonstrated by plotting the scores of taster 6 on those of taster 8. Figure 1.3 shows the paired scores and the fitted least squares line. In fact the 55 correlations are all over a large range, varying from a low of -0.299 to a high of 1.00.

It might be interesting to regress the scores of taster 6 on those of taster 8 using the following WinBUGS code.

BUGS CODE 1.1

```
model;
{
for(i in 1:N) {
# is the simple linear regression model
y[i] ~ dnorm(mu[i], precy);
                mu[i] < - beta[1] + beta[2]*x[i];
                }
# prior distributions − non-informative priors for regression coefficients
                for(i in 1:P) {
                beta[i] ~ dnorm(0, 0.000001);
                }
# non informative prior for precision about regression line
# the coefficients are given normal distributions with mean 0 and precision
0.00001
                precy ~ dgamma(.00001,.00001)
                sigma < - 1/precy
                }
# N is number of wines
# P is number of tasters
```

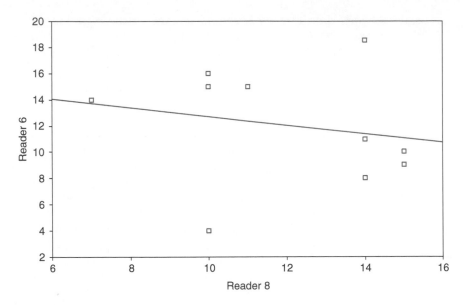

FIGURE 1.3
Taster 6 versus taster 8.

list(N = 10, P = 2, y = c(18.50,15.00,11.00,15.00,10.00,9.00,8.00,14.00,16.00,4.00),
x = c(14.00,11.00,14.00,10.00,15.00,15.00,14.00,7.00,10.00,10.00))
list(beta = c(0,0), precy = 1)

the last list statement gives the starting values for the regression coefficients and the precision about the regression line

The posterior analysis for the regression analysis is given by Table 1.12.

Thus, the intercept is estimated with a posterior mean of 16.18, a standard deviation 7.73, a lower 2½ percentile of 0.7948, a median of 16.12, and an upper 2½ percentile of 31.93. The error of estimation (this is explained in Appendix B) provided by MCMC is 0.2656. Sigma is the standard deviation about the regression line. A 95% credible interval for the intercept is (0.7948, 31.93) and that for the slope is (− 1.621,0.9181), giving evidence that there is very poor agreement between tasters 6 and 8. Good agreement is indicated by a line through the origin with a slope of 1. The Bayesian regression analysis gives results very similar to the classical least squares fit shown in Figure 1.3, which is to be expected, because the prior distributions for the slope and intercept are noninformative.

The software also provides some plots for the posterior analysis, including the posterior density of the slope presented as Figure 1.4. The MCMC simulation generated 30,000 observations from the joint posterior distribution of the three parameters, the slope, intercept, and the standard deviation about the regression line. The refresh was 100 and the sequence began with observation 1001.

TABLE 1.12

Regression of Reader 6 Scores Versus Reader 8

Parameter	Mean	SD	Error	2½ %	Median	97½ %
Intercept	16.18	7.73	0.2656	0.7948	16.12	31.93
Slope	−0.3439	0.6291	0.0215	−1.621	−0.3376	0.9181
Sigma	27.62	18.9	0.2401	9.537	22.74	75.17

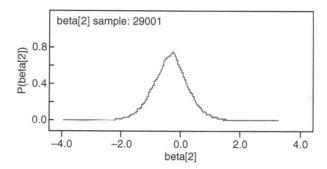

FIGURE 1.4
Posterior density of slope.

Further investigation shows poor agreement with many pairs of observers and also reveals a large range of correlations, thus it would be unwise to use the one-way random model to estimate the intraclass correlation. Such a model is quite restrictive in that it assumes all pairs of tasters have a common correlation. This clearly is not the case.

To the novice user of WinBUGS, the above code probably is somewhat mysterious, but to the more experienced, it is an example of a quite simple case. The explanation of the various lines of code are indicated by a #. Appendix B is provided to give the reader a good introduction to using the software, it is invaluable to a Bayesian practitioner.

What should be concluded from this wine tasting experiment? Since there is some disagreement between tasters, the only "safe" conclusion is to take the comparison between Californian and French wines with a grain of salt. To declare one country or the other as the winner on the basis of this particular wine tasting would be to go out on a limb.

1.4.5 Forgery: Robinson versus Mandell

A famous nineteenth century case of forgery involved the two signatures:

As reported by Devlin and Lorden[26], the evidence was examined by the Pierce family of father and son, two famous mathematicians of the time. The father, Benjamin Pierce, was a Harvard professor of mathematics, and his son a professor of mathematical logic, who also worked for the Coast and Geodetic Survey. He was a founder of the school of American pragmatism, and was well known for his many contributions to philosophy. How did forgery come into play?

Sylvia Ann Howland left an estate of $2,000,000 which was contested by her niece, Hetty Howland Robinson. Her aunt left her only a portion of the estate, but she objected by suing the executor, Thomas Mandell, giving as evidence a purported prior will, that bequeathed her the total estate. The signatures on the two wills appear above and were used by Mandell, who claimed that this was a case of forgery. Mandell employed the Pierces as expert witnesses, claiming the two signatures were identical, that one had been traced to produce the other. This is a different twist on forgery, where usually one wants to demonstrate the signatures are not the same.

A statistical approach to agreement was taken by father and son as expert witnesses for the defense. They identified 30 downstrokes for a signature, and declared the two signatures are a match if the 30 paired downstrokes are the same for both. Fortunately, the defense had 42 genuine signatures of Sylvia Ann Howland, giving 861 pairs of signatures, and consequently 861×30 total downstroke pairs. Among the 861×30 downstroke pairs, they found 5325 coincidences, thus, they reasoned, that the probability that a downstroke pair will coincide is about 0.206156. Next, they assumed that coincidences among downstroke pairs are independent and that a Bernoulli model was appropriate. The probability of observing 30 coincidences among 30 down-stroke pairs is $0.206156^{30} = 1$ in 375 trillion. On the basis of this analysis and other testimony, it was decided the two signatures were identical. Among two Sylvia Ann Howard signatures, one would expect about one in five coincidences among downstoke pairs, and one would expect about six downstroke pair coincidences among 30 between two authentic signatures. Two signatures of the same person are usually not identical or at least not as alike as the two above.

Devlin and Lorden[26] are authors of *The Numbers behind Numbers3RS*, a companion book about the episodes appearing on the hit CBS TV series **NUMB3RS**, to which they are mathematical consultants. They hint that it has been shown that the assumption of a Bernoulli model in not appropriate and that a non parametric approach to the problem should be taken. The Howland forgery case is further examined in the book *Statistics for Lawyers*[27].

1.5 Sources of Information

In order to present a complete account of the subject, many sources of information from diverse areas are needed. This book will draw on several sources of data, including: (1) information from various studies at MD Anderson Cancer center, including many taken from studies in the Division of Diagnostic Imaging. There are some dozen investigations providing data used in this book. These comprise studies with Drs. Munden, Erasmus, Gayed, Marom, Markey, Truong, Ng, Tamm, Kundra, and Das Gupta; (2) examples from the books by Shoukri[1], Von Eye and Mun[2], Liebetrau[3], Broemeling[22], Pepe[28], Zhou et al.[29], and Dobson[30]. The examples provided by Shoukri and Von Eye and Mun are quite valuable and this book will give Bayesians alternatives to their analyses; (3) information taken from the statistical literature, including the journals: *Biometrics, Communications in Statistics, The Statistician, The Canadian Journal of Statistics, Biometrika, Journal of the American Statistical Association, Applied Statistics,* and *Statistics in Medicine*; (4) information from the scientific literature, including the journals: *Journal of Thoracic Imaging, Journal of Clinical Oncology, The Breast, Academic Radiology, Journal of Nuclear Medicine, Journal of Ultrasound Medicine, The British Journal of Radiology*; and (5) some information from the internet, including the wine tasting example of Lindley at http://www.liquidasset.com/lindley.htm, and the valuable information on test accuracy and agreement selected from the datasets of Pepe found at http://www.fhcrc.org/labs/pepe/book.

1.6 Software and Computing

As mentioned several times, this book takes advantage of the software package WinBUGS, which can be downloaded at www.mrc-bsu.cam.ac.uk/bugs/. The package contains a manual and three pages of examples and provides the user a way to perform a posterior analysis for models that can be quite complex. The use in this book will vary from very simple examples, such as computing Kappa from 2×2 tables, to analysis with more sophisticated models, such as logistic linear with covariates and latent variable regressions. MCMC methods generate samples from the joint posterior distribution of the unknown parameters, and based on these, inferential procedures such as hypothesis testing, estimation, and prediction can be carried out. Graphical techniques are also part of the package, providing plots of posterior densities and plots of diagnostic processes that monitor the convergence of the generated sequences.

For the novice, Appendix B will provide the necessary information to understand the code provided as examples in this book and for the student to complete the exercises at the end of each chapter. Woodworth's[31] book is a good introduction to Bayesian statistics and an excellent reference for WinBUGS. Other software packages, such as SAS®, S-Plus®, and SPSS®, are necessary in order to provide good graphics, descriptive statistics, and database management chores.

On occasion, specialized software from MD Anderson Cancer Center is available (at http://Biostatisics/mdanderson.org/SoftwareDownload) and will be utilized, including implementation of Bayesian stopping rules in Phase II clinical trials.

1.7 A Preview of the Book

The book begins with an introductory chapter that introduces the reader to several interesting examples from various fields, including sports, cancer research, wine tasting, and the law. A championship fight between Lennox Lewis and Evander Holyfield with disagreement between the three judges, causing much excitement and agitation in the audience, is presented. The authors of the article, Lee et al., approach the problem in three ways: two are somewhat conventional, while the third is Bayesian with a latent variable regression model that purports to show the American judge is biased. Cancer research is an area, where disagreement between radiologists can occur. For example in Phase II clinical trials, radiologists decide whether a therapy is promising or not, and the example demonstrates the effect of unavoidable subjectivity on the decisions of the people making crucial calls about the efficacy of cancer treatments. Such subjectivity is exemplified in the Erasmus study, where the variation between and within five radiologists measuring tumor size is examined. The implications of the study for cancer research are important and ominous and portray the inherent subjectivity in making medical diagnoses. On a lighter note, the wine tasting example of Lindley is fun to analyze. The data is taken from the internet and the Lindley approach is summarized, and a Bayesian regression analysis used to show the degree of disagreement between two tasters. The last example involves a famous case of fraud, where the authenticity of a second will is questioned. This is one of the first examples with a formal statistical analysis for agreement, where two famous Harvard mathematicians, the father and son team of Benjamin and Charles Pierce give testimony in favor of the defense in Robinson versus Mandell. Matching two signatures is a different twist on agreement studies, and a Bayesian analysis is forthcoming.

The reader of this book is expected to have had courses in introductory probability and mathematical statistics and methods courses in regression and the analysis of variance, but not necessarily in Bayesian theory. Therefore, an introduction to Bayesian inferential techniques appear in Appendix A. Of course, the foundation for the book is Bayes theorem, which is given for continuous random vector variables and unknown vector parameters, for a parametric approach to the subject. The components of Bayes theorem are defined: the prior density, the likelihood function, and the resulting posterior density. Diagnostic techniques for model checking, based on the Bayesian predictive distribution are explained. Special emphasis is focused on hypothesis testing, estimation of parameters, and forecasting future observations. Hypothesis testing is introduced with a sharp null hypothesis, where a prior probability of the null is specified, while the remainder of the prior information is expressed as a density. The null hypothesis is rejected if the posterior probability of the alternative is "large". For estimating parameters the posterior mean, median, standard deviation, lower and upper 2½ percentiles are computed. Along with graphical presentations, the user will have a good idea of what the data and prior information imply about the present state of the parameters.

Many examples illustrate the Bayesian inferential ideas. For instance, the binomial and multinomial populations along with non informative and conjugate priors (Dirichlet), demonstrate Bayesian inferences. Hypothesis testing and estimation techniques for the normal population with the normal-gamma prior (or a Lhoste-Jeffrey non informative prior) are also presented. The t-distribution is derived as the marginal posterior distribution of the mean, and the gamma for the marginal posterior distribution of the precision (the inverse of the variance) is derived analytically. Based on the posterior distribution, the predictive distribution of future observations is also derived as a multivariate t.

Appendix A continues with a section on sample size for binomial and normal populations. The approach is Bayesian, in that the rejection region is based on Bayes theorem, but the sample size is selected using power and alpha levels as constraints. That is, one specifies the null and alternative hypotheses, the probability of a type I error, and the power one wants at a specified value of the parameters under the alternative hypothesis, and then the sample size is estimated. The rejection region itself is based on the posterior probability of rejecting the null in favor of the alternative, thus prior information is available for decision making. Examples involving binomial and normal populations make the ideas clear. A Bayesian stopping rule for a Phase II clinical trial also clarifies the concept of selecting an optimal sample size.

Appendix A continues the discussion of Bayesian methods with direct and indirect methods of sampling from the joint posterior distribution. There are two ways to specify the posterior distribution of the parameters. One is analytical, where the various marginal and conditional distributions can be derived by well-known analytical integrations, such as the mean for a normal population when a conjugate prior is employed. Or to take another example for the Dirichlet posterior distribution, which arises when sampling from a multinomial population, the marginal and conditional distributions are well known, as are its posterior moments. Alternatively, one could generate samples, with suitable software, from the marginal posterior distribution of the mean of a normal population, namely the t-distribution, to provide inferential statements. Another example is the Dirichlet posterior distribution for the probabilities of a 2×2 table. As stated earlier, the marginal moments are well-known, but one could just as well generate large enough samples from the Dirichlet distribution with S-PLUS or WinBUGS, and Bayesian inferences easily determined.

On the other hand, there are more complex cases, such as a longitudinal study with repeated observations on individuals across time, where the response of an individual is expressed with a liner regression and the regression coefficients and regression precision are given normal and gamma prior densities, respectively. Each individual has their own regression line, and the posterior distribution of all the parameters is difficult, if not impossible, to derive analytically; however, using the MCMC sampling techniques of WinBUGS, or other packages such as S-Plus, the posterior distributions can be determined to a high degree of accuracy, and Bayesian inferences implemented. This is an example of what I call indirect sampling, and for other examples of MCMC refer to the example section on the tool bar of the opening page of WinBUGS.

Appendix A concludes with a brief description of Gibbs sampling for generating samples from the posterior distribution of multiparameter models. Gibbs is used as part of the MCMC routines in WinBUGS and is explained with an example from Box and Tiao[12] of estimating the common mean of several normal populations. Also, included are some examples detailing the analyses of diagnostic test accuracy found in Broemeling[22].

Chapter 2 introduces the major topic of this book and stresses the early work in the field. The topics covered begin with the early work of Rogot and Goldberg[31] and Goodman and Kruskall[32], who developed some precursors to the Kappa coefficient. These measures of agreement were not corrected for chance agreement. Fleiss[33] presents some of the earlier contributions and modifies the Rogot and Goldberg indices for chance and show how they are related to the Kappa coefficient. Fleiss also describes the development of Kappa by Scott[9] and Cohen[10]. Both derived a Kappa coefficient, but the assumptions of the two are different. Chapter 2 devotes much space to the history and use of the Kappa coefficient, then continues on to describe one of the first Bayesian contribution namely that of Basu et al.[19]. Next generalization of Kappa to stratified Kappa and to when the scores are nominal, and the index is referred to as a weighted Kappa. Also discussed are the conditional Kappa coefficient, and the intraclass Kappa. For Kappa and each version of Kappa, a Bayesian analog is developed with a description of the prior information, the sample information, and the application of Bayes theorem. This is followed by the posterior and predictive analyses, and all examples are computed with WinBUGS.

Some problems with Kappa are outlined, namely its dependence on disease prevalence. When agreement studies are done in conjunction with measures of test accuracy, it is possible to estimate the dependence of Kappa on disease prevalence. There are some other indices similar to Kappa, such as the G-coefficient, which is not chance corrected, but are useful in some circumstances. Is an index of agreement also a measure of association? Yes, in some ways, but not in others. This is the focus of a discussion on the interface between the McNemar test for association and the Kappa coefficient. Also mentioned in this chapter is the comparison of raters via model based methods, such as multinomial and log-linear models.

Chapter 3 is an extension of the ideas in Chapter 2, whereby the various indices of agreement such as Kappa, weighted Kappa, intraclass Kappa, stratified Kappa, and the G-coefficient are generalized to multiple raters and scores. Remember that originally, Kappa was for two raters and binary nominal scores. This extension presents new problems for the study of agreement, because the more raters and scores there are, the higher the chance of disagreement among them. One would expect more disagreement between five radiologists than two in the tumor size study of Erasmus et al.[8]. If there is disagreement between a larger number of raters, one should determine the rater with the most discrepant scores (for example, with the largest intra rater variability), and the one with the least discrepant scores. It is here where the Lennox Lewis–Evander Holyfield fight is given a Bayesian analysis, and this type of analysis is repeated for the wine tasting example of Lindley. Lastly, the subject of consensus is introduced. This arises in medical diagnoses, where although disagreeing, the radiologists or pathologists must declare a decision about the extent of disease. Consider the Phase II clinical trials where each patient receives one of the following declarations about disease progress. A complete response, a partial response, stable disease, or progressive disease. One of these must be chosen. How are disagreements resolved: Another case of consensus, is a boxing match, where fighter A wins, or fighter B wins, or it is a draw. The scoring system will always declare one of these decisions. The DeGroot[6] consensus formulation is discussed and applied to several examples.

Chapter 4 is motivated by the following scenario. Two or more raters, using a binary score, rate the same pair of eyes. Since they are rating the same pair of eyes, the raters scores are expected to be highly correlated. One way to formulate the problem is to record each pair as $R+$ or $-$ and $L+$ or $-$, meaning the right eye receives a plus or minus score and the left a plus or minus score by each rater, giving 16 possible results in a 4×4 table. See Shoukri[1], who describes the approaches of Oden[34] and Schouten[35], who in turn give different solutions to the problem. The former, proposes a pooled Kappa, one Kappa for the agreement of the two raters for the left eye, the other Kappa for the right eye. On the other hand, Shouten proposes a weighted Kappa, where the weights for the pair of raters depends on their agreement between the left and right eyes. If they agree on both eyes, the weight is 1, and if agree on only one eye, the weight is ½, otherwise the weight is 0, and the usual weighted Kappa can be computed. Thus, Chapter 4 is a generalization of Chapter 3 to more complex situations that induce correlation between the raters. The challenge is to generalize this to more than two raters and to categorical data (nominal or ordinal) with more than two scores. A Bayesian analysis is provided, and it is at this point that a modeling foundation is taken with a multinomial type regression model.

The study of agreement entails much more than a single coefficient such as Kappa, which has limitations. For example, one measure would have a lack of information about the joint distribution of the raters. A modeling approach that takes into account factors such as subject covariates and other experimental settings, like interactions among raters and between raters and other experimental factors, is a more appropriate way to study agreement. Raters can be compared more easily with a modeling approach. Chapter 5 begins with an introduction to the area of modeling and continues by stressing the importance of subject covariates that can affect inter and intra rater agreement. The first model to be considered is the logistic, with binary scores for agreement. The logistic was first discussed in the context of the Lewis–Holyfield fight. Dobson[30] is a good elementary introduction to logistic and ordinal regression models, including log-linear models. The latent variable regression model, as formulated by Johnson and Albert[36], is a strictly Bayesian approach to agreement that is used for our analysis of the Lewis–Holyfield bout and some other examples. However, the majority of the chapter is devoted to the logistic linear model. Tanner and Young were one of the first to employ the log-linear model, but it is Von Eye and Mun[2], who give a complete account of the use of the log linear model to study agreement. Their contribution is explained in detail, and a Bayesian alternative using their modeling methodology is adopted. Briefly, they use the Tanner–Young model and adopt a three phase strategy by first introducing a base model, with a response variable that expresses rates of agreement and assumes independence between the raters. The base model is expanded to include other factors such as interaction terms and covariates in order to better explain the agreement. The last phase is model selection that chooses the model with the best fit. The base model is based on the IxI agreement table for two raters and is

$$Log(m_{ij}) = \lambda_0 + \lambda_i^A + \lambda_j^B + \varepsilon_{ij}$$

where λ_i^A is the i-th effect of rater A, and λ_j^B is the j-th effect of rater B, $i,j = 1,2,\ldots,I$, that is each rater gives a score from 1 to I for each experimental unit, and the last term is the residual. For example, the two raters may be radiologists assigning scores from 1 to 5, to a mammogram, where $i = 1$ means benign, $i = 2$ means perhaps benign, $i = 3$ neither benign or malignant, $i = 4$ signifies probably malignant, and $i = 5$ indicates malignant. The base model is then expanded to

$$Log(m_{ij}) = \lambda_0 + \lambda_i^A + \lambda_j^B + \delta_{ij} + \varepsilon_{ij}$$

where δ assigns weights to the main diagonal, when two raters agree, and 0 is assigned to the off-diagonal elements of the agreement table. Of course, more complicated models can be proposed, such as those with subject covariates and those containing interaction terms, etc. It all depends on the information available and the design of the study. A nested sequence of plausible models are fitted and then one is selected on the basis of the likelihood ratio chi-square test. This book will adopt the strategy by fitting a sequence of hierarchical models, then selecting an optimal model alternative on the basis of a Bayesian selection criterion. Observation from the Bayesian predictive distribution are generated by WinBUGS and compared to the actual observations of the agreement table, as is observed by Petit[37].

The previous chapters are mostly devoted to categorical responses, but Chapter 6 focuses on continuous observations. Some examples of quantitative responses were briefly discusses in earlier sections of this preview, including the tumor size agreement study of Erasmus. When dealing with continuous observations, the usual statistical tools, such as regression and the analysis of variance are often the most functional. These methods are given a Bayesian flavor. For example, consider a situation of five radiologists estimating the tumor size in a Phase II trial. Consider the two-way random model

$$y_{ijk} = \lambda + \theta_i + \phi_j + e_{ijk}$$

where $\lambda \sim n(\mu, \sigma_\lambda^2)$, the $\theta_i \sim nid(0, \sigma_\theta^2)$, the $\phi_j \sim nid(0, \sigma_\phi^2)$, and the $e_{ijk} \sim nid(0, \sigma^2)$, where $i = 1, 2, \ldots, r, j = 1, 2, \ldots, c$, and $k = 1, 2, \ldots, n$.

This is a model with two major factors. For our use, $r = 40$ is the number of images, $c = 5$ is the number of radiologists, and $n = 2$ the number of replications, that is, each radiologist reads each image twice. With this model, the intra observer variability can be measured by

$$\rho_{\text{intra}} = corr(y_{ijk}, y_{ijk'}), \quad k \neq k' \tag{1.4}$$

$$= (\sigma_\lambda^2 + \sigma_\theta^2 + \sigma_\phi^2)/(\sigma_\lambda^2 + \sigma_\theta^2 + \sigma_\phi^2 + \sigma^2)$$

and the inter observer variability expressed as

$$\rho_{\text{inter}} = corr(y_{ijk}, y_{ij'k}), \quad j \neq j' \tag{1.5}$$

Both inter and intra observer variability is estimated in a Bayesian manner by putting prior densities on the parameters μ, σ_λ^2 and the variance components σ_θ^2 and σ_ϕ^2, and generating observations from the joint posterior density via, WinBUGS.

Other methods to assess agreement and to compare raters are regression and correlation, where the scores of one rater are regressed on the scores of another and applying the Bland–Altman mode. Graphical techniques of scatter plots and boxplots greatly add to the analysis. The total variance is partitioned into several components of variation, or variance components, and the percentage attributable to each source of variation can be computed. This is also valuable information for studies of agreement.

Chapter 7 takes us in another direction, that of estimating the sample size. When designing an agreement study, how many raters and how many subjects should be utilized so that the various hypotheses can be tested in a reliable manner? The non Bayesian way to formulate the problem is to state the null and alternative hypotheses in terms of the unknown parameters, then to choose a test statistics or rejection region based on the probability of a Type I error and the power one desired for a setting of the parameter value at an alternative value. The Bayesian can approach the problem in much the same way, however if the null hypothesis is sharp, a prior probability value must be assigned to it and the remaining probability expressed as density over the remainder of the parameter space. The rejection region is then defined as the posterior probability of the alternative hypothesis, if, which is large enough, the null is rejected. Thus, the Bayesian can act as a frequentist by relying on the power and type I error concepts to define the rejection region.

On the other hand, the Bayesian can ignore the frequentist sampling ideas and reject the null hypothesis when the posterior probability of the alternative is "large enough." Both ideas are explored in Chapter 7. Another alternative considered, is to estimate the sample size based on classical formulas, which of course in turn are based on Type I and II considerations. The advantage of doing this is that these formulas are functions of the parameters of the model, thus the sample size formula has a posterior distribution. Sample sizes for Kappa and for log-linear models are illustrated with many proposed agreement studies.

The last part of the chapter summarizes the book and looks to the future for agreement studies by proposing some problems that have to date not been solved. As mentioned earlier, there is an appendix that is an introduction to WinBUGS.

Exercises

1. Refer to Bayes theorem Equation 1.2 and derive the posterior distribution of θ, a Bernoulli parameter, where X is a random sample of size n from a Bernoulli population. Assume the prior distribution for this parameter is Beta with parameters α and is β.

2. Refer to Equation 1.3 and the previous exercise, and suppose $Z = (z_1, z_2, \ldots, z_m)$ is a future sample of size m from the Bernoulli population with parameter θ, derive the predictive distribution of Z.

3. Verify that kappa is estimated as 0.32 for Table 1.3, using a standard software package like SAS or SPSS. What is the standard error of the estimate?

4. Referring to Table 1.8 and Table 1.9, verify the descriptive statistics for Table 1.10. What does Lindley mean by the column labeled SHRUNK?

5. Reproduce the boxplots given in Figure 1.2 of the wine tasting example. What are the interquartile ranges for the French and Californian Chardonnay wines?

6. Compute the Pearson correlations for all pairs of 11 tasters for the Cabernet wines. Are there any obvious cases of disagreement between the tasters? What are the implications for the comparison of French and Californian wines?

7. Verify Figure 1.3 with a scatter plot of the taster 6 versus taster 8 scores. Can you verify the values for the slope and intercept of the fitted line? Why does the plot imply the two tasters disagree?

8. Execute the WinBUGS statements given before Table 1.12 and verify Table 1.12.

9. Devlin and Lorden[26] imply the Bernoulli model of Benjamin and Charles Pierce is not appropriate. Discuss why this may be true. Is the probability of coincident downstrokes the same for all pairs of downstrokes? Experimentally, how would you test the assumptions of a Binomial model?

10. What is a Bland–Altman analysis for agreement between raters? How would a Bayesian implement the Bland-Altman methods? What graphical procedures should accompany a Bland-Altman analysis?

References

1. Shoukri, M. M. 2004. *Measures of interobserver agreement*. Boca Raton, London, New York: Chapman and Hall/CRC.
2. Von Eye, A., and E. Y. Mun. 2005. *Analyzing rater agreement, manifest variable methods*. Mahwash, NJ, London: Lawrence Erlbaum Associates.
3. Liebetrau, A. M. 1983. *Measures of association, quantitative applications in the social sciences*. Thousand Oaks, CA: Sage.
4. Fleiss, J. L., B. Levin, and M. C. Paik. 2003. *Statistical methods for rates and proportions*. 3rd ed. New York: John Wiley and Sons.
5. Agresti, A. 1990. *Categorical data analysis*. New York: John Wiley and Sons.
6. DeGroot, M. H. 1974. Reaching a consensus. *J. Am. Stat. Assoc.*, 69(345), 118.
7. Birkelo, C. C., W. E. Chamberlin, P. S. Phelphs, P. E. Schools, D. Zacks, and J. Yerushalmy. 1947. Tuberculosis case-finding: A comparison of effectiveness of various roentgen graphic and photoflurorographic methods. *J. Am. Med. Assos.* 133, 359.
8. Erasmus, J. J., G. W. Gladish, L. Broemeling, B. S. Sabloff, M. T. Truong, R. S. Herbst, and R. F. Munden. 2003. Interobserver variability in measurement of non-small cell carcinoma of the lung lesions: implications for assessment of tumor response. *J. Clin. Oncol.* 21, 2574.
9. Scott, W. A. 1955. Reliability of content analysis: the cases of nominal scale coding. *Public Opinion Quart.* 19, 321.

10. Cohen, J. 1960. A coefficient of agreement for nominal scales. *Educ. Psychol. Measmt.* 20, 37.
11. Hopkins, W. G. 2004. Bias in Bland–Altman but not regression validity analyses. *Sportscience* 8, 42.
12. Box, G. E. P. 1980. Sampling and Bayes inference in scientific modeling and robustness. *J. Roy. Stat. Soc., Series A.* 143, 383.
13. Zellner, A. 1971. *An introduction to Bayesian inference in econometrics.* New York: John Wiley and Son.
14. Box, G. E. P., and G. C. Tiao. 1973. *Bayesian inference in statistical analysis.* Reading, MA: Addison Wesley.
15. Broemeling, L. D. 1985. *Bayesian analysis of linear models.* New York: Marcel-Dekker.
16. Congdon, P. 2001. *Bayesian statistical modelling.* New York: John Wiley and Sons.
17. Congdon, P. 2003. *Applied Bayesian modelling.* New York: John Wiley and Sons.
18. Congdon, P. 2005. *Bayesian models for categorical data.* New York: John Wiley and Sons.
19. Basu, S., M. Banerjee, and A. Sen. 2000. Bayesian inference for Kappa from single and multiple studies. *Biometrics* 56(2), 577.
20. Lee, H. K. H., D. L. Cork, and D. J. Algranati. 2002. Did Lennox Lewis beat Evander Holyfield? Methods for analyzing small samples interrater agreement problems. *The Statistician* 51(2), 129.
21. Padhani, A. R., and L. Ollivier. 2001. The RECIST criteria: Implications for diagnostic radiologists. *Br. J. Radiol.* 74, 1983.
22. Broemeling L. D. 2007. *Bayesian biostatistics and diagnostic medicine.* Boca Raton: Chapman and Hall/CRC.
23. Thiese, P., L. Ollivier, D. DiStefano-Louieheau, S. Negrier, J. Savary, K. Pignard, C. Lesset, and B. Escudier. 1997. Response rate accuracy in oncology trials: Reasons for interobserver variability. *J. Clin. Oncol.* 15, 3507.
24. Hopkins, W. G. 2004. Bias in Bland–Altman but not regression validity analyses. *Sportscience.* 8, 42.
25. Lindley, D. The wine tasting example, taken from http://www.liquidasset. com/lindley.htm. March 2008.
26. Devlin, K., and G. Lorden. 2007. *The numbers behind numbers NUMB3RS,* PLUME. New York: Penguin.
27. Finklestein, M. O., and B. Levin. 2007. *Statistics for lawyers.* 2nd ed. Berlin: Springer Verlag.
28. Pepe, M. S. 2003. *The statistical evaluation of medical tests for classification and predicition.* Oxford, UK: Oxford University Press.
29. Zhou, H. H., D. K. McClish, and N.A. Obuchowski. 2002. *Statistical methods for diagnostic medicine,* New York: John Wiley and Sons.
30. Dobson, A. J. 2002. *An introduction to generalized linear models.* 2nd ed. Boca Raton, New York, London: Chapman and Hall/CRC.
31. Rogot, E., and I. D. Goldberg. 1966. A proposed idea for measuring agreement in test re-test studies. *J. Chron. Dis.* 19, 991.
32. Goodman, L. A., and W. H. Kruskal. 1954. Measures of association for cross classification. *J. Am. Stat. Assoc.* 49, 732.
33. Fleiss, J. L. 1975. Measuring agreement between two judges on the presence or absence of a trait. *Biometrics* 31(3), 641.

34. Oden, N. 1991. Estimating Kappa from binocular data, *Stat. Med.* 10, 1303.
35. Schouten, H. J. A. 1993. Estimating Kappa from binocular data and comparing marginal probabilities. *Stat. Med.* 12, 2207.
36. Johnson, V. E., and J. H. Albert. 1999. *Ordinal data modeling.* New York: Springer-Verlag.
37. Petit, L. I. 1986. Diagnostics in Bayesian model choice. *The Statistician* 35, 183.

2

Bayesian Methods of Agreement for Two Raters

2.1 Introduction

The previous chapter and Appendix A lay the foundation for the Bayesian approach to agreement. Chapter 1 gives a brief history of the subject, presents a preview of topics to be discussed in future chapters, identifies the literature sources for the book and, goes into some detail about several examples. The examples included wine tasting in a Parisian setting, where both Californian and French wines were rated by several judges, and an example from a cancer research study, where several radiologists estimate the size of lung cancer lesions. Appendix A introduces the theory necessary to understand the Bayesian methods that will analyze data from agreement studies, and introduces WinBUGS as a platform for Bayesian type computations. Special emphasis was put on techniques to check model adequacy and methods to display information.

Chapter 2 is the first encounter with the main topic of the book, Bayesian methods for the analysis of agreement. A brief introduction to elements of a good study design are presented, but much of the chapter centers on precursors of Kappa, Kappa itself, and some generalization of Kappa. The generalizations include conditional Kappa, weighted Kappa, Kappa and stratification, and intraclass Kappa. Kappa plays the central role as a single measure of agreement and most modern developments stem from Kappa. Responses from the raters are restricted to nominal and ordinal variables. The use of Kappa in diagnostic testing studies, where gold standards are available, is also discussed, along with the interface of Kappa to other measures of statistical association. Additional measures of agreement, such as the G coefficient and the Jacquard index are explained, and finally, the chapter is concluded with some model (e.g. logistic model and log linear) based approaches. Each topic of the chapter contains interesting examples taken from cancer research, diagnostic imaging, wine tasting, and editing and reviewing of scientific articles.

2.2 The Design of Agreement Studies

This section is one of the most important in the book, because how the agreement study is to be designed is of the essence. First and foremost, the objectives of the study must be stated. It should be recognized that agreement studies are often included as a subset of a larger study. For example, with diagnostic studies in medicine, the agreement between readers is just one part of a larger plan. On the other hand, agreement is often the main goal of the investigation, such as with judging a boxing match or diving competition. The design of study should include the identification of the raters and subjects and how they are related. Is randomization of the readers to the subjects a part of the design? Consensus is often part of the study, where after the raters assign scores to the subjects, a consensus is reached as to the outcome of the judging. For example in boxing, a consensus is reached via the rules of the contest, as it is in Figure skating, where the rules dictate a winner must emerge after the scoring of the judges. An important feature of any study is to carefully describe the major endpoints, which are to be assigned to the subject, and lastly a description of the proposed analysis is usually detailed, along with an estimate of the sample size.

Objectives to the study describe what is to be accomplished in the study and may contain statistical hypotheses to be tested or parameters to be estimated. For example, one may want to test the null hypothesis that the Kappa value is below some value versus the alternative hypothesis that it exceeds a value. Questions of scientific interest are to be addressed in the objectives of the study plan.

It is important to recognize two basic scenarios in agreement studies, namely: (1) studies without a gold standard; and (2) studies with a gold standard. A good example of the former is in Figure skating, where the scores of the judges are the only ones of importance in the contest. This is in contrast to medical diagnostic testing, where a gold standard is usually present. For example, in lung cancer screening, a CT image indicates a patient might have the disease, but to be sure, a biopsy (the gold standard) of lesion tissue is examined by pathology for a definitive diagnosis. There may be many radiologists who are reading the same image, and the degree of agreement is important to measure. In addition, the scores of the radiologists are to be compared to the gold standard, thus, there is agreement between radiologists and agreement between radiologists and the lesion pathology. The true disease incidence does affect the value of Kappa between the radiologists.

How are the subjects to be selected? At random from some population or in consecutive order as in some clinical studies? No matter how the subjects are selected and from what set of subjects, the selection process should be described in detail. In medical studies, the protocol defines the way the patients are selected from a set of so called eligibility criteria and non-eligibility criteria, which is a list of symptoms and other patient demographics. On the other hand, the situation is completely different in some agreement studies.

Consider a boxing match, then there are only two subjects and the judges are selected by the state boxing commission. There are cases where complex random sampling (probability sampling) schemes are taken to select a sample of subjects in a survey sample over a given time period, say a year. Of interest is the degree of agreement between the respondents of one year versus those of the following or previous years, and is identified as a repeated measures type of agreement study.

How are the readers to be selected? This varies considerably with the type of study. In Figure skating, the number of judges depends on their previous record and the rules of the skating association in charge of the event. The same applies to gymnastics and diving. On the other hand in medical studies, the radiologists chosen to evaluate the patients of a phase II clinical trial are unique to the health care institution and departmental rules. The work schedule of the doctors, nurses, surgeons, and pathologists, as well as the patients' schedules all play a role in determining the assignment of raters to a particular protocol.

After selecting the raters and subjects, the relation between the two groups should be described as part of the design. Do all the judges score all the subjects? This is the case that will occur frequently in this book, however there will be a few exceptions. For example, in some imaging studies, different groups of readers will be assigned to different groups of subjects. Another important question is the order the raters will score the subject. In Figure skating for a particular category, the skaters are scored three times, and the order the skaters appear is different for the three events. The judges may also change from event to event. The rules of the sport explain in detail just how this is to take place.

The design will explain the major responses of the study. Are they binary, discrete, or continuous? When and how will they measure the subjects? We have seen that for some medical studies, the responses are continuous, as for example in measuring the lesion size of lung cancer patients with several radiologists, but with a boxing contest the scores assigned are three for each round of the match. Either fighter A or B wins, or the round is judged a draw by the three judges.

Data collection and data analysis go hand in hand. The study information might be put into a spreadsheet by a research nurse or in the case of a national sample survey, the data will be collected from a number of sources and sent to a central data center, responsible for the total study. In contrast to this, the scores of judges in a boxing contest, are scored by hand, and given to another person at the end of each round. For Figure skating, the scores are entered into a database and displayed for all the audience to see. Depending on the type of study, the design will give the particular facts about how to collect and store the information.

For medical studies, the protocol will describe the way the study information is to be analyzed. It is here that the objectives of the study will be stated in statistical language and phrased as a test of hypothesis. For example, the Kappa value might be a subject of a hypothesis for binary outcomes, or the intraclass correlation coefficient estimated for a continuous outcome.

For other kinds of studies, the analysis would consist of using the scores to arrive at a consensus. The rules of the association would determine the way the scores are to be analyzed to select a "winner." With sports, the design stays the same, however for others, such as with scientific studies, the design will change depending on the context. A good example where the design stays the same, is with the review process for a scientific journal, where there is one paper to judge and several referees recommend acceptance, rejection, or revision to the editor. Obviously, the degree of agreement between referees influence the final decision of the editor, but the design of the process is specified by the journal's editorial policy.

Lastly, if the design has a statistical section as such, the sample size should be justified by a power analysis that determines the size in terms of the probability of a type I error, the power of the test at several alternatives, and length of the difference between the null and alternative values of some measure of agreement like Kappa. The sample size refers to the number of subjects and the number of raters. Some studies have obvious sample sizes: in boxing there are twelve rounds, with two fighters, and three judges, and with a jury trial, there is usually one defendant and 12 jurors!

This brief introduction to study design should be supplemented by reading Zhou, Obuchowski, and McClish[1], and Pepe[2].

2.3 Precursors of Kappa

Our study of agreement begins with some early work before Kappa, but first remember that Kappa was defined in Chapter 1, where Kappa is introduced with the 2×2 Table 1.1 giving a binary score to n subjects. Each subject is classified as either positive, denoted by $X=1$, or negative, denoted by $X=0$, by reader 1 and $Y=0$ or 1, by rater 2.

Let θ_{ij} be the probability that rater 1 gives a score of i and rater 2 a score of j, where $i,j=0$ or 1, and let n_{ij} be the corresponding number of subjects. The experimental results have the structure of a multinomial distribution. Obviously, the probability of agreement is the sum of the diagonal probabilities $\theta_{00} + \theta_{11}$, however this measure is usually not used. Instead, the Kappa parameter is often employed as an overall measure, and is defined as

$$\kappa = [(\theta_{00} + \theta_{11}) - (\theta_{0.}\theta_{.0} + \theta_{1.}\theta_{.1})]/[1 - (\theta_{0.}\theta_{.0} + \theta_{1.}\theta_{.1})]$$

where the dot indicates summation of the θ_{ij} over the missing subscript, and the probability of a positive response for rater 2 is $\theta_{.1}$ Kappa is a so-called chance corrected measure of agreement.

Before Kappa, there were some attempts to quantify the degree of agreement between two raters, and what follows is based on Fleiss[3], account of the early history of the subject, beginning with Goodman and Kruskal[4]. They assert that the raw agreement measure $\theta_{00} + \theta_{11}$ is the only sensible measure, however as Fleiss stresses, other logical measures have been adopted. This presentation closely follows Fleiss, who summarizes the early publications of indices of agreement and the two origins of Kappa, then presents his own version of chance corrected early indices and how they relate to Kappa.

One of the earliest measures is from Dice[5], who formulated

$$S_D = \theta_{11}/[(\theta_{.1} + \theta_{1.})/2] \tag{2.1}$$

as an index of agreement. First select a rater at random, then S_D is the conditional probability that that judge has assigned a positive rating to a subject, given the other judge has assigned a positive rating. This seems reasonable, if the probability of a negative rating is greater than that of a positive rating. In a similar way, define $S_D' = \theta_{00}/[(\theta_{.0} + \theta_{0.})/2]$, then based on the Dice approach, Rogot and Goldberg[6] defined

$$A_2 = \theta_{11}/(\theta_{1.} + \theta_{.1}) + \theta_{00}/(\theta_{0.} + \theta_{.0}) \tag{2.2}$$

as a measure of agreement, with the desirable property that $A_2 = 0$ if there is complete disagreement and is equal to 1 if there is complete agreement.

According to Fleiss, Rogot and Goldberg another measure of agreement was defined as:

$$A_1 = (\theta_{00}/\theta_{.0} + \theta_{00}/\theta_{0.} + \theta_{11}/\theta_{1.} + \theta_{11}/\theta_{.1})/4 \tag{2.3}$$

which is an average of the four conditional probabilities. This index has the same properties as A_1, namely $0 \le A_2 \le 1$, but in addition, $A_2 = \frac{1}{2}$ when the raters are giving independent scores, that is, when $\theta_{ij} = \theta_{i.}\theta_{.j}$ for all i and j.

Suppose two raters assign scores to 100 subjects as outlined in Table 2.1. Then what are the estimated values of A_1 and A_2? It can be verified that $\tilde{A}_1 = (22/55 + 22/37 + 30/45 + 30/63)/4 = (0.4 + 0.59 + 0.68 + 0.47)/4 = 0.53$, and $\tilde{A}_2 = 30/(63 + 45) + 22(55 + 37) = 0.52$, where both indicate fair agreement; remember that the maximum value of both is 1. The first index is very close to $\frac{1}{2}$, does this indicate independence between the two readers? Does

TABLE 2.1

Hypothetical Example of 100 Subjects

Rater 1	Rater 2		
	$Y = 0$	$Y = 1$	
$X = 0$	22	15	37
$X = 1$	33	30	63
	55	45	100

22/100=22/55*22/37? or does 0.22=0.23? This is evidence that the raters are acting independently in their assignment of scores to the 100 subjects.

What is the Bayesian analysis? Some assumptions are in order, therefore suppose the 100 subjects are selected at random from some well defined population and that the structure of the experiment is multinomial, that the assignment of scores to subjects are such that the probability of each of the four outcomes stays constant from subject to subject. Initially, a uniform prior is assigned to the four mutually exclusive outcomes, and the posterior distribution of $(\theta_{00}, \theta_{01}, \theta_{10}, \theta_{11})$ is Dirichlet (23,16,34,31). What is the posterior distribution of A_1 and A_2?

Using 25,000 observations generated from BUGS, with a burn in of 1000 observations, and a refresh of 1000, the posterior analysis is displayed by Table 2.2.

A revision of the BUGS code in Section 7.2 of Appendix A produces the output appearing in Table 1.1. The parameter g is

$$g(\theta) = \sum_{i,j=0}^{i,j=1} (\theta_{ij} - \theta_{i.}\theta_{.j})^2$$

and can be used to investigate the independence of the two raters. From Table 2.3, it appears that independence is a tenable assertion. Note that g is non negative and has a median of 0.0015 and a mean of 0.00308, and that independence is also implied by the posterior mean of $A_2=0.5138$. The role of independence of the two raters is important for chance corrected measures such as Kappa. Fleiss reports two additional indices proposed by Armitage[7] and Goodman and Kruskal, but will not be presented here.

TABLE 2.2

Posterior Distribution of A_1 and A_2

Parameter	Mean	SD	2½	Median	97½
A_1	0.5324	0.0482	0.4362	0.5326	0.6254
A_2	0.5138	0.0487	0.4180	0.5138	0.6088
g	0.00308	0.0040	3.437*10⁻⁶	0.0015	0.01457

TABLE 2.3

Posterior Distribution of Kappa

Parameter	Mean	SD	2½	Median	97½
Kappa	−0.0905	0.1668	−0.4131	−0.0916	0.2437
I_r	0.5196	0.0486	0.4239	0.5199	0.6148
I_c	0.4882	0.0135	0.458	0.4896	0.5126

2.4 Chance Corrected Measures of Agreement

Fleiss continues with a review of the chance corrected indices of agreement, which have a general form of

$$M = (I_r - I_c)/(1 - I_c) \tag{2.4}$$

where I_r is a measure of raw agreement and I_c a measure of agreement by chance. Scott[8] first introduced such a measure called Kappa, which was followed by Cohen[9], who gave another version. Recalling the formula for Kappa as

$$\kappa = [(\theta_{00} + \theta_{11}) - (\theta_{0.}\theta_{.0} + \theta_{1.}\theta_{.1})]/[1 - (\theta_{0.}\theta_{.0} + \theta_{1.}\theta_{.1})]$$

where Scott assumed both raters have the same marginal distribution, Cohen did not however make such an assumption. The Cohen version of Kappa will be used in this book.

From the information in Table 2.1 and the WinBUGS code used to produce Table 2.2, it can be shown that the posterior distribution of Kappa and its components are given in Table 2.3.

The value of Kappa indicates very poor agreement, because the raw agreement estimate of 0.5196 and the estimated chance agreement of 0.4882 are quite close. This demonstrates the effect of chance agreement on Kappa.

Fleiss continues the presentation by correcting the two precursors of Kappa A_1 and A_2 for chance agreement using Equation 2.4, thus the chance corrected A_2 index is

$$M(A_2) = 2(\theta_{11}\theta_{00} - \theta_{01}\theta_{10})/(\theta_{1.}\theta_{.0} + \theta_{.1}\theta_{0.}) \tag{2.5}$$

$$= \kappa$$

where $I_c = E(A_2)$, assuming independent raters.

In a similar fashion, the chance corrected value of S'_D, (2.1), can be shown to be κ, however, this is not true for A_1, that is, the chance corrected value of A_1 is not Kappa.

Kappa has been embraced as the most popular index of agreement and has been extended to ordinal scores and to multiple raters and scores, and its use is ubiquitous in the social and medical sciences. Kappa gives an idea of the overall agreement between two raters with nominal scores, but once the value is estimated, it is important to know why Kappa has that particular value, and this will entail investigating the marginal and joint distribution of the raters. Performing an agreement analysis is much like doing an analysis of variance, where if at the first stage the null hypothesis of equal means is rejected, is followed by a multiple comparison procedure. In the case no agreement is indicated by Kappa, a look at the disagreement between the two readers at the off-diagonal cells would be informative. On the other hand if Kappa indicates strong agreement, a comparison between the two raters at the on diagonal cells can be fruitful.

2.5 Conditional Kappa

The conditional Kappa for rating category i is

$$\kappa_i = (\theta_{ii} - \theta_{i.}\theta_{.i})/(\theta_{i.} - \theta_{i.}\theta_{.i}) \tag{2.6}$$

$$= (\theta_{ii}/\theta_{i.} - \theta_{i.}\theta_{.i}/\theta_{i.})/(1 - \theta_{i.}\theta_{.i}/\theta_{i.})$$

and is the conditional probability that the second rater assigns a score of i to a subject, given the first rater scores an i, adjusted for conditional independence between the raters. Since

$$P[Y=i/X=i] = \theta_{ii}/\theta_{i.}, \quad i = 0,1$$

and assuming conditional independence between the two raters, Equation 2.6 is apparent. Thus, if Kappa overall gives an indication of strong agreement, then conditional Kappa will identify the degree of agreement between the two raters for each possible category, 0 or 1. The Bayesian approach allows one to determine the posterior distribution of conditional Kappa. For the example from Table 2.1, BUGS gives 0.0925 for the posterior mean of κ_0 and a posterior standard deviation of 0.1391. Also, a 95% credible interval for κ_0 is (−0.1837,0.3646). This is not a surprise since overall Kappa is only −0.0905. One would expect a value near 0 for the posterior mean of κ_1. Conditional Kappa should be computed after the value of overall Kappa is known, because such estimates provide additional information about the strength of agreement. The concept of conditional Kappa (sometimes referred to as partial Kappa) was introduced first by Coleman[10] and later by Light[11] and is also described by Von Eye and Mun[12] and Liebetrau[13].

For k nominal scores Kappa is defined as

$$\kappa = \left[\sum_{i=1}^{i=k} \theta_{ii} - \sum_{i=1}^{i=k} \theta_{i.}\theta_{.i}\right] \Bigg/ \left[1 - \sum_{i=1}^{i=k} \theta_{i.}\theta_{.i}\right] \tag{2.7}$$

where θ_{ij} is the probability that raters 1 and 2 assign scores i and j, respectively to a subject, where $i,j = 1,2,...k$, and k is an integer at least 3. With more than two scores, the agreement between raters becomes more complex. There are more ways for the two to agree or disagree.

Von Eye and Mun[12] examine the agreement between two psychiatrists who are assigning degrees of depression to 129 patients. Consider Table 2.4 where they report Kappa is estimated as 0.375, and the scores are interpreted as: 1 = "not depressed", 2 = "mildly depressed", and 3 = "clinically depressed".

If one adopts a Bayesian approach with a uniform prior density for the θ_{ij}, i and j = 1, 2, 3, then the parameters have a Dirichlet (12,3,20,2,4,4,1,9,83) posterior distribution and the resulting posterior analysis is outlined in Table 2.5.

The above description of the posterior analysis is based on 24,000 observations with a burn in of 1000 observations, and a refresh of 100. It shows that the raw agreement probability is estimated as 0.7174 with the posterior mean. The chance agreement probability has a posterior mean of 0.5595, which determines the posterior mean of Kappa as 0.3579, and the difference d between the probability of raw agreement and the probability of chance agreement has a posterior mean of 0.1578. Overall agreement is fair but not considered strong.

A plot of the posterior density of conditional Kappa for the score of 1 (not depressed) has a mean of 0.2639 with a standard deviation of 0.074, again implying only fair agreement for the not depressed category (see Figure 2.1). Von Eye and Mun[12] report an estimated κ_1 of 0.276 and estimated standard deviation 0.0447, thus the conventional and Bayesian are about the same.

Can I confirm the analysis for overall κ compared to Von Eye and Mun? Their estimate of Kappa is 0.375 with a standard error of 0.079, compared to a posterior mean of 0.357, where the standard deviation of the posterior distribution of Kappa is 0.0721. I used SPSS, which gave 0.375 (0.079 is the

TABLE 2.4

Agreement for Depression

Psychiatrist 1	Psychiatrist 2			
	1	2	3	Total
1	11	2	19	32
2	1	3	3	7
3	0	8	82	90
Total	12	13	104	129

TABLE 2.5

Posterior Analysis of Depression

Parameter	Mean	SD	2½	Median	97½
Kappa	0.3579	0.0721	0.219	0.3577	0.4995
Kappa1	0.2639	0.0740	0.1326	0.26	0.4194
agree	0.7174	0.0382	0.62	0.7184	0.7889
cagree	0.5595	0.03704	0.489	0.559	0.6335
d	0.1578	0.0358	0.0914	0.1566	0.232

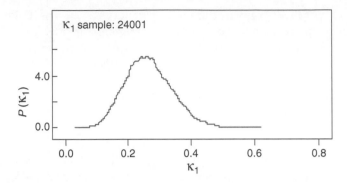

FIGURE 2.1
Posterior density of κ_1.

asymptotic standard error) as an estimate of Kappa, which confirms the Von Eye and Mun analysis.

It is of interest to compare the posterior mean and standard deviation of θ_{11} with its known mean and standard deviation, which are 0.086956 and 0.0238995, respectively. The posterior mean and standard deviation based on 24,000 observations generated by BUGS are 0.08699 and 0.02398, respectively, thus the simulation is accurate to four decimal places (truncated) for estimating the mean and to three places for the standard deviation of θ_{11}.

The BUGS code for these calculations and graphics is:

BUGS CODE 2.1

model

```
{
g[1,1]~dgamma(12,2)
g[1,2]~dgamma(3,2)
g[1,3]~dgamma(20,2)
g[2,1] ~dgamma(2,2)
g[2,2] ~dgamma(4,2)
g[2,3]~dgamma(4,2)
g[3,1]~dgamma(1,2)
g[3,2]~dgamma(9,2)
g[3,3]~dgamma(83,2)
h <- sum(g[,])
for( i in 1 : 3 ) {for( j in 1 :3 ){ theta[i,j] <-g[i,j]/h }}
theta1. <-sum(theta[1,])
theta.1 <-sum(theta[,1])
theta2. <-sum(theta[2,])
theta.2 <-sum(theta[,2])
```

```
theta3. <-sum(theta[3,])
theta.3 <- sum(theta[,3])
kappa <- (agree-cagree)/(1-cagree)
agree <-theta[1,1] + theta[2,2] + theta[3,3]
cagree <-theta1.*theta.1 + theta2.*theta.2 + theta3.*theta.3
d <-agree-cagree
Kappa1 <-(theta[1,1]-theta1.*theta.1)/(theta1.-theta1.*theta.1)
}

list( g = structure(.Data = c(2,2,2,2,2,2,2,2,2), .Dim = c(3,3)))
```

This is a revision of **BUGS CODE A3** that appears in Appendix A.

The raw agreement by itself indicates a strong association between psychiatrists, however the chance agreement is fairly strong also, which reduces the overall agreement to only fair agreement, estimated as 0.357, via a Bayesian approach, and to 0.375, via maximum likelihood by Von Eye and Mun. The Bayesian and conventional analyses of Kappa and conditional Kappa (for the not depressed category) agree quite well, but it is difficult to assign a Kappa value!

2.6 Kappa and Stratification

Stratification is a way to take into account additional experimental information that will be useful in estimating agreement between observers. Up to this point, the only information brought to bear on estimating Kappa are the scores of the two raters, however other information is often available, including age, sex, and other subject demographics. In medical studies, in addition to patient demographics, other information on previous and present medical history is available and might have an influence on the way raters disagree. Stratification is the first stage on the way of utilizing other experimental information that influences Kappa. Later this will be expanded to model based approaches that will incorporate covariate information in a more efficient way.

Consider a hypothetical national trial that compares X-ray and CT for detecting lung cancer in high risk patients. The study is to be conducted at three different sites, under the same protocol. One is primarily interested in the agreement between X-ray and CT, where both images are taken on each patient.

The design calls for a total enrollment of 2500 patients from two major cancer centers enrolling 1000 patients each, and with a smaller cancer center enrolling 500. The idea of estimating Kappa between X-ray and CT is one of the major issues being studied, along with the usual one of estimating and comparing the diagnostic test accuracies of the two images. For the time being this information and patient covariate information, as well as other experimental factors, will not be considered. The major objective is to estimate an overall Kappa and to compare the individual Kappas of the three sites (Tables 2.6 through 2.8). It is assumed the patients are enrolled according to strict eligibility and ineligibility criteria, which describes a well defined population of subjects.

The scores assigned are 1, 2, 3, and 4, where 1 indicates "no evidence of disease", 2 indicates "most likely no evidence of disease", 3 implies it is "likely there is evidence of disease", and 4 indicates "evidence of a malignant lesion". Obviously, a team of radiologists assign the scores, however the details are not considered relevant. Having both images on the same person is somewhat unique in this type of trial, usually a patient receives only one type of image.

A posterior analysis based on BUGS with 25,000 observations generated from the joint posterior distribution, with a burn in of 1000 observations, and a refresh of 100 are reported in Table 2.9. A uniform prior density is used for the parameters.

The parameters K12 is the difference between κ_1 and κ_2, where κ_1 is the Kappa for site 1, and the overall Kappa is denoted by κ, where

$$\kappa = (\kappa_1\omega_1 + \kappa_2\omega_2 + \kappa_3\omega_3)/(\omega_1 + \omega_2 + \omega_3), \tag{2.8}$$

TABLE 2.6

CT Versus X-ray: Site 1

X-ray	CT				
	1	2	3	4	Total
1	320	57	22	15	414
2	44	197	18	10	269
3	20	26	148	5	199
4	11	7	3	73	94
Total	395	287	191	103	976

TABLE 2.7

CT Versus X-ray: Site 2

X-ray	CT				
	1	2	3	4	Total
1	297	44	26	12	379
2	37	210	16	2	265
3	20	26	155	7	208
4	12	6	1	65	84
Total	366	286	198	86	936

TABLE 2.8

CT Versus X-ray: Site 3

X-ray	CT				
	1	2	3	4	Total
1	185	20	11	4	220
2	15	101	16	2	134
3	8	20	80	7	115
4	8	1	1	15	25
Total	216	142	108	28	494

TABLE 2.9

Posterior Analysis for X-ray and CT

Parameter	Mean	SD	2½	Median	97½
K12	−0.0296	0.02761	−0.0836	−0.0297	0.0244
K13	0.0008812	0.03367	−0.0643	0.000677	0.0676
κ	0.6508	0.01245	0.6266	0.6509	0.675
κ_1	0.6391	0.0197	0.6004	0.6393	0.677
κ_2	0.6688	0.0194	0.6298	0.6691	0.7059
κ_3	0.6382	0.0274	0.5828	0.6385	0.6908
D1	0.4464	0.0146	0.4175	0.4466	0.4748
D2	0.4689	0.0144	0.44	0.4691	0.4965
D3	0.4323	0.0191	0.3941	0.4324	0.4964

and ω_i is the inverse of the variance of the posterior distribution of κ_i, $i=1, 2, 3$. This particular average is one of many that could have been computed, but this issue will be explored in a moment.

D1, D2, and D3 are the difference between the raw agreement probability and the chance agreement probability for sites 1, 2, and 3, respectively. The characteristics of the posterior distribution across the three sites are very consistent, as portrayed by the posterior mean and median of K12 and K13. Figure 2.2, illustrates a plot of the posterior density $\kappa_1 - \kappa_2$, where this parameter has a 95% credible interval of (−0.0836, 0.0244). The weighted average of 0.65 is a fairly strong agreement between the two imaging modalities.

The posterior analysis for the partial Kappas of the four scores at each site were not computed, but are left as an exercise.

As seen, the posterior analysis for overall Kappa was based on the weighted average, where the weights were the reciprocals of the posterior standard deviations of the individual site Kappas. This is somewhat of an arbitrary choice, and Shoukri[14] states three alternatives for the weighted average:

A. Equal weights,

B. $\omega_i = n_i$, the number of observations in stratum i, and

C. $\omega_i = [var(\kappa_i/data)]^{-1}$.

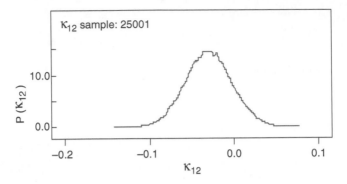

FIGURE 2.2
Posterior density of $\kappa_1 - \kappa_2$.

TABLE 2.10

MRI and Ultrasound Stratified by Lesion Size

Lesion Size (mm)	Missed by Both (0,0)	Seen by MRI Only (0,1)	Seen by Ultrasound Only (1,0)	Seen by Both (1,1)	Total
1–5	40	15	10	20	85
6–10	29	14	10	70	123
11–15	10	7	7	12	36
16–20	3	2	3	24	32
21–25	0	1	1	10	12
>25	1	2	1	8	12
Total	83	41	32	144	300

Source: Shoukri, M., M., 2004. *Measures of Interobserver Agreement.* Boca Raton: Chapman & Hall/CRC.

The var(κ_i/data) are variances of the posterior distribution of the κ_i, and now there is the problem of choosing a particular weighting scheme. Which one should be used for a particular problem? From a frequentist viewpoint, the bias as well as the variance of the estimated (by the posterior mean) should be taken into account, and for more on this see Barlow[15].

Shoukri[14], using hypothetical data, analyzes the association between MRI and ultrasound as a function of lesion size. The usual approach is to use a weighted Kappa parameter with weights chosen according to several schemes. The weighted Kappa parameter is given by Equation 2.8, where the sum of the weights is unity, κ_i is the usual Kappa parameter for the i-th stratum, and w_i is the weight assigned to the i-th stratum.

Consider the information presented in Table 2.10 on the agreement between MRI and ultrasound, stratified by lesion size and involving 300 images. This information is taken from Table 3.6 of Shoukri[14].

The analysis will use the sample size as weights for the weighted Kappa (Equation 2.8). I grouped the data into three strata: the 1–10 mm group, the

TABLE 2.11

Weighted Kappa for Stratified Analysis Weighted by Sample Size

Parameter	Mean	Median	SD	95% Credible Interval
Kappa 1	0.515	0.516	0.057	(0.398,0.621)
Kappa 2	0.350	0.353	0.111	(0.117,0.575)
Kappa 3	0.195	0.184	0.187	(−0.133,0.589)
Weighted Kappa	0.452	0.453	0.049	(0.350,0.545)

11–20 mm group, and the ≥ 21 mm group, with stratum sizes of 208, 68, and 24, respectively. The posterior analysis of the three Kappa parameters is done with Minitab and 1000 generated observations from the posterior distribution, assuming uniform prior information. The posterior distribution of the Kappa parameter of the three strata and the overall weighted Kappa appears in Table 2.11.

There is a large variation between the three Kappas!

2.7 Weighted Kappa

A weighted Kappa was introduced earlier to perform a stratified analysis, however a weighted Kappa is important when ordinal scores are being assigned to subjects. The basic idea is to assign a weight to each cell of the Table where the weight signifies the importance attached to the degree of difference assigned by the raters. Consider the example of the two psychiatrist assigning a degree of depression to patients (Table 2.12).

When the two agree a larger weight is assigned to that cell compared to a cell where they disagree, that is, the Kappa value should account for the weights assigned to the cells as:

$$\kappa_w = \sum_{i=1}^{i=k}\sum_{j=1}^{j=k}\omega_{ij}(\theta_{ij}-\theta_{i.}\theta_{.j})/(1-\sum_{i=1}^{i=k}\sum_{j=1}^{j=k}\omega_{ij}\theta_{i.}\theta_{.j}) \qquad (2.9)$$

where ω_{ij} is the weight assigned to the ij-th cell of the Table. When Cohen[15] introduced this index, he required

$$\text{(a) } 0\leq\omega_{ij}\leq 1$$

and $\qquad\qquad (2.10)$

$$\text{(b) } \omega_{ij} \text{ be a ratio.}$$

TABLE 2.12

Agreement for Depression

Psychiatrist 1	Psychiatrist 2			Total
	1	2	3	
1	11 (1)	2 (0.5)	19 (0)	32
2	1 (0.5)	3 (1)	3 (0.5)	7
3	0 (0)	8 (0.5)	82 (1)	90
Total	12	13	104	129

TABLE 2.13

Posterior Analysis for Weighted Kappa

Parameter	Mean	SD	2½	Median	97½
agree	0.7826	0.0314	0.7177	0.7838	0.8405
cagree	0.6451	0.0310	0.5854	0.6447	0.7068
d	0.1375	0.0310	0.0799	0.1363	0.2014
κ_w	0.3866	0.0764	0.2366	0.387	0.5354

By the latter it means that if one weight is given a value of 1, then a weight of ½ is a value that is half that of the former in value. Also, note that when $\omega_{ii} = 1$ for all i, and $= 0$ otherwise, the weighted Kappa reduces to the usual Kappa parameter.

Von Eye and Mun assign weights enclosed in parentheses in Table 2.12, thus if the two psychiatrists agree the assigned weight is 1, while if they disagree by exactly one unit, the assigned weight is 0.5, and for all other disagreements, the assigned weight is 0.

In order to execute a Bayesian analysis with BUGS, I used a burn in of 5000 observations, generated 50,000 observations from the posterior distribution of the parameters, and used a refresh of 100. The posterior analysis is given in Table 2.13.

The posterior mean of 0.3866 differs from the estimated Kappa of 0.402 stated by Von Eye and Mun, however they are still very close. Also, the posterior mean of the weighted Kappa differs from a value of 0.375 for the unweighted Kappa of Von Eye and Mun. The BUGS statements below will perform the posterior analysis reported in Table 2.13.

BUGS CODE 2.2

Model;

```
{
g[1,1]~dgamma(12,2); g[1,2]~dgamma(3,2); g[1,3]~dgamma(20,2);
g[2,1] ~dgamma(2,2); g[2,2] ~dgamma(4,2); g[2,3]~dgamma(4,2);
g[3,1]~dgamma(1,2); g[3,2]~dgamma(9,2); g[3,3]~dgamma(83,2);
h <- sum(g[,]);
```

```
for( i in 1 : 3 ) {for( j in 1 :3 ){ theta[i,j] < -g[i,j]/h }}
theta1. < - sum(theta[1,]);  theta.1 < -sum(theta[,1]);
theta2. < - sum(theta[2,]);  theta.2 < -sum(theta[,2]);
theta3. < -sum(theta[3,]);  theta.3 < - sum(theta[,3]);
kappa < - (agree-cagree)/(1-cagree);
agree < -
```

w[1,1]*theta[1,1]+w[1,2]*theta[1,2]+w[1,3]*theta[1,3]+w[2,1]*theta[2,1]+
w[2,2]*theta[2,2]+w[2,3]*theta[2,3]+w[3,1]*theta[3,1]+w[3,2]*theta[3,2]+
w[3,3]*theta[3,3];

cagree < -

w[1,1]*theta1.*theta.1 + w[1,2]*theta1.*theta.2 + w[1,3]*theta1. *theta.3 +
w[2,1]*theta2.*theta.1 + w[2,2]*theta2.*theta.2 + w[2,3] *theta2.*theta.3 +
w[3,1]*theta3.*theta.1 + w[3,2]*theta3.*theta.2 + w[3,3]* theta3.*theta.3;

```
d <-agree-cagree ;

}
list(g = structure(.Data = c(2,2,2,2,2,2,2,2,2), .Dim = c(3,3)))
list(w = structure(.Data = c(1,.5,0,.5,1,.5,0,.5,1), .Dim = c(3,3)))
```

Note that the first list statement contains the initial values of the *g* vector of gamma values, while the second include the data for the vector of weights *w*.

Obviously, the assignment of weights in Equation 2.10 is somewhat arbitrary, resulting in many possible estimates of weighted Kappa for the same data set. One set of weights advocated by Fleiss and Cohen[16] are

$$\omega_{ij} = 1 - (i-j)^2/(k-1)^2, \quad i, j = 1, 2, \dots, k \tag{2.11}$$

where *k* is the number of scores assigned by both raters. With this assignment of weights to the cells, weighted Kappa becomes an intraclass Kappa, a measure of agreement that will be explained in the following section, along with some additional assumptions that are needed to define the index.

2.8 Intraclass Kappa

The intraclass correlation for the one-way random model is well known as an index of agreement between readers assigning continuous scores to subjects, however, the intraclass Kappa for binary observations is not as well known. Intraclass Kappa is based on the following probability model:

Let

$$P(X_{ij} = 1) = \pi \tag{2.12}$$

for all $i = 1, 2,..., c$, and $j = 1, 2,..., n_i$, $i \neq j$, where X_{ij} is the response of the j-th subject in the i-th group, and suppose the responses from different groups are independent; and assume the correlation between two distinct observations in the same group is ρ.

Thus, the intraclass correlation measures a degree of association. Other characteristics of the joint distribution are:

$$P(X_{ij} = 1, X_{il} = 1) = \theta_{11} = \pi^2 + \rho\pi(1-\pi)$$

$$P(X_{ij} = 0, X_{il} = 0) = \theta_{00} = (1-\pi)[(1-\pi) + \rho\pi] \qquad (2.13)$$

$$P(X_{ij} = 0, X_{il} = 1) = P(X_{ij} = 1, X_{il} = 0) = \pi(1-\pi)(1-\rho) = \theta_{01} = \theta_{10}$$

where $j \neq l$, $i = 1, 2,..., c$, and $j = 1, 2,..., n_i$.

Note this model induces some constraints on the joint distribution, where the two probabilities of disagreement are the same, and the marginal probabilities of all raters are the same. For this model, the subjects play the role of raters, and there may be several raters, and the raters may differ from group to group. Also, all pairs of raters for all groups have the same correlation! Table 2.14 shows the relation between the θ_{ij} parameters and π and ρ determined by Equation 2.13.

Under independence $\rho = 0$, the probability of chance agreement is $\pi^2 + (1-\pi)^2 = 1 - 2\pi(1-\pi)$, thus a Kappa type index is given by

$$\kappa_I = \{[1 - 2\pi(1-\pi)(1-\rho)] - [1 - 2\pi(1-\pi)]\} / \{1 - [1 - 2\pi(1-2\pi)]\} \qquad (2.14)$$

$$= \rho$$

The likelihood function as a function of $\theta = (\theta_{00}, \theta_{01}, \theta_{11})$ is

$$L(\theta) \propto \theta_{00}^{n_{00}} \theta_{01}^{(n_{01} + n_{10})} \theta_{11}^{n_{11}} \qquad (2.15)$$

where the n_{ij} are the observed cell frequencies. This is the form of a multinomial experiment with four cells, but with three cell probabilities, because $\theta_{01} = \theta_{10}$.

As an example consider the study in Table 2.15 with five groups and five subjects per group. What is the common correlation between the subjects? The corresponding Table is Table 2.16.

TABLE 2.14

Cell Probabilities for Intraclass Kappa

	0	1
0	$(1-\pi)^2 + \rho\pi(1-\pi) = \theta_{00}$	$\pi(1-\pi) + (1-\rho) = \theta_{01}$
1	$\pi(1-\pi) + (1-\rho) = \theta_{10}$	$\pi^2 + \rho\pi(1-\pi) = \theta_{11}$

TABLE 2.15

Five Groups and 15 Subjects

Group	Response
1	1,1,1,1,1
2	0,1,0,1,0
3	1,1,0,0,1
4	1,1,1,0,0
5	0,0,1,1,1

TABLE 2.16

Cell Frequencies

Intraclass Kappa	$X=0$	$X=1$	Total
$X=0$	6	11	17
$X=1$	13	20	33
Total	19	31	50

The Table follows by counting the number of pairs within a group where $x_{ij}=k$, $x_{il}=m$, where k, $m=0$ or 1, $i=1, 2, 3, 4, 5$, and j, $l=1, 2, 3, 4, 5$ and $j\neq l$. By definition

$$\kappa_I = \text{cov}(X_{ij}, X_{il})/\sqrt{\text{Var}(X_{ij})\text{Var}(X_{il})} \tag{2.16}$$

$$\text{Cov}(X_{ij}, X_{il}) = E(X_{ij}X_{il}) - E(X_{ij})E(X_{il})$$

$$= \theta_{11} - \pi^2$$

Also,

$$\text{Var}(X_{ij}) = \pi(1-\pi)$$

Thus,

$$\kappa_I = (\theta_{11} - \pi^2)/\pi(1-\pi) \tag{2.17}$$

Using a uniform prior for the cell probabilities, the following BUGS CODE 2.3 presents the posterior distribution of the interclass Kappa, and the statement for Kappa is from Equation 2.17.

BUGS CODE 2.3
model

```
{
phi~dbeta(1,1)
for( i in 1 : 5) {
for( j in 1 : 5 ) {
Y[i , j] ~ dbern(phi)}}
theta11~dbeta(20,30)
kappa < -(theta11-phi*phi)/phi *(1-phi)
}
list(Y= structure( .Data=c(1,1,1,1,1,
0,1,0,1,0,1,1,0,0,1,1,1,1,0,0,
0,0,1,1,1), .Dim=c(5,5)))
list(theta11=.5 )
```

The posterior mean of the intraclass Kappa is 0.02546 (0.0995), median −0.00137, and a 95% credible interval of (−0.0853, 0.2883), indicating a weak agreement between pairs of subjects of the groups. The posterior mean (SD) of π is 0.6295 (0.0911). The first list statement is the data from Table 2.15, while the second list statement defines the starting values of θ_{11}. Note that θ_{11} is given a beta(20,30) posterior density, which implies the prior for θ_{11} is beta(0,0), namely

$$g(\theta_{11}) = \theta_{11}^{-1}(1-\theta_{11})^{-1}$$

an improper distribution. 25,000 observations are generated from the posterior distribution, with a burn in of 1000, and a refresh of 100.

This is a hypothetical example with equal group sizes, often, however, the groups differ in size, such as in heritability studies, where the groups correspond to litters and the groups consist of litter mate offspring that share common traits. The binary outcomes designate the presence or absence of the trait.

The development for the intraclass Kappa is based on the constant correlation model described by Mak[17]. The constant correlation model is a special case of the beta binomial model, which will be used for the following example. See Kupper and Haseman[18] and Crowder[19] for some non Bayesian approaches using the beta binomial model as the foundation. Other sources for information about intraclass Kappa type indices are Banerjee, Capozzoli, McSweeney, and Sinha[20], and Bloch and Kraemer[21].

Paul[22] presents an example where the intraclass Kappa can be used to estimate the agreement between pairs of fetuses affected by treatment.

There are four treatment groups, and several litters are exposed to the same treatment, where each litter consists of live Dutch rabbit fetuses, among which a number respond to the treatment. What is the correlation between the binary responses (yes or no) to treatment? For each treatment group, an intraclass Kappa can be estimated, then the four compared. In order to do this, two models will be the basis for the analysis: (a) the constant correlation model seen earlier; and (b) the beta-binomial model, which is a generalization of the constant correlation model. Paul reported Table 2.17 for the study.

The 1 designates the control group, $2 =$ low dose, $3 =$ medium dose, and $4 =$ high dose. In the high dose group, there are 17 litters, where the first litter has nine live fetuses and one responding to treatment, while the second litter of ten has none responding. For this treatment group: $n_{00} = 135$, $(n_{01} + n_{10}) = 78$, $n_{11} = 17$, for a total of 230.

The **BUGS CODE 2.3** was revised and executed to give Table 2.18, which provides the posterior analysis of the intraclass Kappa for the high dose treatment group.

Thus, there is a low correlation between responses (yes or no) of live fetuses in the same litter, with a 95% credible interval of $(-0.0752, 0.3429)$, and the posterior mean of π estimates the probability that a fetus will respond to a high dose as 0.22 (0.04).

The Bayesian results can be compared to the conventional values calculated from Equation 2.4 in Shoukri[14], and are 0.081525 for the estimated intraclass correlation, with an estimated standard deviation of 0.071726. Also, the conventional estimate of π is 0.2434, thus the Bayesian results differ, but there is very little difference. In fact the posterior median of 0.0766 is very close to the conventional estimate.

2.9 Other Measures of Agreement

The two standard references on measures of agreement are Shoukri[14] and Von Eye and Mun[12], and their presentations have a lot in common, but they also differ is some respects. Shoukri gives a very complete account of the many measures of agreement in Chapter 3, while Von Eye and Mun put heavy emphasis on the log-linear model, which is barely mentioned by Shoukri. The majority of the material in this section is based mostly on Shoukri, with the exception of the Kendall measure (*W*) of concordance for ordinal scores, and our presentation will be based on Von Eye and Mun. Of course, their developments are not Bayesian, but they give many examples, from which we draw on to do the Bayesian counterpart.

The *G* coefficient, Jacquard index, the concordance ratio, and the Kendall concordance coefficient *W* will be the subject of this section. The *G* coefficient

TABLE 2.17

Response to Therapy of Dutch Rabbit Fetuses

Group		Values
1	x	1 1 4 0 0 0 0 0 1 0 2 0 5 2 1 2 0 1 0 0 0 0 3 2 4 0
	n	12 7 6 8 10 7 8 6 11 7 9 8 9 7 2 9 7 11 10 4 8 10 12 8 7 8
2	x	0 1 1 0 2 0 3 0 0 1 0 0 3
	n	5 11 7 12 1 3 0 1 6 0 5 0 3
3	x	2 3 2 1 2 3 0 4 0 6 0 6
	n	4 9 1 7 8 0 4 6 6 5 17 4 13 8 6 3 11 1
4	x	1 0 1 0 1 2 0 4 1 1 4 3 1 6
	n	9 10 7 5 6 3 8 6 4 4 3 8 6 8

Source: Paul, S. R. 1982. Analysis of proportions of affected fetuses in teratological experiments. *Biometrics.* 38, 361.

TABLE 2.18

Posterior Analysis for Dutch Rabbit Fetuses

Parameter	Mean	SD	2½	Median	97½
κ_I	0.09075	0.1063	−0.0752	0.0766	0.3429
π	0.2262	0.0403	0.1521	0.2245	0.3103
θ_{11}	0.074	0.0172	0.0435	0.0727	0.1110

and the Jacquard index are not well known, and are very rarely used in agreement studies, but do have some attractive properties. Of the four, the Kendall concordance W is probably the most recognized, while the concordance ratio is also not well known. Very little is known about the Jacquard coefficient, but is given by Shoukri as

$$J = \theta_{11}/(\theta_{11} + \theta_{01} + \theta_{10}) \tag{2.18}$$

an index that ignores the (0,0) outcome, and is the proportion of (1,1) outcomes among the three cell probabilities. I cannot find any references to this coefficient, but it has the property that $0 \le J \le 1$, and if there is perfect agreement $J = 1$, but if there is perfect disagreement $J = 0$.

Shoukri reports that J is estimated by

$$\tilde{J} = n_{11}/(n_{11} + n_{01} + n_{10}) \tag{2.19}$$

with an estimated variance of

$$\text{Var}(\tilde{J}) = \tilde{J}^2(1 - \tilde{J})/n_{11} \tag{2.20}$$

Another adjusted index is the G coefficient defined as

$$G = (\theta_{00} + \theta_{11}) - (\theta_{01} + \theta_{10}) \tag{2.21}$$

and is estimated by

$$\tilde{G} = [(n_{00} + n_{11}) - (n_{01} + n_{10})]/n, \quad -1 \le G \le 1 \tag{2.22}$$

with an estimated variance of $(1 - \tilde{G}^2)/n$.

Also note that $G = 1$ if the agreement is perfect and if there is perfect disagreement $G = -1$.

There are many ways of measuring agreement, and still another is the concordance ratio

$$C = (1/2)[P(X_2 = 1/X_1 = 1) + P(X_1 = 1/X_2 = 1)]$$

$$= [\theta_{11}/(\theta_{10} + \theta_{11})] + \theta_{11}/(\theta_{01} + \theta_{11})]/2, \quad 0 \le C \le 1 \tag{2.23}$$

which is estimated by

$$\tilde{C} = [n_{11}/(n_{11} + n_{01}) + n_{11}/(n_{10} + n_{11})/2 \qquad (2.24)$$

C is the average of two conditional measures of agreement, and $C=1$ if there is perfect agreement and is equal to 0 if $\theta_{11}=0$. Such measures of agreement are not chance corrected like Kappa and are not as popular, but they take on values that are easy to interpret. The G coefficient is easy to generalize to more than two scores, however it is not obvious how to do it for the Jacquard and concordance ratio.

The estimates and their estimated variances depend on the sampling plan of the 2×2 Table 2.19, which is assumed to be multinomial, that is, the n subjects are selected at random from a well defined population and fall into the four categories with the probabilities indicated in the cells.

Suppose the example of Table 2.1 is reconsidered, and that the agreement between the two raters is estimated with Kappa, the Jacquard and G coefficients, the concordance ratio, and their Bayesian counterparts.

The posterior analysis for these four measures of agreement are contained in Table 2.20. These can be compared with the non Bayesian estimates: For J, the estimate is 0.3846 with a standard error of 0.0549 via Equation 2.19 and Equation 2.20, while 0.04 is the estimate of the G coefficient with a standard error of 0.0999, via Equation 2.22. The conventional estimate of the concordance ratio is 0.5713 compared to the posterior mean of 0.5714, the same to three decimal places.

TABLE 2.19

2×2 Outcomes

	Rater 2		
Rater 1	$X_2=0$	$X_2=1$	
$X_1=0$	(n_{00}, θ_{00})	(n_{01}, θ_{01})	$(n_{0.}, \theta_{0.})$
$X_1=1$	(n_{10}, θ_{10})	(n_{11}, θ_{11})	$(n_{1.}, \theta_{1.})$
	$(n_{.0}, \theta_{.0})$	$(n_{.1}, \theta_{.1})$	$(n, 1)$

TABLE 2.20

Posterior Analysis of Four Indices of Agreement

Parameter	Mean	SD	2½	Median	97½
G	0.0325	0.0469	−0.0612	0.033	0.123
J	0.3845	0.0550	0.2802	0.3828	0.4955
C	0.5714	0.0559	0.4573	0.5726	0.6767
Kappa	0.0636	0.0919	−0.1199	0.0643	0.2415

As for Kappa, the estimate is 0.0637, which is almost the same as the posterior mean of 0.0636. The Bayesian and conventional estimates are approximately the same, but they differ because the four measure different properties of agreement. The G and Kappa indicate very poor agreement, but Kappa is adjusted for chance agreement and G adjusts raw agreement by non agreement.

The Bayesian analysis is based on the following:

BUGS CODE 2.4

```
g00~dgamma(a00,b00); g01~dgamma(a01,b01);
g10~dgamma(a10,b10); g11 ~dgamma(a11,b11);
h<- g00+g01+g10+g11; theta00 <-g00/h;
theta01 <-g01/h; theta10 <-g10/h; theta11 <-g11/h;
theta0. <-theta00+theta01; theta.0 <-theta00+theta10;
theta1. <-theta10+theta11; theta.1 <-theta01+theta11;
kappa <- (agree-cagree)/(1-cagree); g <-agree-cagree;
c<- (1/2)*(theta11/(theta11+theta01) +theta11/(theta11+theta10));
J<-theta11/(theta11+theta10+theta01); agree <-theta11+theta00;
                 cagree <-theta1.*theta.1+theta0.*theta.0;

}
list( a00=22, b00=2, a01=15, b01=2, a10=33, b10=2, a11=30, b11=2)
list( g00=2, g01=2, g10=2, g11=2)
```

which is executed with 25,000 observations, 1000 for the burn in, and a refresh of 100. An improper prior

$$g(\theta) \propto \prod_{i,j=1}^{i,j=2} \theta_{ij}^{-1} \tag{2.25}$$

is also implemented for the Bayesian analysis, which implies that the Bayesian and conventional analyses will provide very similar estimates and standard errors.

2.10 Agreement with a Gold Standard

When a gold standard is present, two or more raters are compared with reference to the gold standard, not to each other. If the raters are judging the severity of disease, the gold standard provides an estimate of disease prevalence, and the effect of prevalence on measures of agreement can be explored, but there are situations where raters are judging disease and no gold standard is available. Judges and readers need to be trained, and frequently the gold standard provides the way to do this efficiently. However, in other cases, such as in gymnastics, judges are trained with reference to more experienced people, because a gold standard is not available.

Consider a case where two radiologists are being trained with ultrasound to diagnose prostate cancer, where they either identify the presence of disease or no disease. The lesion tissue from the patient is sent to pathology where the disease is either detected or declared not present. The pathology report serves as the gold standard. The same set of 235 lesions is imaged by both, then how should the two be compared? One approach is to estimate a measure of agreement between the radiologist and pathology for both radiologists and compare them, as in Table 2.21a and b.

With the improper prior density (Equation 2.25), the Bayesian analysis based on the two Tables (Table 2.21a and b), and executed with WinBUGS, the output is reported in Table 2.22.

There is an obvious difference in the agreement between the two radiologists relative to pathology based on Kappa and the G coefficient. The second radiologist is a beginner and is a very inexperienced resident in diagnostic radiology.

Another approach is to estimate the diagnostic accuracy of each radiologist with the sensitivity, specificity, or ROC area (area under the radio operating characteristic curve) of ultrasound, where pathology is the gold standard. If disease detected is the threshold for diagnosis, the sensitivity of ultrasound

TABLE 2.21a

Radiologist 1: Prostate Cancer with Ultrasound

Radiologist 1	Pathological Stage		
	No disease = 0	Disease = 1	Total
No disease = 0	85	25	110
Disease = 1	15	110	125
Total	100	135	235

TABLE 2.21b

Radiologist 2: Prostate Cancer with Ultrasound

Radiologist 2	Pathological stage		
	No disease = 0	Disease = 1	Total
No disease = 0	2	85	87
Disease = 1	98	50	148
Total	100	135	235

TABLE 2.22

Posterior Analysis for Two Radiologists

Parameter	Mean	SD	2½	Median	97½
κ_1	0.6555	0.0492	0.5541	0.6575	0.7458
κ_2	−0.6156	0.04101	−0.3389	−0.2971	−0.2527
G_1	0.3237	0.0246	0.2731	0.3245	0.3692
G_2	−0.2968	−0.0221	−0.3389	−0.2972	−0.2527

with the first radiologist is 0.814 and 0.370 with the second, and one would conclude that the first radiologist is doing a much better job than the second. In fact, the first radiologist is a very experienced diagnostician. What is the effect of disease incidence on Kappa?

Shoukri[14] provides a value of Kappa, the agreement between the two radiologists, based on disease incidence, sensitivity, and specificity. The dependence of Kappa on these measures of test accuracy were formulated by Kraemer[23] and later by Thomson and Walter[24] and is given by

$$\kappa_{tw} = 2\theta(1-\theta)(1-\alpha_1-\beta_1)(1-\alpha_2-\beta_2)/[\pi_1(1-\pi_1)+\pi_2(1-\pi)] \qquad (2.26)$$

where θ, as determined by the gold standard, is the proportion with disease, π_i the proportion having disease according to rater i, $1-\alpha_i$ the sensitivity of rater i, and $1-\beta_i$ the specificity of rater i. Based on Tables 2.21a and b, a quick calculation for an estimate of κ_{tw} is −0.65667, thus the two radiologists have very poor agreement (relative to each other), which in view of Table 2.22 (they had poor agreement relative to the gold standard) is not a surprise. It should be noted that Equation 2.26 is often not relevant because a gold standard is not always available.

The dependence of Kappa on disease incidence shows that the two raters can have high sensitivity and specificity, but nevertheless Kappa can be small. It all depends on the disease incidence, which if small, can produce a low value of κ_{tw}. A similar phenomenon occurs in diagnostic testing, when a test can have high sensitivity and specificity, but because of low disease incidence, the positive predictive value is low.

How should the Bayesian analysis for estimating κ_{tw} be conducted? One approach is to treat the scores of the two radiologists as independent and find its posterior distribution based on Equation 2.26, where θ is the incidence of disease estimated from either Table, π_i the proportion having disease according to rater i. Also, $1-\alpha_1$, the sensitivity of rater 1, is estimated by θ_{11}/θ_1 (see Table 2.21a) The specificity of rater 2, $1-\beta_2$, is estimated by $\phi_{10}/\phi_{.0}$ etc. The posterior distribution of all the terms in Equation 2.26 and consequently for κ_{tw}, is determined from the following:

BUGS CODE 2.5

model;

```
{

    g00~dgamma(a00,b00);  g01~dgamma(a01,b01);    g10~dgamma(a10,b10);
    g11 ~dgamma(a11,b11);

    h <- g00+g01+g10+g11 ; theta00 <-g00/h;

    theta01 <-g01/h; theta10 <-g10/h; theta11 <-g11/h;

    theta0. <-theta00+theta01; theta.0 <-theta00+theta10;

    theta1. <-theta10+theta11; theta.1 <-theta01+theta11;

    p00~dgamma(c00,d00);  p01~dgamma(c01,d01);    p10~dgamma(c10,d10);
    p11 ~dgamma(c11,d11);

    q <-p00+p01+p10+p11;  phi00 <-p00/q;

    phi01 <-p01/q;  phi10 <-p10/q;  phi11 <-p11/q;

    phi0. <-phi00+phi01;  phi.0 <-phi00+phi10;

    phi1. <-phi10+phi11;  phi.1 <-phi01+phi11 ;

    kappatw <-k11/k12;   k11 <-2*(theta.1)*(1-theta.1)*(1-theta01/theta.1-
    theta10/theta.0)*(1-phi01/phi.1-phi10/phi.0);

    k12 <-theta1.*(1-theta1.)+ phi1.*(1-phi1.);

}
```

list(a00=85, b00=2, a01=25, b01=2, a10=15, b10=2, a11=110, b11=2,

    c00=2, d00=2, c01=85, d01=2, c10=98, d10=2, c11=50, d11=2)

list(g00=2, g01=2, g10=2, g11=2, p00=2, p01=2, p10=2, p11=2)

The theta parameters refer to Table 2.21a, while the phi parameters correspond to the entries of Table 2.21b, and the Kappa parameter κ_{tw} is referred to as "kappatw" in the above statements. I computed the posterior mean of κ_{tw} as −0.4106 (0.04013), a median of −0.4105, and a 95% credible interval of (−0.49, −0.3325), indicating very poor agreement between the two radiologists. Utilizing 35,000 observations generated from the posterior distribution of κ_{tw}, with a burn in of 1000, and a refresh of 100, produces Figure 2.3, a graph of the posterior density of κ_{tw}.

2.11 Kappa and Association

Obviously agreement is a form of association between raters, and the word association is ubiquitous in the statistical literature. It can be argued that the goal of most, if not all, statistical methods is to demonstrate an association or

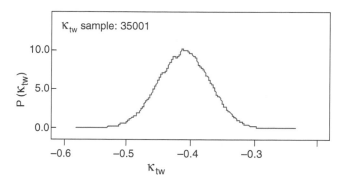

FIGURE 2.3
Posterior density of κ_{tw}.

lack thereof between various experimental factors. By itself, the term association is often used in an informal way. In the context of a 2×2 Table, the way to test for no association is to test the hypothesis that the marginal probabilities of the raters are the same. Our last version of Kappa is in the context of association, when the marginal distributions of the two raters are not the same as represented in Table 2.23. Where, $\theta_1 = P(X_1 = 1)$, $\theta_2 = P(X_2 = 1)$, and $\phi_i = 1 - \theta_i$, $i = 1, 2$. It can be shown that:

$$\gamma = \rho\sqrt{\theta_1\theta_2\phi_1\phi_2} \tag{2.27}$$

where ρ is the correlation between X_1 and X_2, and the chance corrected Kappa under the above distribution for the two raters is

$$\kappa_a = 2\rho\sqrt{\theta_1\theta_2\phi_1\phi_2}/[\theta_1\phi_2 + \theta_2\phi_1] \tag{2.28}$$

This derivation of κ_a is found in Shoukri[14], who references Bloch and Kraemer[21], who give a quite complete account of measures of agreement and association.

Note that

$$\rho = (\lambda_{11} - \theta_1\theta_2)/\sqrt{\theta_1\phi_1\theta_2\phi_2} \tag{2.29}$$

where $\theta_1 = \lambda_{1.}$, $\theta_2 = \lambda_{.1}$.

TABLE 2.23

Cell Probabilities of Two Raters

	Rater 2	
Rater 1	$X_2 = 0$	$X_2 = 1$
$X_1 = 0$	$\phi_1\phi_2 + \gamma = \lambda_{00}$	$\phi_1\theta_2 - \gamma = \lambda_{01}$
$X_1 = 1$	$\phi_2\theta_1 - \gamma = \lambda_{10}$	$\theta_1\theta_2 + \gamma = \lambda_{11}$

Under a multinomial sampling scheme for the 2×2 Table of outcomes, the cell probabilities have a Dirichlet $(n_{00}+1, n_{01}+1, n_{10}+1, n_{11}+1)$ distribution under a uniform prior density. This induces a prior distribution for the θ_i and hence for the correlation coefficient ρ and also for γ, thus these distributions should be determined and examined to see if they are compatible with the investigator's prior information about these particular parameters.

Suppose the outcomes for two raters are represented by Table 2.24, where the probability of a success is estimated with rater 2 as $25/125 = 0.2$ and $90/125 = 0.72$ for rater 1, an obvious case of no association between the two raters. What is the posterior distribution of κ_a? I generated 15,500 observations from the joint posterior distribution of κ_a and ρ, and a burn in of 500, with a refresh of 500 and computed Table 2.25.

In order to determine if there is an association between the two raters, the posterior distribution of $d = \lambda_{1.} - \lambda_{.1}$ should be determined. From Table 2.25, the posterior means of κ_a and ρ imply a weak agreement between the two readers, and the posterior analysis reveals that there is no or very little association between the two. It can be verified that the posterior distribution of the usual Kappa is the same as κ_a.

Shoukri also states the MLE of κ_a is

$$\kappa_a = 2(n_{11}n_{00} - n_{10}n_{01})/[(n_{11}+n_{10})(n_{00}+n_{10}) + (n_{11}+n_{10})(n_{00}+n_{10})],$$

thus, using the information from Table 2.24 determines an estimate of 0.0506, which is almost the same as the posterior mean.

TABLE 2.24

Kappa in the Context of Association

	Rater 2		
Rater 1	0	1	Total
0	30	5	35
1	70	20	90
Total	100	25	125

TABLE 2.25

Posterior Analysis for κ_a

Parameter	Mean	SD	2½	Median	97½
κ_a	0.0503	0.0472	−0.0476	0.0516	0.1406
ρ	0.0883	0.0799	−0.0839	0.0932	0.2294
d	0.5205	0.0509	0.4183	0.522	0.6168

2.12 Consensus

Many agreement situations involve a method to come to a conclusion. For example, in boxing the rules of the contest determine a definite conclusion about the outcome of the contest. If the bout goes the limit without a knock-out or technical knockout, one of three things can occur: (1) fighter A wins; or (2) fighter B wins; or (3) the contest is ruled a draw. Or, as in a jury trial, each juror has an opinion as to guilt or innocence, and one of three outcomes will happen: (a) the defendant is guilty; or (b) the defendant is innocent; or (c) there is hung jury.

Still another example is Figure skating, where the skaters perform at three events (compulsory Figures, the short program, and the long) and for each the skater receives a score from a panel of judges. Then finally each receives a final score and the winner is the one with the highest total score. How they are scored is governed by the rules of skating association, and they can be quite involved, however; there is a mechanism to come to a consensus.

Reviewing and editing papers submitted to a scientific journal involves coming to a consensus about the acceptance of the paper for publication. The consensus revolves around the editor who receives input from a panel of reviewers, and sometimes other editors, but somehow a final decision is made. The paper can be accepted with or without revision, rejected, etc., depending on the rules set down by the journal's editorial board.

In Phase II clinical trials, a panel of radiologists declare for each patient that the response to therapy is in one of the four categories: a complete response, a partial response, progressive disease, or stable disease. How does this panel resolve the individual differences between the radiologists? That is a good question!

There are some statistical approaches that study the way people come to a consensus. For example, DeGroot[25] addresses this issue and is briefly described as follows. A group of individuals must work jointly and each has their own subjective probability distribution about an unknown parameter. DeGroot describes how the group might come to a decision about a common probability distribution for the parameter by combining their individual opinions. The process is iterative, where each individual revises their own subjective probability distribution by taking into account the opinions of the other judges. The author describes how the iterative process might converge, how a consensus is reached, and how the process is related to the Delphi technique.

For additional information about statistical consensus methods, see Winkler[26]. Is the DeGroot method applicable to the Phase II trial example given above? If applied, apparently a common probability distribution among the panel of radiologists would be agreed upon. This distribution would place a probability on each of the four possible outcomes: a complete response, a partial response, progressive disease, or stable disease. This could be very helpful, because if one outcome had a high probability, presumably the panel would agree to label that particular outcome to the patient. On the other hand, if the consensus distribution is uniform, another mechanism is necessary in order to label the patients's outcome.

Exercises

1. Use the BUGS CODE 2.1 of Section 2.5 to confirm Table 2.5.
2. Use the **BUGS CODE 2.1** of Section 2.5 to find the posterior mean and standard deviation of the two conditional Kappas for responses 2 and 3. Use 25,000 observations, with a burn in of 1000.
3. Refer to Table 2.1 and suppose the results of a previous related experiment are given as Table 2.26. Combine this prior information with Table 2.1 and perform the appropriate analysis and compare with the posterior analysis of Table 2.2. Use 35,000 observations, with a burn in of 5000 and a refresh of 100. Also, plot the posterior densities of A_1 and A_2.

TABLE 2.26

Two Raters

Rater 1	Rater 2	
	$Y=0$	$Y=1$
$X=0$	2	1	3
$X=1$	3	22	25
	5	23	28

4. Show the chance corrected value of S_D (Equation 2.1) is Kappa.
5. Refer to Section 2.5 and determine the posterior distribution of the conditional Kappas κ_2 and κ_3. What are the mean, standard deviation, median, and upper and lower 2½ percentiles of both? Use 50,000 observations, with 10,000 as a burn in with a refresh of 500. Employ a uniform prior density.
6. Refer to Table 2.6, Table 2.7, and Table 2.8 and perform a Bayesian analysis by estimating Kappa with a weighted average based on: (a) the sample size of the three sites; and (b) equal weights. Use 35,000 observations generated from the joint posterior distribution, and 2500 as a burn in, with a refresh of 100. How accurate is the posterior mean of the two weighted averages? You choose the prior density, but carefully specify it.
7. Refer to Table 2.9 for the posterior analysis of three Kappas corresponding to three lesion size ranges, and their weighted average, using the sample sizes as weights. The computations are done with Minitab: (a) Verify this Table with BUGS using 50,000 observations, with a burn in of 5000, a refresh of 100, and a uniform prior density;

(b) compute a weighted average based on the inverse of the posterior variance of each Kappa; (c) What weighted average is preferred and why? and; (d) Are your computations more accurate than those of Minitab? Why?

8. Refer to Table 2.9 and compare the three Kappas of the three strata. Test the hypothesis that $\kappa_1 = \kappa_2 = \kappa_3$ and justify your decision as to the outcome. If this hypothesis is rejected, why is it rejected? On the other hand if it is not rejected, why isn't it? If it is rejected, what is you conclusion about the three? In order to do the calculations, provide the BUGS code and execute the program with a burn in of 1000 observation, generate 20,000 observations from the joint posterior distribution, with a refresh of 100. Use a non-informative prior density.

9. Using **BUGS CODE 2.2**, verify Table 2.13.

10. Modify the **BUGS CODE 2.2**, and reanalyze the depression data of Table 2.12, but use the weights from Equation 2.11. Do not use a uniform prior, but instead assume the prior density is

$$g(\theta) \propto \prod_{i=1}^{i=3} \prod_{j=1}^{j=3} \theta_{ij}^{-1}$$

which is an improper prior density. Use a burn in of 1000 and generate 20,000 observations from the joint posterior distribution with a refresh of 50. Find the posterior mean, standard deviation, lower and upper 2½ percentiles, and the median. How accurate is your simulation? Compare your results to Table 2.13. Any comments?

11. How would one define a weighted conditional Kappa? Give a definition for the three responses of the depression information in Table 2.12 and estimate the three conditional kappas. What do they tell us about the agreement between the two psychologists? Use the prior density in problem 10, and modify the **BUGS CODE 2.2** accordingly, and state exactly what you did to execute WinBUGS!

12. Show that the intraclass correlation as defined by the constant correlation model, is indeed given by Equation 2.14.

13. Would it be appropriate to have the same raters for all groups in order to estimate interclass kappa? Why?

14. Verify the likelihood function, Equation 2.15.

15. Verify Equation 2.17.

16. Is there a lower bound to intraclass kaapa given by Equation 2.17? If so, what is it?

17. The BUGS statements that follow are based on the beta-binomial model. With 25,000 observations generated from the joint posterior distribution of theta11, π, and κ_I, with a burn in of 1000, and a refresh of 100, execute the statements below and estimate the three parameters. How do your results compare with those based on the constant correlation model? See Table 2.18.

BUGS CODE 2.7

model

  { for (i in 1:17) {phi[i]~dbeta(1,1)}

for(i in 1 : 17) {

for(j in 1 : 10) {

Y[i , j] ~ dbern(phi[i])}}

theta11~dbeta(17,213)

kappa < -(theta11-pi*pi)/pi*(1-pi)

 pi < -mean(phi[])   }

list(Y = structure(.Data =

c(1,0,0,0,0,0,0,0,0,NA,0,0,0,0,0,0,0,0,0,0,

1,0,0,0,0,0,0,NA,NA,NA,0,0,0,0,0,NA,NA,NA,NA, NA,

1,0,0,0,NA,NA,NA,NA, NA,NA,0,0,0,0,0,0,NA,NA, NA,NA,

1,0,0,NA,NA,NA,NA, NA,NA,NA,1,0,0,0,0,0,0,0,NA,NA,

1,1,0,0,0,0,0,0,NA,NA,0,0,0,0,NA,NA,NA, NA,NA,NA,

1,1,1,1,NA,NA,NA, NA,NA,NA,1,0,0,0,0,NA,NA, NA,NA,NA,

1,0,0,NA,NA,NA, NA,NA,NA,NA,1,1,1,1,0,0,0,0,NA,NA,

1,1,0,0,0,0,NA,NA,NA,NA,1,1,1,0,0,0,0,0,NA,NA,

1,0,0,0,0,0,NA,NA,NA,NA), .Dim = c(17,10)))

list(theta11 = .5)

18. Using the above program, estimate the intraclass kappa for the four treatment groups of Table 2.17. Do the four groups differ with respect to the correlation between fetuses, across groups. Explain in detail how the BUGS code is executed and carefully explain the posterior analysis for comparing the four groups. If there is no difference in the four kappas, how should they be combined in order to estimate the overall intraclass correlation?

19. Refer to Table 2.20 and give an interpretation about the degree of agreement to the four indices. What is your overall conclusion about the degree of agreement? Should only one index be used, or should all four be reported? Kappa is the usual choice as an index of agreement. In view of Table 2.20, is this a wise choice?

20. Refer to Table 2.22, and verify the Table, and determine the posterior distribution of the difference in the two kappas? Is there a difference between them? Based on the G coefficient, what is your interpretation of the degree of agreement between the radiologists and the gold standard? Test for a difference in $G1$ and $G2$.

21. In view of Equation 2.26, show that two raters can have high sensitivity and specificity, but κ_{tw} can be small.

22. Verify Equation 2.26. Refer to Kraemer[22] and Thompson and Walter[23], if necessary.
23. Verify Equation 2.28 and Equation 2.29.
24. If $\theta_1 = \theta_2$, what is the value of κ_a (Equation 2.28)?
25. Verify Table 2.25 with your own BUGS code. Verify the posterior distribution of the usual Kappa (Equation 1.1) is the same as that of κ_a, given by Table 2.25.

References

1. Zhou, H. H., D. K. McClish, N. A. Obuchowski. 2002. *Statistical models for diagnostic medicine.* New York: John Wiley & Sons.
2. Pepe, M. S. 2003. *The statistical evaluation of medical tests for classification and prediction.* Oxford UK: Oxford University Press.
3. Fleiss, J. L. 1975. Measuring agreement between two judges on the presence or absence of a trait. *Biometrics* 31(3), 641.
4. Goodman, L. A., and W. H. Kruskal. 1954. Measures of association for cross classification. *J. Am. Stat. Assoc.* 49, 732.
5. Dice, L. R. 1945. Measures of the amount of ecologic association and disagreement. *Ecology* 26, 297.
6. Rogot, E., and I. D. Goldberg. 1966. A proposed idea for measuring agreement in test re-test studies. *J. Chron. Dis.* 19, 991.
7. Armitage, P., L. M. Blendis, and H. C. Smyllie. 1966. The measurement of observer disagreement in the recording of signs. *J. Royal Stat. Soc., Series A* 129, 98.
8. Scott, W. A. 1955. Reliability of content analysis: The cases of nominal scale coding. *Public Opinion Quart.* 19, 321.
9. Cohen, J. 1960. A coefficient of agreement for nominal scales. *Educ. Psychol. Measmt.* 20, 37.
10. Coleman, J. S. 1966. Measuring concordance in attitudes, unpublished manuscript. Department of Social Relations, Johns Hopkins University.
11. Light, R. J. 1969. Analysis of variance for categorical data with applications for agreement and association. Dissertation. Harvard University, Department of Statistics.
12. Von Eye, A., and E. Y. Mun. 2005. *Analyzing rater agreement, manifest variable methods.* Mahwah, NJ, London: Lawrence Erlbaum Associates.
13. Liebetrau, A. M. 1983. *Measures of association.* Newbury Park: Sage.
14. Shoukri, M., M., 2004. *Measures of interobserver agreement.* Boca Raton: Chapman & Hall/CRC.
15. Barlow, W., M. Y. Lai, and S. P. Azen. 1991. A comparison of methods for calculating a stratified Kappa. *Statistics in Medicine* 10, 1465.
16. Fleiss, J. L., and J. Cohen. 1973. The equivalence of the weighted kappa and the interclass kappa. *Psychological Bulletin* 72, 232.
17. Mak, T. K. 1988. Analyzing intraclass correlation for dichotomous variables. *Applied Statistics* 37, 344.

18. Kupper, L. L., and J. K. Haseman. 1978. The use of correlated binomial model for the analysis of certain toxicological experiments. *Biometrics* 25, 281.
19. Crowder, M. J. 1979. Inferences about the intraclass correlation coefficient in the beta-binomial ANOVA for proportions. *J. R. Stat. Soc.* B 41, 230.
20. Banerjee, M., M. Capozzoli, L. McSweeney, and D. Sinha. 1999. Beyond Kappa: A review of interrater agreement measures. *Can. J. Stat.* 27, 3.
21. Bloch, D. A., and H. C. Kraemer. 1989. 2 x 2 Kappa coefficients: Measures of agreement or association. *Biometrics* 45, 269.22.
22. Paul, S. R. 1982. Analysis of proportions of affected fetuses in teratological experiments. *Biometrics* 38, 361.
23. Kraemer, H. C. 1979. Ramifications of a population model for kappa as a coefficient of reliability. *Psychometrika* 44, 461.
24. Thompson, W. D., and S. D. Walter. 1988. A reappraisal of the kappa coefficient. *J. Clin. Epi.* 41(10), 949.
25. DeGroot, M. H. 1974. Reaching a consensus. *J. Am. Stat. Assoc.* 69, 118.
26. Winkler, R. L. 1969. The consensus of subjective probability distributions. *Management Science* 15, B61.

3

More than Two Raters

3.1 Introduction

With more than two raters, the determination of agreement becomes much more complicated. How should one measure agreement among three or more raters? One could use one overall measure like kappa in order to estimate perfect agreement, corrected for chance, or, one could estimate all pairwise agreements with kappa, then take an average of them. And as seen in the last chapter, one could estimate a common intraclass correlation between all the raters. Still another possibility with multiple raters, is to introduce the idea of partial agreement, say among six raters, and measure the agreement between five, four, and three raters. What is the role of conditional type kappa measures with multiple raters? When the scores are ordinal, how should the weights be defined for a weighted kappa?

It is obvious in some situations how to generalize the kappa for a 2×2 table to multiple scores and raters, but not so obvious in other cases.

The Bayesian approach to measure agreement is continued, and the chapter begins with several interesting examples. For example, six pathologists studying Crohn's disease are asked to identify (yes or no) lesions in 68 biopsy specimens. This example illustrates the complexity involved when there are multiple raters. Several issues not addressed earlier are considered with this example, including: (1) is there an overall general measurement of agreement for six readers? (2) how is the problem of partial agreement to be met? For example, how does one measure partial agreement between five of the six pathologists? (3) can one use a weighted kappa by averaging all pairwise kappas, and what is the optimal weighting scheme? (4) should a modeling approach be taken? The example is based on a study by Rogel, Boelle, and Mary[1], who cite Theodossi et al.[2] with an earlier analysis of the data.

Another example with several raters and a binary response serves to exemplify the complexity of analyzing such information. Shoukri[3], in his Table 4.4, presents an example of four veterinary students from a college in Ontario who are asked to identify foals that have a cervical vertebral malformation. They each examine 20 X-rays and score each as either affected, designated by 1, or not affected scored as a 0. Shoukri's analysis consists of testing for inter rater bias via Cochran's test and estimating the interclass correlation coefficient with the one-way ANOVA. The Bayesian approach will be compared to his conventional analysis and the approach extended in order to answer questions about partial agreement and kappa in the context of association.

A third example from Williams[4] involves multiple raters and outcomes taken on by the College of American Pathologists who conduct a reliability test with four laboratories and three classifications for each pathology specimen. How reliable is the testing of these laboratories? The labs are the raters, and the analysis is presented by Shoukri[3], who based his examination on Fleiss[5], who developed a kappa type index $\kappa(m, c)$, which was further studied by Landis and Koch[6] for its relation to intraclass kappa.

Chapter 3 continues with presentations about weighted kappa, conditional kappa, agreement with many raters and ordinal scores, stratified kappa, the intraclass kappa, a generalization of the G coefficient, agreement indices in the context of association, and an introduction of a model based approach to comparing raters with logistic and multinomial regression models. Thus Chapter 4 is seen as a generalization of Chapter 3, which greatly expands the scope of the Bayesian approach to the subject.

Finally, it should be pointed out that it is just as important to focus on disagreement between raters. Often more emphasis is placed on agreement, however in any given real problem, one will usually find more disagreement than agreement, especially so when multiple raters and multiple outcomes are encountered. What is a good index of disagreement?

3.2 Kappa with Many Raters

There are several ways to define a measure of overall agreement between multiple raters, and two approaches will be taken: (1) A generalization of kappa from 2×2 tables; and (2) an average of the kappas for all pairs of 2×2 tables. The details of both approaches and the issues that are raised follow.

For the first approach, suppose each of m raters assign scores to all of n subjects using a nominal response with c outcomes labeled 1, 2,..., c, then an overall kappa can be defined for $m = c = 3$ as

$$\kappa(3,3) = \left[\sum_{i=1}^{i=3} \theta_{iii} - \sum_{i=1}^{i=3} \theta_{i..} \theta_{.i.} \theta_{..i} \right] \bigg/ \left[1 - \sum_{i=1}^{i=3} \theta_{i..} \theta_{.i.} \theta_{..i} \right], \tag{3.1}$$

which is a chance corrected index for all three agreeing simultaneously at the three possible outcomes. Of course, it is obvious how to extend kappa to $\kappa(m, c)$ in general for m raters and c scores, where m is a positive integer at least three, and c is a positive integer greater than or equal to 2. It should be pointed out that this is just one of the possible ways to define an agreement measure for multiple raters, and this will be addressed in a later section. Note that θ_{ijk} is the probability that rater 1 assigns a score of i, rater 2 a score of j, and rater 3 a score of k, where $i,j,k = 1,2,3$. Assume for the moment that the n subjects are selected at random from a population so that the study has the structure of a multinomial experiment.

The first example to be presented is the assessment of cervical vertebral malformation of foals detected by four veterinary students who on the basis of 20 X-rays report either a 1 for not affected and a 2 for affected. These responses are reported in Table 3.1, and the appropriate kappa is

$$\kappa(4,2) = \left[\sum_{i=1}^{i=2} \theta_{iiii} - \sum_{i=1}^{i=2} \theta_{i...}\theta_{.i..}\theta_{..i.}\theta_{...i} \right] / \left[1 - \sum_{i=1}^{i=2} \theta_{i...}\theta_{.i..}\theta_{..i.}\theta_{...i} \right]. \quad (3.2)$$

The frequencies that are contained in the 16 categories are represented in Table 3.1.

Assuming an improper prior for the θ_{ijkl}, the probability that student A assigns an i, B a j, C a k, and D an l, the posterior distribution of these parameters is Dirichlet (3,0,2,0,3,0,0,0,0,1,1,0,2,1,0,7), and the posterior analysis for $\kappa(4, 2)$ is given in Table 3.2.

The program is executed with BUGS CODE 3.1 using 25,000 observations with a burn in of 1000 and a refresh of 100. The analysis estimates $\kappa(4, 2)$ with a posterior mean (SD) of 0.4777(0.1171) and an upper 97.5 percentile of 0.7010, and its posterior density is portrayed in Figure 3.1. The raw agreement of the

TABLE 3.1

Agreement of Four Students

Students				
A	B	C	D	Frequencies
1	1	1	1	3
1	1	1	2	0
1	1	2	1	2
1	2	1	1	0
2	1	1	1	3
1	1	2	2	0
1	2	1	2	0
2	1	1	2	0
1	2	2	1	0
2	1	2	1	1
2	2	1	1	1
2	2	2	1	0
2	2	1	2	2
2	1	2	2	1
1	2	2	2	0
2	2	2	2	7

TABLE 3.2

Posterior Analysis for Cervical Vertebral Malformation

Parameter	Mean	SD	2½	Median	97½
agree	0.498	0.1086	0.2884	0.4978	0.7086
cagree	0.0369	0.0185	0.0105	0.0342	0.0796
κ(4, 2)	0.4777	0.1171	0.2497	0.4792	0.7010
D	0.461	0.1168	0.2344	0.4619	0.6857

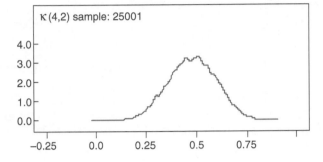

FIGURE 3.1
Posterior density of κ(4, 2).

four students, agreeing at the outcomes (1,1,1,1) or (2,2,2,2) is 0.498(0.1086) with the posterior mean (SD). A quick calculation based on Table 3.2 gives a raw agreement of $10/20 = 0.5$. The only way they can perfectly agree with the same score is at those two outcomes. The posterior mean (SD) of theta (1,1,1,1) = 0.1486 (0.0763) of theta(2,2,2,2) = 0.3486(0.1038), and of theta(1,1,1,2) is 0.0005297(0.00486). The latter is an empty cell and was given a posterior probability of 0.0005297. The simulation uses all 16 cells of the multinomial, see the BUGS code later, and an improper prior distribution for the parameters.

$$g(\theta) \propto \prod_{i,j,k,l=1}^{i,j,k,l=2} \theta_{ijkl}^{-1} \tag{3.3}$$

All 16 cells were used for the analysis, however there were eight empty cells, and the analysis could have been done with the eight non empty cells employing an uninformative prior such as Equation 3.3.

The parameter d is the difference between the probability of agreement and the probability of agreement by chance (assuming the four raters are independent) and implies that there is a difference in the two probabilities that comprise kappa 42. This parameter is defined as (agree − cagree)/(1 − cagree) from Equation 3.2.

What does this tell us about the agreement between the four? The majority of their agreement occurs at (2,2,2,2), where they designate seven images as indicating cervical vertebral malformation, which has a posterior mean (SD) = 0.3486(0.1038), while they next agree the most three times at (1,1,1,1) with a posterior mean (SD) = 0.1486(0.0767).

The BUGS code 3.1 is self explanatory. The gamma vector consists of gamma random variables and *h* is their sum, thus theta is the corresponding gamma divided by the sum *h*. The resulting vector theta has the required Dirichlet distribution. The list statement at the end of the code, gives the starting values for the gamma vector; and the κ(4, 2) of Equation 3.2 is listed as kappa42, the raw probability of agreement is listed as agree, while the chance probability of agreement between the four students is denoted by cagree, and their difference is listed as *d* below.

BUGS CODE 3.1

```
model;
  {
    # the following assumed an improper prior
    # instead of a 0 for the first scale parameter a .01 is assigned
    g[1,1,1,1] ~ 75dgamma(3,2); g[1,1,1,2] ~ dgamma(.01,2);
    g[1,1,2,1] ~ dgamma(2,2); g[1,2,1,1] ~ dgamma(.01,2);
    g[2,1,1,1] ~ dgamma(3,2); g[1,1,2,2] ~ dgamma(.01,2);
    g[1,2,1,2] ~ dgamma(.01,2); g[2,1,1,2] ~ dgamma(.01,2);
    g[1,2,2,1] ~ dgamma(.01,2); g[2,1,2,1] ~ dgamma(1,2);
    g[2,2,1,1] ~ dgamma(.01,2); g[2,2,2,1] ~ dgamma(.01,2);
    g[2,2,1,2] ~ dgamma(2,2); g[2,1,2,2] ~ dgamma(1,2);
    g[1,2,2,2] ~ dgamma(.01,2); g[2,2,2,2] ~ dgamma(7,2);
    h <-sum(g[,,,])
    for(i in 1:2){ for (j in 1:2) {for (k in 1:2){for (l in 1:2)
    {theta [i,j,k,l] <- g[i,j,k,l]/h}}}}
    atheta2... <-sum(theta[2,,,]); atheta.2.. <-sum(theta[,2,,]);
    atheta..2. <-sum(theta[,,2,]); atheta...2 <-sum(theta[,,,2]);
    atheta1... <-sum(theta[1,,,]); atheta.1.. <-sum(theta[,1,,]);
    atheta..1. <-sum(theta[,,1,]); atheta...1 <-sum(theta[,,,1]);

    kappa42 (agree-cagree)/(1-cagree)
    agree <- theta[1,1,1,1] + theta[2,2,2,2]
    cagree <- atheta1...*atheta.1..*atheta..1.*atheta...1+
              atheta2...*atheta.2..*atheta..2.*atheta...2
    d <-agree-cagree
  }
list(g = structure(.Data = c(2,2,2,2,2,2,2,2,2,2,2,2,2,2,2,2), .Dim = c(2,2,2,2)))
```

TABLE 3.3

Pairwise Agreement Between Four Students

Parameter	Mean	SD	2½	Median	97½
agree					
AB	0.7366	0.0990	0.5196	0.7463	0.9019
AC	0.5993	0.1068	0.3856	0.602	0.7985
AD	0.7495	0.0948	0.5419	0.7575	0.9069
BC	0.6509	0.1034	0.439	0.6567	0.8358
BD	0.8994	0.0652	0.7414	0.9121	0.9871
CD	0.7509	0.0948	0.5439	0.7598	0.9097
cagree					
AB	0.526	0.0702	0.4098	0.5145	0.6881
AC	0.5284	0.0606	0.4175	0.5214	0.6628
AD	0.5115	0.0591	0.412	0.5005	0.6532
BC	0.5073	0.0277	0.4584	0.5022	0.5782
BD	0.5191	0.0287	0.4902	0.5086	0.5997
CD	0.5119	0.0287	0.4687	0.5042	0.5885
kappa					
AB	0.4513	0.1641	0.1466	0.4472	0.7697
AC	0.1528	0.1868	−0.2052	0.148	0.5314
AD	0.4939	0.1585	0.1884	0.495	0.7993
BC	0.2937	0.1973	−0.1016	0.3009	0.6573
BD	0.7919	0.132	0.4766	0.8157	0.9732
CD	0.492	0.1841	0.0994	0.5051	0.8085

A second approach to overall agreement is now presented where an average kappa of the six kappas of all 2×2 tables corresponding to pairs of raters will provide another index of agreement (Table 3.3).

Consider the pair BD, then from Table 3.1 it can be confirmed that the four cells have frequencies: 1–1(9), 1–2(1), 2–1(1), and 2–2(9), that is the 1–1 cell has nine images that the two students agree are not affected, while the 2–2 cell also has nine images, indicating the two students B and D agree of evidence of cervical vertebral malformation, and so on, which allows for the determination of the posterior analysis of Table 3.3. The computations are based on 15,000 observations generated by WinBUGS, with a burn in of 1000, and a refresh of 100. A non-informative improper type prior density similar to Equation 3.3 was assumed for the parameters of the 2×2 tables for all pairs of students, and BUGS CODE 3.2 provides for the calculation of the posterior analysis.

The pairwise kappa varies from a low of 0.1528 for the AC pair to 0.7919 for the BD pair, which tells something about partial agreement, which will be explored in the next section. An overall kappa based on an average of the six pairwise values is another approach to the problem, but what

TABLE 3.4

Homogeneity of Four Students

Parameter	Mean	SD	2½	Median	97½
$p[1]$	0.7269	0.0933	0.5263	0.7341	0.8857
$p[2]$	0.5457	0.104	0.3384	0.5481	0.7423
$p[3]$	0.4632	0.1029	0.2584	0.4518	0.6554
$p[4]$	0.5004	0.1049	0.2958	0.510	0.7013
$g[1]$	0.1812	0.1389	−0.0938	0.1835	0.4495
$g[2]$	0.2737	0.1394	−0.01101	0.2777	0.5334
$g[3]$	0.0454	0.1471	−0.2415	0.0466	0.329
Average	0.5566	0.0512	0.4555	0.5567	0.6557

average should be employed? Based on Table 3.3, the arithmetic average is 0.4459, and a weighted average using the inverse of the posterior variance is 0.4962.

What is the probability of a 1 (or 2) for each student, and are they the same for all four? Student A scores 2 15/20=0.75, followed by 0.55, 0.45, and 0.50 for students B, C, and D, respectively. Are these different, that is, is there inter rater bias? A Bayesian counterpart to the Cochran Q test is contained in Table 3.4, where $p[i]$ is posterior mean of the probability that student i, $i=1,2,3,4$, scores a 2, and where it is assumed the prior distribution of these four probabilities is a beta(1,1) or uniform. The $g[i]$ are contrasts, among the $p[i]$, where $g[1]=p[1]-p[2]$, $g[2]=p[1]-p[3]$ and $g[3]=p[2]-p[3]$. The credible intervals among the contrasts give some evidence, but not very much, that there is a difference between students A and C, however to reject outright the hypothesis of homogeneity appears to be risky. The evidence is not that strong! The Cochran Q test, as reported by Shoukri[3], does not reject the hypothesis of homogeneity.

The computations for inter rater bias revealed in Table 3.4 are executed with 25,000 observations generated from the posterior distribution of the parameters and based on BUGS CODE 3.2.

BUGS CODE 3.2
```
model
{
for(i in 1:4) {for(j in 1:20) { y[i,j] ~ dbern(p[i])}}
for(i in 1:4){ p[i] ~ dbeta(1,1)}
for(i in 1:4){ h[i] <- (p[i]-mean(p[]))*(p[i]-mean(p[])) }
g[1] <- p[1]-p[2];g[2] <- p[1]-p[3]
g[3] <- p[2]-p[4];all <- mean(p[])
```

test1 <-sum(h[]);test2 <-sum(g[])}
list(y=structure(.Data=c(
1,1,1,1,1,1,1,1,1,1,1,1,1,1,1,0,0,0,0,0,
1,1,1,1,1,1,1,1,1,1,1,0,0,0,0,0,0,0,0,0,
1,1,1,1,1,1,1,1,0,0,0,0,0,0,0,0,0,0,0,0,
1,1,1,1,1,1,1,1,1,0,0,0,0,0,0,0,0,0,0,0), .Dim=c(4,20)))
list(p=c(.5,.5,.5,.5))

3.3 Partial Agreement

Another multirater example with binary scores is presented to demonstrate the Bayesian approach to estimating partial agreement between six
pathologists, who are examining the presence of epithelioid granuloma in
68 intestinal biospy lesions of patients with Crohn's disease. The study was
carried out by Rogel, Boelle, and Mary[1], using a log linear model approach.
For the time being, a more basic descriptive approach is proposed. Consider
Table 3.5 where 1 designates presence and 2 absence of epithelioid granuloma. An overall measure of agreement for the six pathologists is

$$\kappa(6,2) = (\text{agree} - \text{cagree})/(1 - \text{cagree}) \tag{3.4}$$

where,

$$\text{agree} = \sum_{i=1}^{i=2} \theta_{iiiiii} \tag{3.5}$$

is the probability of agreement among the six. The probability of agreement
by chance is

$$\text{cagree} = \sum_{i=1}^{i=2} \theta_{i.....}\theta_{.i....}\theta_{..i...}\theta_{...i..}\theta_{....i.}\theta_{.....i} , \tag{3.6}$$

where θ_{ijklmn} is the probability that pathologist 1 assigns a score of i, and
pathologist 6 assigns a score of n, etc., where $i, j, k, l, m, n = 1, 2$. It can be
verified that the posterior distribution of a kappa type index $\kappa(6, 2)$ of overall
agreement is given in Table 3.6.

The analysis is based on BUGS CODE 3.3 and executed with 75,000 observations, with a burn in of 5000, and a refresh of 100.

TABLE 3.5

Observed Frequencies of Epithelioid Granuloma

	Pathologist						
Case	A	B	C	D	E	F	Number
1	1	1	1	1	1	1	15
2	1	1	1	1	1	2	0
3	1	1	1	1	2	1	2
4	1	1	1	1	2	2	0
5	1	1	1	2	1	1	2
6	1	1	2	2	2	1	1
7	1	1	2	1	1	1	0
8	1	1	2	1	2	1	0
9	1	1	2	1	2	2	0
10	1	1	2	2	1	1	1
11	1	2	1	1	1	1	0
12	1	2	1	1	1	2	0
13	1	2	1	1	2	1	0
14	1	2	1	1	2	2	1
15	1	2	1	2	2	1	0
16	1	2	2	2	2	2	1
17	1	2	2	1	1	1	0
18	1	2	2	1	1	2	1
19	1	2	2	1	2	1	0
20	1	2	2	1	2	2	0
21	1	2	1	2	2	2	0
22	2	1	1	1	1	1	1
23	2	1	1	1	1	2	0
24	2	1	1	1	2	1	0
25	2	1	1	2	1	1	1
26	2	1	1	2	2	1	1
27	2	1	2	2	2	2	1
28	2	1	2	1	1	1	0
29	2	1	2	1	2	1	0
30	2	1	2	1	2	2	0
31	2	1	1	2	2	2	0
32	2	2	1	1	1	2	0
33	2	2	1	1	2	1	0
34	2	2	1	1	2	2	0
36	2	2	1	2	2	1	2
37	2	2	1	2	2	2	2
38	2	2	2	1	1	2	0
39	2	2	2	1	2	1	0
40	2	2	2	1	2	2	0

(continued)

TABLE 3.5 (*continued*)

Case	A	B	C	D	E	F	Number
			Pathologist				
41	2	2	2	2	1	2	2
42	2	2	2	2	2	1	1
43	2	2	2	2	2	2	30

TABLE 3.6

Agreement of Six Pathologists

Parameter	Mean	SD	2½	Median	97½
θ_{111111}	0.2206	0.0498	0.1307	0.2178	0.3257
θ_{222222}	0.4414	0.0597	0.3267	0.4408	0.5595
agree	0.6615	0.0571	0.5461	0.6627	0.7696
cagree	0.0631	0.0268	0.0308	0.0587	0.1316
$\kappa(6, 2)$	0.6382	0.0598	0.5183	0.6392	0.7521
agree5	0.8088	0.0473	0.7076	0.8118	0.8918
cagree5	0.2072	0.0502	0.1389	0.1984	0.327
$\kappa_5(6, 2)$	0.758	0.0611	0.6266	0.7623	0.8644
GC	0.3241	0.1143	0.0918	0.3269	0.5382
GC5	0.6178	0.0948	0.4146	0.6235	0.7854

BUGS CODE 3.3

model

```
{
# g1 corresponds to the first case

g1 ~ dgamma(15,2) g3 ~ dgamma(2,2)
g5 ~ dgamma(2,2) g6 ~ dgamma(1,2)
g10 ~ dgamma(1,2)
g14 ~ dgamma(1,2) g16 ~ dgamma(1,2)
g18 ~ dgamma(1,2) g22 ~ dgamma(1,2)
g25 ~ dgamma(1,2)  g26 ~ dgamma(1,2)
g27 ~ dgamma(1,2) g35 ~ dgamma(3,2)
g36 ~ dgamma(2,2) g37 ~ dgamma(2,2)
g41 ~ dgamma(2,2) g42 ~ dgamma(1,2)
g43 ~ dgamma(30,2)

h <- g1 + g3 + g5 + g6 + g10 + g14 + g16 + g18 + g22 + g25 + g26 + g27 + g35 + g36
   + g37 + g41 + g42 + g43
```

a1 is the probability of the outcome for case 1

a1 < -g1/h; a3 < -g3/h; a5 < -g5/h; a6 < -g6/h; a10 < -g10/h;
a14 < -g14/h ; a16 < -g16/h; a18 < -g18/h; a22 < -g22/h;
a25 < -g25/h ; a26 < -g26/h; a27 < -g27/h; a35 < -g35/h;
a36 < -g36/h; a37 < -g37/h; a41 < -g41/h; a42 < -g42/h;
a43 < -g43/h ;

agree < -a1 + a43

GC < -agree -(1-agree)

a1….. is the probability that pathologist A assigns a 1 to a lesion

a1….. < -a1 + a3 + a5 + a6 + a10 + a14 + a16 + a18
a.1…. < -a1 + a3 + a5 + a6 + a10 + a22 + a25 + a26 + a27
a..1… < -a1 + a3 + a5 + a14 + a22 + a25 + a26 + a35 + a36 + a37
a…1.. < -a1 + a3 + a14 + a18 + a22
a….1. < -a1 + a5 + a10 + a18 + a22 + a25 + a35 + a41
a…..1 < -a1 + a3 + a5 + a6 + a10 + a22 + a25 + a26 + a35 + a36 + a42

a2….. < -a22 + a25 + a26 + a27 + a35 + a36 + a37 + a41 + a42 + a43
a.2…. < -a14 + a16 + a18 + a35 + a36 + a37 + a41 + a42 + a43
a..2… < -a6 + a10 + a16 + a18 + a27 + a41 + a42 + a43
a…2.. < -a5 + a6 + a10 + a16 + a25 + a26 + a27 + a35 + a36 + a37 + a41 + a42 + a43
a….2. < -a3 + a6 + a14 + a16 + a26 + a27 + a36 + a37 + a42 + a43
a…..2 < -a14 + a16 + a18 + a27 + a37 + a41 + a43

a…..2 is the probability that reader F assigns a score of 2 to a lesion

cagree < - (a1…..*a.1….*a..1…*a…1..*a….1.*a…..1) + (a2…..*a.2….*a..2…*a…2..*a….2.
           *a…..2)

kappa is the index that all six agree

kappa < -(agree-cagree)/(1-cagree)
al5 < -a1 + a3 + a5 + a22 + a37 + a41 + a42 + a43

#al5 is the probability at least five agree

cal5 < -a1…..*a.1….*a..1…*a…1..*a….1.*a…..1 + a1…..*a.1….*a..1…*a…1..*a….2.
*a…..1 + a1…..*a.1….*a..1…*a…2..*a….1.*a…..1 + a2…..*a.1….*a..1…*a…1..*a….1.
*a…..1 + a2…..*a.2….*a..1…*a…2..*a….2.*a…..2 + a2…..*a.2….*a..2…*a…2..*a….1.*a…..2
+ a2…..*a.2….*a..2…*a…2..*a….2.*a…..1 + a2…..*a.2….*a..2…*a…2..*a….2.*a…..2

cal5 is the probability at least 5 agree by chance

kappa5 is the index that at least 5 agree

kappa5 < - (al5-cal5)/(1-cal5)

GC5 < - al5 -(1-al5)

  }

list(g1 = 2,g3 = 2,g5 = 2,g6 = 2, g14 = 2, g16 = 2, g18 = 2, g22 = 2, g25 = 2, g26 = 2, g27 = 2, g35 = 2, g36 = 2,g37 = 2, g41 = 2, g42 = 2,g43 = 2).

In the above code, a3 denotes the probability of the outcome corresponding to case 3 of Table 3.6.

The analysis show a fairly good overall agreement between the six pathologists based on the kappa $\kappa(6, 2)$. The most agreement is when all six concur 30 times of the absence of a epithelioid granuloma, which has a posterior mean of 0.441 and a 95% credible interval of (0.326, 0.559), while they concur 15 times that a epithelioid granuloma is present, with a posterior mean of 0.2206 and a 95% credible interval of (0.1307, 0.3257). Also, there is very little evidence that the pathologists are presenting independent scores.

As the number of raters increase, the chances of overall agreement decrease and it is important to measure partial agreement. For example, when do at least five of the pathologists agree on either the absence or presence of the condition?

Suppose the agreement of at least five agree is measured by a chance type kappa parameter

$$\kappa_5(6, 2) = (al5 - cal5)/(1 - cal5) \tag{3.7}$$

where al5 = al + a3 + a5 + a10 + a22 + a37 + a41 + a42 + a43 and cal5 = a1.....
*a.1....*a..1...*a...1..*a....1.*a.....1 + a1.....*a.1....*a..1...*a...1..*a....2.*a.....1 + a1.....
*a.1....*a..1...*a...2..*a....1.*a.....1 + a2.....*a.1....*a..1...*a...1..*a....1.*a.....1 + a2.....
*a.2....*a..1...*a...2..*a....2.*a.....2 + a2.....*a.2....*a..2...*a...2..*a....1.*a.....2 + a2.....
*a.2....*a..2...*a...2..*a....2.*a.....1 + a2.....*a.2....*a..2...*a...2..*a....2.*a.....2.

Note, al1... is the probability that pathologist A assigns a score of 1 to a lesion, while a....2 is the probability that pathologist F assigns a score of 2 to a lesion.

The al5 parameter is the raw probability that at least five agree and cal5 is the probability that at least five agree, assuming all six are giving independent scores. Table 3.6 also shows the posterior analysis for the agreement of at least five pathologists. Exactly five agree (at either 1 or 2) at cases 3, 5, 10, 22, 41, and 42 of Table 3.6, with frequencies 2, 2, 1, 1, 2, and 1, respectively. All six agree for cases 1 and 43 with frequencies 15 and 30, respectively. A quick calculation shows a raw probability of 8/68 = 0.1117 for the probability that exactly five agree and a probability of 0.794 that at least five agree. These results agree very well with the corresponding posterior means of Table 3.6.

TABLE 3.7

Posterior Analysis for Split Agreement

Parameter	Mean	SD	2½	Median	97½
GCS	−0.7942	0.0729	−0.9142	−0.8019	−0.6311
agrees	0.1029	0.0364	0.0428	0.0990	0.1844
cagrees	0.0723	0.0114	0.0473	0.0734	0.0915
kappas	0.03292	0.0389	−0.0331	0.0294	0.1178

Another approach to partial agreement is to ask: When do the pathologists disagree the most? Of course, when they split 3–3, i.e. when three pathologists agree for absence of epithelioid granuloma and when the other three agree for the presence of granuloma or vice versa. This occurs at cases 6, 14, 18, 26, and 35, of Table 3.5, with frequencies 1, 1, 1, 1, and 3, respectively. See Table 3.5, where a quick calculation reveals that the probability of such a split is $7/68 = 0.1029$, while the posterior analysis for a kappa-like index for this event appears in Table 3.7.

As one would expect, the analysis shows very poor agreement between the six pathologists, when they split 3–3. The probability of split agreement is 0.1029 and the corresponding chance agreement is 0.0723. When discussing kappa, the probability of agreement and of chance agreement of the event should always be stated.

Of course, the analysis of partial agreement can be expanded, by determining the kappa in the context of association and the intraclass kappa. This will be considered in the exercise section. One could examine the following situations for partial agreement: (a) when exactly five of the six agree, or a 5–1 split, (either way for absence or presence of granuloma); and (b) when exactly four agree (a 4–2 split). The case where exactly three agree with a 3–3 split and pairwise agreement has already been analyzed. See Table 3.6 for the latter case, and Table 3.7 for the former.

3.4 Stratified Kappa

The example of a national lung cancer screening trial of Chapter 2 (see Table 2.6, Table 2.7, and Table 2.8), is continued, but another modality is added, namely MRI, that is, each subject receives three images: One from X-ray, one from CT, and lastly one from MRI, and three hospital sites are involved. The study was designed to have 1000 subjects in the first two sites, and 500 in the third, but the final assignment of patients was 976 and 936 for sites 1 and 2, respectively, and 494 for the third.

For the first site, the outcomes are displayed in Table 3.8, where the cell entries are interpreted as follows: 1 indicates no evidence of disease, 2 indicates that most likely no disease is present, 3 denotes that it is likely disease is present, and 4 that there is a malignant lesion. Thus for case 64 of site 1, the three images agree 50 times that there is a malignant lesion.

Obviously, this is the most complex example introduced so far. It involves multiple raters, the three imaging modalities, four ordinal outcomes, and

TABLE 3.8

X-ray, CT, and MRI

Case	X-ray	CT	MRI	Total
Site 1				
1	1	1	1	200
2	1	1	2	55
3	1	1	3	40
4	1	1	4	25
5	1	2	1	25
6	1	2	2	15
7	1	2	3	10
8	1	2	4	7
9	1	3	1	9
10	1	3	2	6
11	1	3	3	4
12	1	3	4	3
13	1	4	1	3
14	1	4	2	1
15	1	4	3	1
16	1	4	4	10
17	2	1	1	10
18	2	1	2	20
19	2	1	3	8
20	2	1	4	6
21	2	2	1	100
22	2	2	2	40
23	2	2	3	35
24	2	2	4	22
25	2	3	1	2
26	2	3	2	3
27	2	3	3	7
28	2	3	4	6
29	2	4	1	1
30	2	4	2	1
31	2	4	3	1
32	2	4	4	7
33	3	1	1	10

(continued)

TABLE 3.8 *(continued)*

Case	X-ray	CT	MRI	Total
34	3	1	2	5
35	3	1	3	3
36	3	1	4	2
37	3	2	1	1
38	3	2	2	3
39	3	2	3	20
40	3	2	4	2
41	3	3	1	15
42	3	3	2	10
43	3	3	3	100
44	3	3	4	23
45	3	4	1	1
46	3	4	2	1
47	3	4	3	2
48	3	4	4	1
49	4	1	1	1
50	4	1	2	1
51	4	1	3	2
52	4	1	4	7
53	4	2	1	1
54	4	2	2	3
55	4	2	3	1
56	4	2	4	2
57	4	3	1	0
58	4	3	2	0
59	4	3	3	2
60	4	3	4	1
61	4	4	1	8
62	4	4	2	6
63	4	4	3	9
64	4	4	4	50
Site 2				
1	1	1	1	150
2	1	1	2	50
3	1	1	3	50
4	1	1	4	47
5	1	2	1	5
6	1	2	2	25
7	1	2	3	10
8	1	2	4	4
9	1	3	1	5
10	1	3	2	5

(continued)

TABLE 3.8 *(continued)*

Case	X-ray	CT	MRI	Total
11	1	3	3	15
12	1	3	4	1
13	1	4	1	3
14	1	4	2	2
15	1	4	3	1
16	1	4	4	6
17	2	1	1	20
18	2	1	2	8
19	2	1	3	6
20	2	1	4	3
21	2	2	1	50
22	2	2	2	100
23	2	2	3	40
24	2	2	4	20
25	2	3	1	3
26	2	3	2	2
27	2	3	3	10
28	2	3	4	1
29	2	4	1	0
30	2	4	2	0
31	2	4	3	0
32	2	4	4	2
33	3	1	1	10
34	3	1	2	5
35	3	1	3	4
36	3	1	4	1
37	3	2	1	2
38	3	2	2	18
39	3	2	3	3
40	3	2	4	3
41	3	3	1	30
42	3	3	2	20
43	3	3	3	90
44	3	3	4	15
45	3	4	1	2
46	3	4	2	1
47	3	4	3	0
48	3	4	4	4
49	4	1	1	6
50	4	1	2	3
51	4	1	3	2

(continued)

TABLE 3.8 *(continued)*

Case	X-ray	CT	MRI	Total
52	4	1	4	1
53	4	2	1	1
54	4	2	2	3
55	4	2	3	1
56	4	2	4	1
57	4	3	1	0
58	4	3	2	0
59	4	3	3	1
60	4	3	4	0
61	4	4	1	7
62	4	4	2	3
63	4	4	3	50
64	4	4	4	5
Site 3				
1	1	1	1	100
2	1	1	2	40
3	1	1	3	22
4	1	1	4	23
5	1	2	1	4
6	1	2	2	10
7	1	2	3	3
8	1	2	4	3
9	1	3	1	2
10	1	3	2	2
11	1	3	3	5
12	1	3	4	2
13	1	4	1	1
14	1	4	2	1
15	1	4	3	0
16	1	4	4	2
17	2	1	1	10
18	2	1	2	2
19	2	1	3	2
20	2	1	4	1
21	2	2	1	20
22	2	2	2	60
23	2	2	3	11
24	2	2	4	10
25	2	3	1	3
26	2	3	2	2
27	2	3	3	10
28	2	3	4	1

(continued)

88

Bayesian Methods for Measures of Agreement

TABLE 3.8 *(continued)*

Case	X-ray	CT	MRI	Total
29	2	4	1	0
30	2	4	2	0
31	2	4	3	1
32	2	4	4	1
33	3	1	1	4
34	3	1	2	2
35	3	1	3	1
36	3	1	4	1
37	3	2	1	2
38	3	2	2	10
39	3	2	3	4
40	3	2	4	4
41	3	3	1	5
42	3	3	2	5
43	3	3	3	60
44	3	3	4	10
45	3	4	1	2
46	3	4	2	1
47	3	4	3	1
48	3	4	4	3
49	4	1	1	4
50	4	1	2	1
51	4	1	3	1
52	4	1	4	2
53	4	2	1	0
54	4	2	2	1
55	4	2	3	0
56	4	2	4	0
57	4	3	1	0
58	4	3	2	0
59	4	3	3	1
60	4	3	4	0
61	4	4	1	3
62	4	4	2	2
63	4	4	3	2
64	4	4	4	8

three sites. Such complexity introduces a myriad of ways that agreement is to be approached. The overall kappa for agreement between the three modalities can be determined, as well as many indices for partial agreement. How should the various indices of agreement of the three strata (sites) be combined and compared?

Consider combining the three kappas for the three sites into one, by a weighted average, where the weights depend on the inverse of the variance of the posterior distribution of kappa of Table 3.9.

TABLE 3.9

Stratified Kappa for Screening Trial

Parameter	Mean	SD	2½	Median	97½
agree1	0.4612	0.0159	0.43	0.4611	0.4925
cagree1	0.0942	0.0047	0.0855	0.0940	0.104
kappa1	0.4051	0.0171	0.3717	0.405	0.4387
agree2	0.3152	0.0151	0.2858	0.3151	0.3451
cagree2	0.0878	0.0030	0.0821	0.0876	0.0942
kappa2	0.2492	0.0158	0.2184	0.2492	0.2806
agree3	0.3803	0.0219	0.3378	0.3801	0.4239
cagree3	0.0987	0.0056	0.0887	0.0983	0.1109
kappa3	0.3125	0.0237	0.2679	0.3122	0.3585
kappa stratified	0.3193	?	?	?	?

The BUGS CODE 3.4 is for the posterior analysis of overall kappa for site 3 and was executed with 30,000 observations generated from the joint posterior distribution of kappa3, agree3, and cagree3, with a burn in of 5000 observations and a refresh of 100. A similar program is executed for the other two sites, giving kappas for the three sites (strata) and the corresponding standard deviations, thus the stratified kappa is easily computed as 0.3193. A non-informative prior was employed, namely,

$$f(\theta) \propto \prod_{i,j,k=1}^{i,j,k=4} \theta_{ijk}^{-1} \tag{3.8}$$

where θ_{ijk} is the probability that X-ray, CT, and MRI assigns a score of i, j, and k, respectively for a subject with i, j, $k = 1, 2, 3$, or 4.

BUGS CODE 3.4

model

```
{
# for site 3
# g[i] corresponds to case i of Table 3.9c.
# .01 was used in lieu of 0 for the first scale parameter of the gamma

g[1] ~ dgamma(100,2)  g[2] ~ dgamma(40,2)  g[3] ~ dgamma(22,2)
g[4] ~ dgamma(23,2)  g[5] ~ dgamma(4,2)  g[6] ~ dgamma(10,2)
g[7] ~ dgamma(3,2)  g[8] ~ dgamma(3,2)  g[9] ~ dgamma(2,2)
g[10] ~ dgamma(2,2)  g[11] ~ dgamma(5,2)  g[12] ~ dgamma(2,2)
g[13] ~ dgamma(1,2)  g[14] ~ dgamma(1,2)  g[15] ~ dgamma(.01,2)
g[16] ~ dgamma(2,2)  g[17] ~ dgamma(10,2)  g[18] ~ dgamma(2,2)
g[19] ~ dgamma(2,2)  g[20] ~ dgamma(1,2)  g[21] ~ dgamma(20,2)
```

g[22] ~ dgamma(60,2) g[23] ~ dgamma(11,2) g[24] ~ dgamma(10,2)
g[25] ~ dgamma(3,2) g[26] ~ dgamma(2,2) g[27] ~ dgamma(10,2)
g[28] ~ dgamma(1,2) g[29] ~ dgamma(.01,2) g[30] ~ dgamma(.01,2)
g[31] ~ dgamma(1,2) g[32] ~ dgamma(1,2) g[33] ~ dgamma(4,2)
g[34] ~ dgamma(2,2) g[35] ~ dgamma(1,2) g[36] ~ dgamma(1,2)
g[37] ~ dgamma(2,2) g[38] ~ dgamma(10,2) g[39] ~ dgamma(4,2)
g[40] ~ dgamma(4,2) g[41] ~ dgamma(5,2) g[42] ~ dgamma(5,2)
g[43] ~ dgamma(60,2) g[44] ~ dgamma(10,2) g[45] ~ dgamma(2,2)
g[46] ~ dgamma(1,2) g[47] ~ dgamma(1,2) g[48] ~ dgamma(3,2)
g[49] ~ dgamma(4,2) g[50] ~ dgamma(1,2) g[51] ~ dgamma(1,2)
g[52] ~ dgamma(2,2) g[53] ~ dgamma(.01,2) g[54] ~ dgamma(1,2)
g[55] ~ dgamma(.01,2) g[56] ~ dgamma(.01,2) g[57] ~ dgamma(.01,2)
g[58] ~ dgamma(.01,2) g[59] ~ dgamma(1,2) g[60] ~ dgamma(.01,2)
g[61] ~ dgamma(3,2) g[62] ~ dgamma(2,2) g[63] ~ dgamma(2,2)
g[64] ~ dgamma(8,2)

```
h <- sum(g[])
  for(i in 1:64){a[i] <- g[i]/h}
agree3 <- a[1] + a[21] + a[43] + a[64]
b1.. <-
a[1] + a[2] + a[3] + a[4] + a[5] + a[6] + a[7] + a[8] + a[9] + a[10]
 + a[11] + a[12] + a[13] + a[14] + a[15] + a[16]
b.1. <-
a[1] + a[2] + a[3] + a[4] + a[17] + a[18] + a[19] + a[20]
 + a[33] + a[34] + a[35] + a[36] + a[49] + a[50] + a[51] + a[52]
b..1 <-
a[1] + a[5] + a[9] + a[13] + a[17] + a[21] + a[25] + a[29]
 + a[33] + a[37] + a[41] + a[45] + a[49] + a[53] + a[57] + a[61]
b2.. <-
a[17] + a[18] + a[19] + a[20] + a[21] + a[22] + a[23] + a[24] +
 a[25] + a[26] + a[27] + a[28] + a[29] + a[30] + a[31] + a[32]
b.2. <-
a[5] + a[6] + a[7] + a[8] + a[21] + a[22] + a[23] + a[24] + a[37]
 + a[38] + a[39] + a[40] + a[53] + a[54] + a[55] + a[56]
b.2 <-
a[2] + a[6] + a[10] + a[14] + a[18] + a[22] + a[26] + a[30] + a[34]
 + a[38] + a[42] + a[46] + a[50] + a[54] + a[58] + a[62]
b3.. <-
a[33] + a[34] + a[35] + a[36] + a[37] + a[38] + a[39] + a[40] + a[41]
 + a[42] + a[43] + a[44] + a[45] + a[46] + a[47] + a[48]
```

b.3. <-

a[9] + a[10] + a[11] + a[12] + a[25] + a[26] + a[27] + a[28] +
a[41] + a[42] + a[43] + a[44] + a[57] + a[58] + a[59] + a[60]

b..3 <-

a[3] + a[7] + a[11] + a[15] + a[19] + a[23] + a[27] + a[31] +
a[35] + a[39] + a[43] + a[47] + a[51] + a[55] + a[59] + a[63]

b4.. <-

a[49] + a[50] + a[51] + a[52] + a[53] + a[54] + a[55] + a[56] +
a[57] + a[58] + a[59] + a[60] + a[61] + a[62] + a[63] + a[64]

b.4. <-

a[13] + a[14] + a[15] + a[16] + a[29] + a[30] + a[31] + a[32] +
a[45] + a[46] + a[47] + a[48] + a[61] + a[62] + a[63] + a[64]

b..4 < − a[4] + a[8] + a[12] + a[16] + a[20] + a[24] + a[28] + a[32]
+ a[36] + a[40] + a[44] + a[48] + a[52] + a[56] + a[60] + a[64]

cagree3 < - b1..*b.1.*b..1 + b2..*b.2.*b..2 + b3..*b.3.*b..3 + b4..*b.4.*b..4

kappa3 < - (agree3-cagree3)/(1-cagree3) }

list(

g = c(2,2,2,2,2,2,2,2,2,2,2,2,2,2,2,2,2,2,2,2,2,2,2,2,2,2,2,2,2,2,2,2,2,2,2,2,2,2,2,2,2,2,
2,2,2,2,2,2,2,2,2,2,2,2,2,2,2,2,2,2,2,2,2,2))

3.5 Intraclass Kappa

This multirater example with binary scores is revisited to demonstrate the Bayesian approach to estimating the intraclass correlation between three pathologists, who are examining the presence of epithelioid granuloma in 68 intestinal biopsy lesions of patients with Crohn's disease. The study was carried out by Rogel, Boelle, and Mary[1], using a log linear model approach. For the time being, a more basic descriptive approach is proposed. Consider Table 3.10, where 1 designates presence and 0 absence of epithelioid granuloma. This example is taken from Table 3.5, but with only the first three pathologists reported.

What is the intraclass correlation coefficient for this information? The model of Bahadur[7] as generalized by George and Bowman[8] to several raters for binary observations and is presented by Altaye, Donner, and Klass[9]. Shoukri[3] presents the account by Altaye, Donner, and Klar. Consider n raters and binary responses and let $X_{ij} = 1$ if the j-th rater scores a success for the i-th subject and let $P(X_{ij} = 1) = \pi$ be the probability of a success for all raters and all subjects. Now let

$$X_i = \sum_{j=1}^{j=n} X_{ij} \tag{3.10}$$

TABLE 3.10

Observed Frequencies of Epithelioid Granuloma

Case	Pathologist			Number
	A	B	C	
1	1	1	1	19
2	1	1	0	2
3	1	0	1	1
4	1	0	0	2
5	0	1	1	3
6	0	1	0	1
7	0	0	1	7
8	0	0	0	33

be the number of successes for the i-th subject, then the correlated binomial model of Bahadur may be represented by

$$P(X_i = x_i) = \text{BC}(n, x_i) \sum_{u=0}^{u=n-x_i} (-1)^u \text{BC}(n - x_i)\lambda_{x_i+u}, \qquad (3.11)$$

where $\text{BC}(n, x_i)$ is the binomial coefficient of x_i from n and $\lambda_k = P(X_{i1} = X_{i2} = \cdots X_{ik} = 1)$ where $\lambda_0 = 1$. One may show that

$$P_0 = P(X_i = 0) = 1 - 3(\lambda_1 - \lambda_2) - \lambda_3,$$

$$P_1 = P(X_i = 1) = 3(\lambda_1 - 2\lambda_2 + \lambda_3), \qquad (3.12)$$

$$P_2 = P(X_i = 2) = 3(\lambda_2 - \lambda_3),$$

and

$$P_3 = P(X_i = 3) = \lambda_3.$$

Also,

$$\lambda_1 = (3P_3 + 2P_2 + P_1)/3 = \pi,$$

$$\lambda_2 = (3P_3 + P_2)/3 \qquad (3.13)$$

and

$$\lambda_3 = P_3.$$

It can also be shown that the intraclass correlation is

$$\rho = (\lambda_2 - \lambda_1^2) / \lambda_1 (1 - \lambda_1), \qquad (3.14)$$

and may be expressed as a kappa type index as

$$\rho = (P_0 - Q)/(1 - Q) \tag{3.15}$$

where,

$$Q = 1 - 3\lambda_1(1 - \lambda_1). \tag{3.16}$$

The layout of such an experiment is exhibited in Table 3.11 and is a multinomial with four categories. From Table 3.10, $m_0 = 33$, $m_1 = 10$, $m_2 = 6$, $m_3 = 19$, thus, assuming an improper prior for P_0, P_1, P_2, P_3, implies the posterior distribution of these parameters is Dirichlet (33,10,6,19) and Bayesian inferences for ρ are easily implemented.

Another approach for estimating ρ is also presented by Altaye, Donner, and Klass[9] and is based on the maximizing the likelihood function (Equation 3.11), which provides the following estimators.

$$\tilde{\lambda}_1 = (3m_3 + 2m_2 + m_1)/3N, \tag{3.17}$$

$$\tilde{\lambda}_2 = (3m_3 + m_2)/3N,$$

and

$$\tilde{\lambda}_3 = m_3/N.$$

The above provide the maximum likelihood estimator of ρ via Equation 3.15, where

$$\tilde{P}_0 = (m_0 + m_3)/N. \tag{3.18}$$

Bayesian inferences for the intraclass correlation and its components are portrayed in Table 3.13, and Figure 3.2 displays the posterior density of ρ.

TABLE 3.11

Intraclass Correlation

Category	Ratings	Frequency	Probability
0	(0,0,0)	m_0	P_0
1	(0,0,1), (0,1,0), (1,0,0)	m_1	P_1
2	(1,1,0), (1,0,1), (0,1,1)	m_2	P_2
3	(1,1,1)	m_3	P_3
Total		N	

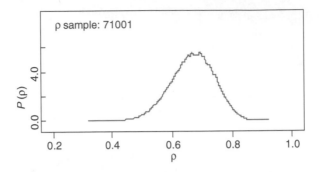

FIGURE 3.2
Posterior density of ρ.

TABLE 3.12

Bayesian Inferences for Intraclass Correlation

Parameter	Mean	SD	2½	Median	97½
λ_1	0.3872	0.0518	0.2884	0.3864	0.4907
P	0.7648	0.051	0.6581	0.7676	0.8568
Q	0.2961	0.0359	0.2507	0.2887	0.3832
ρ	0.6654	0.0726	0.5134	0.669	0.7964

A WinBUGS program was executed with 70,000 observations generated from the joint posterior distribution of the above four parameters of Table 3.12 with a burn in of 1000, and a refresh of 100.

3.6 The Fleiss Generalized Kappa

Fleiss[10] proposed a generalized kappa for measuring agreement among multiple raters and his method is illustrated by Shoukri[3] with an example from Williams[4] with information about the evaluation program of the College of American Pathologists, and the results are presented in Table 3.13 where four laboratories assign nominal scores as follows to each specimen: 1 designates reactive, 2 denotes borderline, and 3 means non reactive. There are 12 specimens where the four laboratories agree that they are reactive, while they all agree that four specimens are non reactive, giving an overall probability of $16/28 = 0.571$ for overall agreement. These results are summarized in Table 3.14.

Thus, there are five specimens where laboratory 1 gives a score of 2 (borderline) and the other three a score of 3 (non reactive). With the Bayesian approach, the analysis gives Table 3.15 where 0.5712 is the posterior mean of the probability that the four laboratories agree, while the chance probability

TABLE 3.13

Laboratory Evaluation for Syphilis Serology of 28 Specimens

Specimen	Laboratory			
	1	2	3	4
1	1	1	1	1
2	1	1	1	1
3	2	3	3	3
4	2	3	3	3
5	2	3	3	3
6	1	1	1	1
7	2	3	3	3
8	1	1	1	1
9	3	3	3	3
10	3	3	3	3
11	1	1	1	1
12	1	1	2	2
13	1	1	1	1
14	1	1	2	2
15	1	1	1	1
16	1	1	3	2
17	1	1	3	2
18	1	1	1	1
19	1	1	1	1
20	2	2	3	3
21	1	1	1	1
22	2	3	3	3
23	2	2	3	3
24	2	2	3	3
25	1	1	1	1
26	3	3	3	3
27	1	1	1	1
28	3	3	3	3

TABLE 3.14

Summary for Syphilis Serology

Case	Laboratory				Frequency
	1	2	3	4	
1	1	1	1	1	12
2	2	3	3	3	5
3	3	3	3	3	4
4	1	1	2	2	2
5	1	1	3	2	2
6	2	2	3	3	3

TABLE 3.15

Posterior Analysis for Syphilis Serology

Parameter	Mean	SD	2½	Median	97½
kappa	0.5347	0.0865	0.3461	0.5353	0.7027
agree	0.5712	0.0922	0.3864	0.573	0.745
cagree	0.0835	0.0426	0.03651	0.0719	0.1957

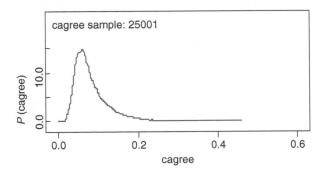

FIGURE 3.3
Posterior density of cagree.

(the four labs are acting independently) the four agree is 0.0835, giving kappa as 0.5347. A small right skewness is detected with the posterior density of the chance probability cagree (see Figure 3.3).

I used WinBUGS with 25,000 observations generated from the posterior distribution, with a burn in of 1000 observations, and a refresh of 100. An improper prior density similar to Equation 3.9 was deemed appropriate.

The Fleiss approach is completely different and in terms of Table 3.11 is defined as follows:

$$\kappa_F = (p_0 - p_1)/(1 - p_1),\tag{3.19}$$

where,

$$p_0 = \left(\sum_{i=1}^{i=k}\sum_{j=1}^{j=c} k_{ij}^2 - nk\right)\bigg/ kn(n-1)\tag{3.20}$$

and,

$$p_1 = \sum_{j=1}^{j=c} p_j,\tag{3.21}$$

and

$$p_j = \sum_{i=1}^{i=k} k_{ij}/nk,$$ (3.22)

where k_{ij} (these are random and the primary information) is the number of raters who assign a score of j to the i-th specimen, c is the number of scores, n the number of raters, and k the number of specimens. Applying this methodology to Table 3.11 produces $p_0 = 0.732$, $p_1 = 0.403$, and the estimated $\kappa_F = 0.18$. See Shoukri[3] for the details of these calculations and the interpretation of p_0, which is as follows: Select a specimen and two laboratories at random, then p_0 is the probability the two laboratories will agree on the score for that specimen. The Fleiss estimate of 0.18 is low compared to the Bayesian kappa posterior mean of 0.5347, indicating fair agreement according to the latter index and fairly poor agreement according to the Fleiss. Of course, they constitute different approaches to measuring overall agreement. The Bayesian kappa is a simple generalization of the original index for binary scores and two raters, while the Fleiss approach focuses on the average pairwise agreement between raters, averaged over all pairs and specimens. According to Woolson[11], the estimated standard error of the estimator of κ_F is

$$\tilde{\sigma}_{\kappa_F} = \left[\left(2/kn(n-1)\left(1 - \sum_{j=1}^{j=c} p_j^2 \right)^2 \right) \right]$$
$$\times \left[\sum_{j=1}^{j=c} p_j^2 - (2n-3)\left(\sum_{j=1}^{j=c} p_j^2 \right)^2 + 2(n-2)\sum_{j=1}^{j=c} p_j^3 \right]$$ (3.23)

3.7 The G Coefficient and Other Indices

The G coefficient is one of the attractive alternatives to kappa and is defined as

$$G = \text{agree} - (1 - \text{agree})$$ (3.24)

where, $\text{agree} = \sum_{i=1}^{i=c} \theta_{iii...i}$, and $\theta_{iii...i}$ is the probability that all k raters score a subject with a score of i, where $i = 1, 2,..., c$. The study is designed as a multinomial with k raters that each assign a score to each of n subjects and the scores range from 1 to c. Since 1–agree is the probability of not agreeing, the G coefficient has a range form –1 to 1, where if $G = 1$, only when all raters agree on their score for all subjects, but if $G = -1$, they never agree unanimously on any subject. Although the G coefficient is not corrected for chance agreement, like Cohen's kappa, its interpretation is easy to understand, which makes it an attractive index.

The first example is a return to Table 3.1 where four veterinary students are scoring 20 images as a 1 or 2, where 1 indicates little evidence of a cervical vertebral malformation, and 2 indicates evidence of such a malformation. What is the G coefficient? There are ten images where the four agree unanimously out of a total of 20, thus the probability of unanimous agreement is estimated as 0.5 and $G=0$. The Bayesian analysis is reported in Table 3.16.

The G coefficient is easy to work with for partial agreement, for example suppose G is estimated for the partial agreement of exactly three of four students agreeing, then the posterior analysis is also reported in Table 3.16.

This is continued for the partial agreement of a 2–2 split among the four students and is also reported in Table 3.17. The G coefficient decreases from −0.0035 for all four agreeing to a value of −0.1985 for exactly three agreeing, to a final value of −0.798 for a 2–2 split. The corresponding probabilities of agreement decrease from a value of 0.4983 for unanimous agreement to a value of 0.4008 that exactly three agree, to a final value of 0.1008 for the 2–2 split.

Based on BUGS CODE 3.5, the posterior analysis of Table 3.16 was executed with 31,000 observations generated from the joint posterior distribution, with a burn in of 1000, and a refresh of 100 observations.

TABLE 3.16

Posterior Analysis for G Coefficient of Agreement Among Four Students

Parameter	Mean	SD	2½	Median	97½
agree	0.4982	0.1088	0.2857	0.4977	0.7094
G	−0.003594	0.2177	−0.4285	−0.004697	0.4187
G3	−0.1985	0.2136	0.5944	0.2052	0.2361
agree3	0.4008	0.1068	0.2028	0.3974	0.6181
G22	−0.7984	0.1315	−0.9742	−0.8248	−0.4777
agree22	0.1008	0.0657	0.0128	0.0876	0.2611

TABLE 3.17

Homogeneity of Four Students Cervical Vertebral Malformation

Parameter	Mean	SD	2½	Median	97½
1 minus 2	0.2776	0.1407	−0.056	0.2296	0.494
2 minus 3	−0.04431	0.148	−0.3317	−0.0449	0.2463
3 minus 4	0.0437	0.147	−0.2472	0.0451	0.3292
$p1$	0.7276	0.0929	0.5278	0.7347	0.887
$p2$	0.5	0.1047	0.2952	0.5007	0.7037
$p3$	0.5443	0.104	0.3393	0.5455	0.7409
$P4$	0.5005	0.1026	0.3016	0.5002	0.7004

BUGS CODE 3.5
model

```
{
g1111 ~ dgamma(3,2)  g1121 ~ dgamma(2,2)
g2111 ~ dgamma(3,2)  g2121 ~ dgamma(1,2)
g2211 ~ dgamma(1,2)  g2212 ~ dgamma(2,2)
g2122 ~ dgamma(1,2)  g2222 ~ dgamma(7,2)
h <- g1111 + g1121 + g2111 + g2121 + g2211 + g2212 + g2222
a1111 <- g1111/h  a1121 <- g1121/h  a2111 <- g2111/h
a2121 <- g2121/h a2211 <- g2211/h  a2212 <- g2212/h
a2122 <- g2122/h a2222 <- g2222/h agree <- a1111 + a2222
G <- agree - (1 - agree)
G3 <- agree3 - (1 - agree3)
agree3 <- a1121 + a2111 + a2212 + a2122
agree22 <- a2121 + a2211
G22 <- agree22 - (1 - agree22)}
list(g1111 = 2,g1121 = 2,g2111 = 2,g2121 = 2,g2211 = 2, g2212 = 2, g2122 = 2,
g2222 = 2)
```

3.8 Kappa and Homogeneity

At this point, it is important to stress the analysis that should follow an analysis of agreement. If there is high agreement between the raters, just why and how do they indeed agree? On the other hand, when they do not agree, it is also important to know why. An analysis of agreement should always be followed by a test for homogeneity between raters. If there is a high degree of agreement, are the mean scores of the raters the same, and if not, how do they differ? This question is considered with the example of Table 3.1, where we see that the overall level of agreement of ½ for the probability that all four students agree, with a G coefficient of −0.003 indicating poor agreement. What is the probability of evidence of a malformation (see Table 3.1 where evidence of a malformation is indicated by 2) for the four students?

The students are assigning the same probability to a malformation as evidence of the three contrasts 1 minus 2, 2 minus 3, and 3 minus 4. The 95% credible interval for the three differences imply very little difference in the probabilities of a malformation among the four veterinary students. The largest probability corresponds to student 1 and has a posterior mean of 0.727, while the same probability has a mean of 0.5005 for student 4, and the average probability is 0.5685. The four probabilities of the four students should be reported together with the index of overall agreement and partial agreement. According to Table 3.16 there is very poor agreement overall and also very poor partial agreement. The computations appearing in Table 3.17 are based on BUGS CODE 3.6 and an improper prior distribution for the eight non empty cell probabilities of Table 3.1 using 35,000 observations with a burn in of 5000 and a refresh of 100.

The Friedman test for homogeneity among the four raters gave an exact *p*-value of 0.134, also confirming homogeneity, as did the Bayesian analysis.

BUGS CODE 3.6

```
Model
# 0 indicates no evidence
# 1 indicates evidence of malformation
model
{
for(i in 1:20) {for(j in 1:4) { y[i,j] ~ dbern(p[j])}}
for(i in 1:4){ p[i] ~ dbeta(1,1)}
for(i in 1:4){ h[i] < - (p[i] - mean(p[]))*(p[i] - mean(p[])) }
g[1] < - p[1] - p[2];
g[2] < - p[2] - p[3]
g[3] < - p[3] - p[4]
all < - mean(p[])
test1 < - sum(h[])
test2 < - sum(g[])
}
list(y = structure(.Data = c(0,0,0,0,0,0,0,0,0,0,0,0,0,0,1,0,
                  0,0,1,0,1,0,0,0,1,0,0,0,1,0,0,0,
                  1,0,1,0,1,1,0,0,1,1,0,1,1,1,0,1,
                  1,0,1,1,1,1,1,1,1,1,1,1,1,1,1,1,
                  1,1,1,1,1,1,1,1,1,1,1,1,1,1,1,1), .Dim = c(20,4)))
list(p = c(.5,.5,.5,.5))
```

3.9 Introduction to Model Based Approaches

An introduction to model based approaches is now presented, but a complete account of the subject appears in a later chapter. Of course, the principal advantage of a model based approach to agreement is that one may include subject and other study covariates as import predictors that might have an effect on agreement. As an introduction to the subject the homogeneity of several raters will be investigated with a logistic mode based on the information of Table 3.1. This study was first examined in Section 3.2, where four veterinary students examine 20 X-ray images for the presence or absence of cervical vertebral malformation. The outcomes are summarized in Table 3.18.

One discerns that for three images, the four agree that there is no evidence of a malformation, while for seven images they all agree that there is evidence of a malformation, giving an estimated probability of total agreement of 10/20=0.5 and a probability of 2–2 split of 2/20=0.10. A formal Bayesian analysis is displayed in Table 3.3. One also observes that the probability of a 1 is estimated as 15/20=0.75 for student A, as 10/20=0.5 for student B, as 11/20=0.55 for C, and for student D as 10/2=0.5. Are the four students estimating the probability of a 1 with the same odds? To answer this question a classical and Bayesian logistic regression is performed. The classical approach with SPSS results in Table 3.19, while the Bayesian is given by Table 3.20.

The analyses are based on the model

$$\text{logit}(s[i]) = \beta_0 + \beta_1 x_1(i) + \beta_2 x_2(i) + \beta_3 x_3(i) + \beta_4 x_4(i), \tag{3.25}$$

where β_0 is a constant and β_1 is the effect on the logit of student A, β_3 is the effect on the logit of student C and $i = 1, 2, ..., 80$. Note that $s[i]$ is the probability of a 1 (presence of malformation) for the i-th image, and the

TABLE 3.18

Four Students and Twenty Images for Cervical Vertebral Malformation

Students				
A	B	C	D	Frequency
0	0	0	0	3
0	0	1	0	2
1	0	0	0	3
1	0	1	0	1
1	1	0	0	1
1	1	0	1	2
1	0	1	1	1
1	1	1	1	7

TABLE 3.19

Logistic Regression for Homogeneity of Four Students

Parameter	B	SE	Wald	df	Sig.	Exp(B)
Constant	0.000	0.447	0.000	1	1.000	1.000
Student A	1.099	0.683	2.586	1	0.108	3.000
Student B	0.201	0.634	0.100	1	0.752	1.222
Student C	0.201	0.634	0.100	1	0.752	1.222

TABLE 3.20

Bayesian logistic Regression for Homogeneity

Parameter	Mean	SD	2½	Median	97½	Exp(B)
Constant	0.002108	0.4544	−0.8930	0.0000237	0.887	1.022
Student A	1.165	0.7012	−0.1745	1.149	2.603	3.205
Student B	0.2107	0.6501	−0.1053	0.2094	1.49	1.234
Student C	0.209	0.6451	−1.046	0.2066	1.482	1.232
D12	0.9541	0.7092	−0.4159	0.9437	2.375	2.596
D23	0.00169	0.6557	−1.295	0.0065	2.198	1.001

analysis is executed with $\beta_4 = 0$, and the $x_j(i)$, $j = 1$, 2, 3, are specified in BUGS CODE 3.7. The parameters D12 and D13 are the differences of the effects of the students. For example, D12 is the effect of student A minus the effect of student B on the logit. Note the classical and Bayesian analyses give similar results, however it should be noted that the Bayesian includes a vague type prior on the parameters of the logistic model. For example the constant and the student effects are given $N(0,0.000001)$ prior distribution, that is the prior mean is 0 and the prior precision is 0.000001 (or a variance of 1,000,000). See the BUGS CODE 3.7 for additional details of the Bayesian analysis.

The Bayesian analysis is based on BUGS CODE 3.7. Both analyses show the effect of the students on the logit are negligible as are the differences in the student effects on the logit.

BUGS CODE 3.7

Model

alpha is a constant

r1 is the effect of student A on the logit

r4 = 0 for the constraint

the 80 by 1 vector y is the data

the 80 by 1 vector x1 identifies the effect of student A

x2 identifies the effect of student B

```
{ for ( i in 1:80){
y[i] ~ dbern(s[i])
logit(s[i]) < - alpha + r1*x1[i] + r2*x2[i] + r3*x3[i]}
alpha ~ dnorm(0.0,.000001)
r1 ~ dnorm(0.0,.000001)
r2 ~ dnorm(0.0,.000001)
r3 ~ dnorm(0.0,.000001)
d12 < - r1 - r2
```

```
d23 <- r2-r3
}
list
x1 = c(1,1,1,1,1,1,1,1,1,1,1,1,1,1,1,1,1,1,1,1,1,1,0,0,0,0,0,0,0,0,0,0,0,0,0,0,0,0,0,0,0,0,0,0,0,0,0,0,
0,0,0,0,0,0,0,0,0,0,0,0,0,0,0,0,0,0,0,0,0,0,0,0,0,0,0,0,0,0,0,0,0,0,0,0,0,0,0,0,0,0),
x2 = c(0,0,0,0,0,0,0,0,0,0,0,0,0,0,0,0,0,0,0,0,0,1,1,1,1,1,1,1,1,1,1,1,1,1,1,1,1,1,1,1,1,1,0,0,0,0,
0,0,0,0,0,0,0,0,0,0,0,0,0,0,0,0,0,0,0,0,0,0,0,0,0,0,0,0,0,0,0,0,0,0,0,0,0,0,0,0,0,0),
x3 = c(0,0,0,0,0,0,0,0,0,0,0,0,0,0,0,0,0,0,0,0,0,0,0,0,0,0,0,0,0,0,0,0,0,0,0,0,0,0,0,0,0,0,1,1,1,
1,1,1,1,1,1,1,1,1,1,1,1,1,1,1,1,1,0,0,0,0,0,0,0,0,0,0,0,0,0,0,0,0,0,0,0,0),
y = c(.00,.00,.00,.00,.00,1.00,1.00,1.00,1.00,1.00,1.00,1.00,1.00,1.00,1.00,1.00,1.00,1.0
0,1.00,1.00,
.00,.00,.00,.00,.00,.00,.00,.00,.00,1.00,1.00,1.00,1.00,1.00,1.00,1.00,1.00,1.00,1.00,1.00,
.00,.00,.00,1.00,1.00,.00,.00,.00,1.00,.00,.00,.00,1.00,1.00,1.00,1.00,1.00,1.00,1.00,1.00,
.00,.00,.00,.00,.00,.00,.00,.00,.00,1.00,1.00,1.00,1.00,1.00,1.00,1.00,1.00,1.00,1.00))
list(alpha = 0, r1 = 0,r2 = 0,r3 = 0)
```

3.10 Agreement and Matching

There is a close connection between matching problems and agreement and this can be seen by the following description of the classical problem of M urns and M balls, where the balls and urns are labeled from 1 to M. Suppose a ball is selected and tossed at random into one of the M urns, then what is the chance that the label of all balls match the label of the urn into which the balls are tossed? The answer is $1/M!$. The first ball has a chance of $1/M$ of landing in the urn that has a label that matches the label of the ball. Then given the first ball matched the urn into which it was tossed, the probability that the second ball lands in a matched urn is $1/(M-1)$, and so on. Thus, for three balls and three urns, the probability is $1/6$.

Another version of matching is when two people are tossing a coin. In n paired tosses, what is the probability that there will be n matches? Assuming independence, the probability of a match on the first toss is θ^2, where θ is the probability of a head on both coins, thus in n tosses, the probability of n matches is $(\theta^2)^n$. Thus, if the two coins are fair, the probability of 10 matches is $(0.25)^{10} = 0.000000954$.

This situation of matching is quite different than the usual problem of agreement between two raters. In the usual case, the two raters are usually judging the same subjects and one would expect some degree of correlation between the two raters, however in the matching problem last described, the two raters are acting independently of each other. The result is as the number of paired tosses increase, the probability of agreement becomes small very rapidly.

Exercises

1. Verify the posterior analysis for Table 3.2 using the BUGS CODE 3.1, and produce a plot of the posterior distribution of $\kappa(4,2)$.
2. What are the posterior mean, standard deviation, upper and lower 2½ percentiles of the 16 components of the theta vector? From this information, identify those images where they disagree the most (see BUGS CODE 3.1).
3. Verify Table 3.3 and present plots of the posterior densities of the six kappa parameters. Repeat the analysis, but assuming a uniform prior density for the parameters of all six 2×2 tables.
4. Refer to Table 3.1 and estimate the intraclass correlation kappa with BUGS CODE 2.3 and compare your answer to Shoukri[3], who gives an estimate of 0.472. How many observations did you generate with WinBUGS, and how many observations were used for the burn in and the refresh? Also, estimate the probability a student will give a score of 2 (the image was affected). What prior distribution did you use?
5. Refer to Table 3.3 and find the posterior mean and standard deviation of the weighted kappa, where the weights are the inverse of the posterior variance of each of the six individual kappas. Use Win-BUGS with 50,000 observations generated from the posterior distribution of the weighted average. Assume an improper prior for the parameters and compare your answer with the posterior mean of $\kappa(4, 2)$ given in Table 3.2.
6. Amend BUGS CODE 3.3 and verify Table 3.7 and perform the posterior analysis for a split agreement between the six radiologists. Generate 75,000 observations from the joint posterior distribution of the parameters, with a bur in of 5000 and a refresh of 100.
7. Amend BUGS CODE 3.3 and determine the posterior distribution of a kappa type index for the following events: (a) exactly five agree, or a 5–1 split between pathologists; and (b) exactly four agree for a 4–2 split. For each of the two events above, find the posterior probability of the event and the posterior probability of the same event, but assuming the six radiologists are acting independently in their assignment of scores to the lesions. Use 125,000 observations with a burn in of 10,000 observations and a refresh of 100 and employ a uniform prior density for the parameters. Provide a plot of the posterior distribution of the kappa type index for both events, (a) and (b).
8. For Table 3.5, find the posterior distribution of the intraclass correlation kappa assuming a constant correlation model, Equation 3.14.
9. Using BUGS CODE 3.4, verify the posterior analysis of Table 3.9. What is the posterior mean and standard deviation of the stratified kappa? Provide the posterior density of stratified kappa and give a 95% credible interval for the same parameter. What is your overall conclusion about the agreement between the three imaging

modalities? Recall that 30,000 observations were generated from the posterior distribution with the improper prior (Equation 3.10).
10. Write a WinBUGS program and verify Table 3.15.
11. Based on Table 3.14, verify the Fleiss kappa value of 0.18. Why is this estimate different than the Bayesian value of 0.5347? What is the estimated standard error of the Fleiss estimate? See Equation 3.23 and Table 3.15.
12. The following table is from Woolson[11] and is a study with seven nosologists of the cause of death of 35 hypertensive patient, where the raters used death certificates to determine the cause of death. Find the Fleiss estimate of κ_F (Equation 3.19) along with its estimated standard error. Note for the Fleiss estimate, the k_{ij} are the key observations and that for the first row of the table, $k_{11}=1$, the number of raters who assigned a score of 1 to the first subject, and $k_{12}=0$, etc. The following scores were assigned to the categories as follows: 1 designates arteriosclerotic disease, 2=cerebrovascular disease, 3=other heart disease, 4=renal disease, and 5=other disease (see Table 3.21).

TABLE 3.21

Primary Cause of Death

	Class of disease				
Decendent	Arteriosclerotic Disease	Cerebrovascular	Other Heart Disease	Renal Disease	Other
1	1		5	1	
2					7
3	1		5	1	
4		3		4	
5	4				3
6		5		2	
7	1		5		1
8		3			4
9	2		3	2	
10		1		6	
11					7
12		1	2	4	
13		2		5	
14	6	1			
15	4			1	2
16			3	4	
17	1	2	1	3	
18	6	1			

(continued)

TABLE 3.21 *(continued)*

	Class of Disease				
Decendent	Arteriosclerotic Disease	Cerebrovascular	Other Heart Disease	Renal Disease	Other
19			1	6	
20	5		2		
21	3				4
22			2	5	
23		6			1
24	5		1	1	
25	7				
26		1	6		
27	1		1	5	
28		2			5
29		7			
30			1	2	4
31		6		1	
32	5		2		
33			1	6	
34			6		1
35				7	

13. Write the BUGS code and verify Table 3.15. Use an improper prior density for (P_0, P_1, P_2, P_3) and generate 70,000 observations from the posterior distribution of ρ with a burn in of 1000 and a refresh of 100.

14. Based on Table 3.14, compute the maximum likelihood estimates of P, Q, λ_1, and ρ. Base your answers on Equations 3.16 through 3.18.

15. Based on Table 3.5 of the agreement of six pathologists, perform the Bayesian analysis for the G coefficient of: (a) unanimous agreement of all six; (b) the agreement of exactly four; and (c) the partial agreement of a 3–3 split. List the posterior mean, median, and a 95% credible interval for the three parameters, using an improper prior for the multinomial cell probabilities, with 55,000 observations generated from the joint posterior distribution, with a burn in of 10,000 and a refresh of 250. In order to execute the analysis, write the necessary code.

16. Based on BUGS CODE 3.6, verify Table 3.17. Also test for homogeneity among the four students with the Friedman test.

17. Perform a posterior analysis that investigates the homogeneity of the four laboratories of Table 3.13, which display the results for 28 specimens, where nominal scores are assigned as follows: 1 designates reactive, 2 denotes borderline, and 3 means non reactive. A summary follows in Table 3.14. Use 5000 observations as a burn in and 40,000 generated from the posterior distribution of all the parameters with a refresh of 100 and a non informative prior. I performed a Friedman

test for homogeneity and got a *P*-value of 0.000, and the descriptive statistics are given by Table 3.22. How do your results compare to the Friedman test?

TABLE 3.22

Homogeneity of Four Laboratories

Lab	Mean	SD	N
1	1.571	0.741	28
2	1.750	0.927	28
3	2.071	0.978	28
4	2.000	0.942	28

18. Table 3.23 presents the scores of six pathologists about local infiltrate among 74 biopsy specimens from patients with Crohn's disease. The information is a summary of Table 1 of Rogel, Boelle, and Mary[1], where 1 denotes presence of local infiltrate and 0 denotes absence. Perform a Bayesian analysis for homogeneity (the probability of a 1) among the six pathologists using 65,000 observations with a burn in of 5000 and a refresh of 100 employing an improper prior density.

TABLE 3.23

Observed Frequencies of Local Inflitrate

	Pathologist						
Case	A	B	C	D	E	F	Number
1	0	0	0	0	0	0	6
2	0	0	0	0	0	1	1
3	0	0	1	1	1	0	6
4	0	0	1	0	1	1	1
5	0	1	0	0	0	0	2
6	0	1	0	0	0	1	1
7	0	1	0	0	1	0	7
8	0	1	0	1	1	0	1
9	0	1	1	0	0	0	1
10	0	1	1	0	1	0	2
11	0	1	1	0	1	1	1
12	0	1	0	1	1	1	6
13	1	1	0	0	1	0	2
14	1	1	0	0	1	1	1
15	1	1	0	1	1	0	1
16	1	1	0	1	1	1	2
17	1	1	1	0	1	0	2
18	1	1	1	0	1	1	1
19	1	1	1	1	1	0	8
20	1	1	1	1	1	1	22

108   *Bayesian Methods for Measures of Agreement*

19. Verify Table 3.20 with BUGS CODE 3.7 using 65,000 observations generated from the joint posterior distribution of the six parameters with a burn in of 5000 and a refresh of 100.

20. From the Bayesian analysis of Table 3.20, estimate the probability that student A assigns a 1 (presence of a malformation) to an image.

21. Three pathologists at a major cancer center are examining lymph node specimens of melanoma patients who have just had their primary tumor removed. The pathologists must determine if the disease has metastasized to the lymphatic system, and they examined 109 specimens with the following results and scored a 1 for presence of metastasis and a 0 for absence of metastasis. See Table 3.24. Write a BUGS program and examine the melanoma results for metastasis above with a logistic model and perform a Bayesian analysis with 55,000 observations generated from the posterior distribution of the parameters of the model, with a burn in of 2500 observations and a refresh of 100. The Bayesian analysis should be in the form of Table 3.20. Also, perform a conventional analysis and compare to the Bayesian. The Bayesian analysis should be based on a prior distribution for the coefficients of the model by using a vague prior $N(0,0.00001)$. See BUGS CODE 3.7! Based on the Bayesian results, what is the probability that pathologist one assigns a 1 (presence of metastasis) to a specimen?

TABLE 3.24

Melanoma Pathology for Metastasis

Case	Pathologist			Frequency
	1	2	3	
1	0	0	0	22
2	1	0	0	4
3	0	1	0	9
4	0	0	1	6
5	0	1	1	17
6	1	0	1	3
7	1	1	0	11
8	1	1	1	37
Total				109

22. Refer to Section 3.10, and suppose two people are tossing two coins, one of which has a probability of θ of landing heads and the other a probability of ϕ of landing heads. A match is said to occur if both coins land heads when the two toss their respective coins. (a) What is the probability of n matches in n paired tosses of the two coins? (b) What is the probability of m matches in n paired tosses of a coin, where $m=0, 1, 2,..., n$?

23. Refer to Section 3.10, and the ball–urn matching problem. Given M balls and M urns, what is the probability of S matches, where $S = 0$, 1, 2,..., M?

24. Refer to Section 3.10, where M people are each tossing one coin, and the coins have probabilities $\theta_1, \theta_2,...,\theta_M$ of landing heads: (a) What is the probability that there are n matches in n paired tosses of the M coins? (b) What is the probability of S matches in n paired tosses of M coins, where $S = 0, 1, 2,..., n$?

References

1. Rogel, A., P. Y. Boelle, and J. Y. Mary. 1998. Global and partial agreement among several observers. *Stat. Med.* 17, 489.
2. Theodossi, A., D. J. Spieglehalter, J. Jass, J. Firth, M. Dixon, M. Leader, D. A. Levison, et al. 1994. Observer variation and discriminatory value of biopsy features of inflammatory bowel disease. *Gut.* 35, 961.
3. Shoukri, M. M. 2004. *Measures of interobserver agreement.* Boca Raton: Chapman & Hall/CRC.
4. Williams, G. W. 1976. Comparing the joint agreement of several raters with one rater. *Biometrics* 32, 619.
5. Fleiss, J. L. 1971. Measuring nominal scale agreement among many raters. *Psychol. Bull.* 76, 378.
6. Landis, J. R., and G. C. Kock. 1977. The measurement of observer agreement for categorical data. *Biometrics* 33, 159.
7. Bahadur, R. 1961. A representation of the joint distribution of responses to n dichotomous items. In *Studies in items analysis and prediction,* ed. H. Soloman, Palo Alto, CA: Stanford University Press, 158–568.
8. George, E. O., and D. Bowman. 1995. A full likelihood procedure for analyzing exchangeable binary data. *Biometrics* 51, 512.
9. Altaye, M., A. Donner, and N. Klar. 2001. Inference procedures for assessing agreement among multiple raters. *Biometrics* 57, 584.
10. Fleiss, J. L. 1981. *Statistical methods for rates and proportions.* 2nd ed. New York: John Wiley & Sons.
11. Woolson, R. F. 1987. *Statistical methods for the analysis of biomedical data.* New York: John Wiley & Sons.

4

Agreement and Correlated Observations

4.1 Introduction

Suppose that two radiologists are reading mammograms and score the images as follows: 1 denotes evidence of a lesion while 0 signifies no evidence of a lesion. They see each image twice with the same scoring system. After scoring the first image each radiologist waits for some two weeks to score the image again, thus there are two replications and it is of interest to measure the degree of agreement between the two radiologists overall, to measure their agreement at each replication, and to compare their agreement between the two replications. One would expect the two scores between replications to be correlated, and it is of interest to explore the many issues that entail this type of agreement study.

Shoukri[1] presents several other examples where the observations are correlated. One very interesting one is from an eye study, where there are two ophthalmologists examining both eyes of a patient for evidence of geographic atrophy, where 0 is scored for absence of the pathology and 1 denotes evidence of atrophy. One would expect the scores from the two eyes to be correlated and this correlation should be taken into account when estimating agreement between the two raters. This study had been analyzed by Oden[2], Schouten[3], and Donner, Shoukri, Klar, and Baretfay[4]. The Oden approach is to estimate the agreement between the two raters by kappa, separately for the two eyes, then combine the two kappas by a weighted average, while Schouten also used a weighted kappa to estimate overall agreement. The methodology of Donner et al. is quite different and is based on a generalized correlation model, a generalization of the common correlation model presented in Section 2.8. The main focus of the Donner et al. approach is to compare the two kappas and two see if they are the same for each eye.

The approach taken here is Bayesian, where Bayesian analogs of the Oden[2], Schouten[3], and Donner et al.[4] methods are developed. For example, for the Oden weighted average, the posterior distribution of the Oden parameter is determined with a uniform prior distribution of the multinomial parameters. A similar way is taken for the Schouten parameter and for the Donner et al. index.

Chapter 4 continues with model-based methods of logistic regression to compare agreement between two raters across subgroups, adjusted for other relevant covariates of the experiment. Consider a 2×2 table of two raters with binary scores assigned to each subject, then the probability of crude agreement (the two diagonal elements of the 2×2 table) is estimated by the logistic model. The estimate of the probability of agreement for the entire study is adjusted for other covariates in the model, and also can be estimated for other interesting subgroups. The logistic approach is extended beyond binary scores to multiple ordinal and nominal scores with a multinomial type regression model. The last part of the chapter describes the log-linear model approach which was pioneered by Agresti[5] and Tanner and Young[6] and is major focus of the book by von Eye and Mun[7]. This is the first glimpse of using regression models to examine agreement and will be the main topic of Chapter 5.

4.2 An Example of Paired Observations

Oden[2], Schouten[3], and Shoukri[1] all base their analysis of dependent data for the eye study example on Table 4.1, where $n_{00} = 800$ is the number of patients scored 0 for the right and left eyes by rater 1 and who scored 0 for the right and left eyes by rater 2. Also, $n_{12} = 0$ is the number of patients scored 0 for the right eye and 1 for the left by rater 1 and 1 for the right and 0 for the left by rater 2, etc. for the remaining frequencies of the table. Thus, 840 patients are selected at random from some population and the right and left eyes of each patient is examined by two ophthalmologists for the presence of geographic atrophy and each eye receives a score of either 0 (for the absence) or 1 (for the presence) of atrophy. From Table 4.1 it is immediately evident that the 2×2 table for the right eye is given by Table 4.2 and for the left eye by Table 4.3.

A glance at Table 4.1 provides an estimate of $0.965 = 811/840$ for the probability of agreement for the two examiners. From Table 4.2 for the right eye, the probability of agreement is 0.982, while for the left eye it is $823/840 = 0.979$, based on Table 4.3. This will be analyzed in a Bayesian way by computing the posterior distribution of kappa for the overall 4×4 Table 4.1, then estimating kappa separately for the right eye and the left eye.

The posterior distribution of kappa for the right eye is based on the truncated distribution of the four parameters corresponding to the right eye, while in the same manner the truncated distribution of the four parameters for the left eye provides the posterior distribution of kappa for the left eye. Note that the posterior distribution of overall kappa has a mean (SD) 0.4225 (0.065) with a 95% credible interval of (0.2945, 0.4413), and for kappar of the right eye, the posterior mean (SD) is 0.4916 (0.088) with a 95% credible interval of (0.3145, 0.6579). Lastly for the left eye, the posterior mean (SD) of kappal is 0.4059 (0.092) with (0.2287, 0.5852) as a 95% credible interval. It

TABLE 4.1

Eye Data for Two Examiners with Schouten Weights

Rater 1	R_0L_0	R_0L_1	R_1L_0	R_1L_1	Total
			Rater 2		
R_0L_0	$n_{00}=800$	$n_{01}=10$	$n_{02}=9$	$n_{03}=2$	$n_{0.}=821$
	δ_{00}	δ_{01}	δ_{02}	δ_{03}	$\delta_{0.}$
	$w_{00}=1$	$w_{01}=.5$	$w_{02}=.5$	$w_{03}=0$	
R_0L_1	$n_{10}=4$	$n_{11}=2$	$n_{12}=0$	$n_{13}=0$	$n_{1.}=6$
	δ_{10}	δ_{11}	δ_{12}	δ_{13}	$\delta_{1.}$
	$w_{10}=.5$	$w_{11}=1$	$w_{12}=0$	$w_{13}=.5$	
R_1L_0	$n_{20}=3$	$n_{21}=0$	$n_{22}=5$	$n_{23}=0$	$n_{2.}=8$
	δ_{20}	δ_{21}	δ_{22}	δ_{23}	δ_{2}
	$w_{20}=.5$	$w_{21}=0$	$w_{22}=1$	$w_{23}=.5$	
R_1L_1	$n_{30}=1$	$n_{31}=0$	$n_{32}=0$	$n_{33}=4$	$n_{3.}=5$
	δ_{30}	δ_{31}	δ_{32}	δ_{33}	$\delta_{3.}$
	$w_{30}=0$	$w_{31}=.5$	$w_{32}=.5$	$w_{33}=1$	
Total	$n_{.0}=808$	$n_{.1}=12$	$n_{.2}=14$	$n_{.3}=6$	$N=840$
	$\delta_{.0}$	$\delta_{.1}$	$\delta_{.2}$	$\delta_{.3}$	

TABLE 4.2

Data for Right Rye

Rater 1	0	1	Total
		Rater 2	
0	816	11	827
	θ_{00}	θ_{01}	
1	4	9	13
	θ_{10}	θ_{11}	
Total	820	20	840

appears that the two ophthalmologists agree more for the right eye than for the left, however, the posterior distribution of the difference in the two kappas has a 95% credible interval of (-0.1246, 0.29), implying there is actually very little difference in the two indices. See Figures 4.1 through 4.3 for the posterior densities of kappar for the right eye, kappal for the left eye, and diff for the difference in the two kappas, respectively. In order to compare the left and right eye kappas, a generalized correlation model due to Donner et al.[4] and Shoukri and Donner[8] will be explored in a forthcoming section of this chapter.

TABLE 4.3

Data for Left Eye

Rater 1	Rater 2		Total
	0	1	
0	817	12	829
	ϕ_{00}	ϕ_{01}	
1	5	6	11
	ϕ_{10}	ϕ_{11}	
Total	822	18	840

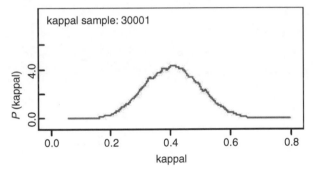

FIGURE 4.1
Posterior density of kappa for left eye.

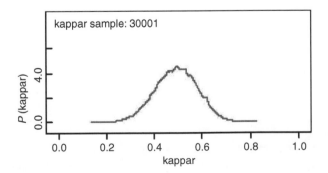

FIGURE 4.2
Posterior density of kappa for right eye.

The posterior analysis is based on BUGS CODE 4.1, where 30,000 observations were generated from the posterior distribution of the parameters, and 5000 observations were used for the burn in with a refresh of 100.

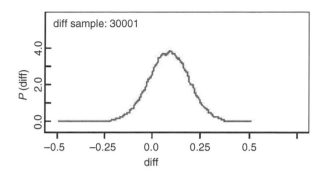

FIGURE 4.3
Posterior density of kappa for right eye minus left eye.

BUGS CODE 4.1

Model;

```
{
g[1] ~ dgamma(801,2)
g[2] ~ dgamma(11,2)
g[3] ~ dgamma(10,2)
g[4] ~ dgamma(3,2)
g[5] ~ dgamma(5,2)
g[6] ~ dgamma(3,2)
g[7] ~ dgamma(1,2)
g[8] ~ dgamma(1,2)
g[9] ~ dgamma(4,2)
g[10] ~ dgamma(1,2)
g[11] ~ dgamma(6,2)
g[12] ~ dgamma(1,2)
g[13] ~ dgamma(2,1)
g[14] ~ dgamma(1,2)
g[15] ~ dgamma(1,2)
g[16] ~ dgamma(5,2)
h <- sum(g[])
for (i in 1 :16) { a[i] <- g[i]/h}
# the a[i] are the Dirichlet parmeters for the 4 × 4 Table 4.1
kappa <- (agree -cagree)/(1-cagree)
# kappa is the overall kappa parameter for the 4 × 4 Table 4.1
```

```
agree <- a[1] + a[6] + a[11] + a[16]

G <- agree - (1-agree)

# G is the g coeffient for overall agreement

cagree <- (a[1] + +a[2] + a[3] + a[4])*(a[1] + a[5] + a[9] + a[13])
            + (a[2] + a[6] + a[10] + a[14])*(a[5] + a[6] + a[7] + a[8])
            + (a[3] + a[7] + a[11] + a[15])*( a[9] + a[10] + a[11] + a[12])
            + (a[13] + a[14] + a[15] + a[16])*(a[4] + a[8] + a[12] + a[16])

a1[1] <- a[1] + a[2] + a[5] + a[6]
a1[2] <- a[3] + a[4] + a[7] + a[8]
a1[3] <- a[9] + a[10] + a[13] + a[14]
a1[4] <- a[11] + a[12] + a[15] + a[16]

g1 <- sum(a1[])

for ( i in 1:4) { b[i] <- a1[i]/g1}

kappar <- (agreer-cagreer)/(1-cagreer)

Gr <- agreer - (1-agreer)

#Gr is the G-coefficient for the right eye
# kappar is the kappa for the right eye

agreer <- b[1] + b[4]
cagreer <- (b[1] + b[2])*( b[1] + b[3]) + (b[2] + b[4])*( b[3] + b[4])

c[1] <- a[1] + a[9] + a[3] + a[11]
c[2] <- a[2] + a[10] + a[4] + a[12]
c[3] <- a[5] + a[13] + a[7] + a[15]
c[4] <- a[6] + a[14] + a[8] + a[16]

g2 <- sum(c[])

for ( i in 1:4){ d[i] <-c[i]/g2}

kappal <- (agreel-cagreel)/(1-cagreel)

Gl <- agreel - (1-agreel)

# kappal is the kappa for the left eye

# Gl is the G coefficient for the left eye

agreel <- d[1] + d[4]

cagreel <- (d[1] + d[2])*( d[1] + d[3]) + (d[2] + d[4])*( d[3] + d[4])

diff <-kappar-kappal

kappap <- ((1-cagreer)*kappar + (1-cagreel)*kappal)/
            ((1-cagreer) + (1-cagreel))
```

kappap is the Oden parameter formula 4.1

p0 <- a[1]+a[2]/2+a[3]/2+a[5]/2+a[6]+a[8]/2+a[9]/2+a[11]+a[12]/2
+a[14]/2+a[15]/2+a[16]

pe <- f1+f2+f3+f4+f5+f6+f7+f8+f9+f10+f11+f12

f1 <- (a[1]+a[2]+a[3]+a[4])*(a[1]+a[5]+a[9]+a[13])

f2 <- (a[1]+a[2]+a[3]+a[4])*(a[2]+a[6]+a[10]+a[14])/2

f3 <- (a[1]+a[2]+a[3]+a[4])*(a[3]+a[7]+a[11]+a[15])/2

f4 <- (a[5]+a[6]+a[7]+a[8])*(a[1]+a[5]+a[9]+a[13])/2

f5 <- (a[5]+a[6]+a[7]+a[8])*(a[2]+a[6]+a[10]+a[14])

f6 <- (a[5]+a[6]+a[7]+a[8])*(a[4]+a[8]+a[12]+a[16])/2

f7 <- (a[9]+a[10]+a[11]+a[12])*(a[1]+a[5]+a[9]+a[13])/2

f8 <- (a[9]+a[10]+a[11]+a[12])*(a[3]+a[7]+a[11]+a[15])

f9 <- (a[9]+a[10]+a[11]+a[12])*(a[4]+a[8]+a[12]+a[16])/2

f10 <- -(a[13]+a[14]+a[15]+a[16])*(a[2]+a[6]+a[10]+a[14])/2

f11 <- -(a[13]+a[14]+a[15]+a[16])*(a[3]+a[7]+a[11]+a[15])/2

f12 <- -(a[13]+a[14]+a[15]+a[16])*(a[4]+a[8]+a[12]+a[16])

kw <- (p0-pe)/(1-p)

kw is the Schouten parameter formula (4.2)

}

list (g=c(1,1,1,1,1,1,1,1,1,1,1,1,1,1,1,1))

How does the Bayesian analysis above compare to the Oden[1] and Schouten[2] approaches to estimating agreement? First consider Oden, who defined a pooled kappa as

$$\kappa_{po} = \frac{\left[\left(1-\sum_{i=0}^{i=1}\sum_{i=0}^{i=1}w_{ij}\theta_{i.}\theta_{.j}\right)\kappa_r + \left(1-\sum_{i=0}^{i=1}\sum_{i=0}^{i=1}w_{ij}\phi_{i.}\phi_{.j}\right)\kappa_1\right]}{\left[\left(1-\sum_{i=0}^{i=1}\sum_{i=0}^{i=1}w_{ij}\theta_{i.}\theta_{.j}\right)+\left(1-\sum_{i=0}^{i=1}\sum_{i=0}^{i=1}w_{ij}\phi_{i.}\phi_{.j}\right)\right]} \quad (4.1)$$

where κ_r is the usual chance corrected kappa of agreement for the two raters for the right eye, and κ_1 is the corresponding index for the left eye. The θ_{ij} (see Table 4.2) is the probability of a score of i assigned by the first rater and a j by the second rater for the right eye, while ϕ_{ij} (see Table 4.3) is the corresponding probability for the left eye, with $i, j = 0, 1$. In addition, Oden assigned weights w_{ij}, which is assigned to raters 1 and 2, when they assign scores of i and j, respectively.

On the other hand, Schouten[2] defines a weighted index as

$$\kappa_w = (P_0 - P_e)/(1 - P_e) \tag{4.2}$$

where,

$$P_0 = \sum_{i=0}^{i=3} \sum_{i=0}^{i=3} w_{ij} \delta_{ij}$$

and

$$P_e = \sum_{i=0}^{i=3} \sum_{i=0}^{i=3} w_{ij} \delta_{i.} \delta_{.j}$$

See Table 4.1 for the definition of δ_{ij} and the values of the weights w_{ij} used by Schouten, where $i, j = 0, 1, 2, 3$.

4.3 The Oden Pooled Kappa and Schouten Weighted Kappa

The pooled kappa of Oden is defined as Equation 4.1, and in order to illustrate the method for the eye data of Table 4.1, the following weights are assigned as: $w_{00} = w_{11} = 1$ and the off-diagonal weights are given a value of zero. As for the Schouten weighted kappa defined as Equation 4.2, the weights are given in Table 4.1. A Bayesian analog of the two indices is presented with BUGS CODE 4.1 and executed with 35,000 observations, a burn in of 5000, a refresh of 100, and a uniform prior density for the 16 multinomial parameters δ_{ij} of Table 4.1. The posterior analysis is reported in Table 4.5.

It is interesting to see that the numerical results for the Oden and Schouten measures of agreement are identical, but remember that the weights chosen for the Oden parameter are 1 for the diagonal elements and 0 elsewhere of Table 4.2 and Table 4.3. The computing formulas used for the two indices are not the same! (see BUGS CODE 4.1). The components P_0 and P_e of the Schouten index are also displayed in Table 4.5 and imply the weighted probability of agreement is about the same as the weighted probability of agreement, assuming independence between raters. A comparison between Table 4.5 and Table 4.4 reveals that the posterior mean of 0.4513 for the Schouten and Oden indices lie between the posterior mean of 0.4059 for the kappa of the left eye and 0.4916 for the kappa of the right eye. Of course, this is to be expected, because the Oden parameter is a weighted average of the right eye and left eye kappas. It is also interesting to compare the posterior mean (SD) of 0.4513

TABLE 4.4

Posterior Distribution of Agreement for Eye Data

Parameter	Mean	SD	2½	Median	97½
agree	0.9499	0.007	0.933	0.950	0.963
cagree	0.913	0.011	0.889	0.913	0.934
kappa	0.4225	0.065	0.2945	0.4225	0.4413
agreer	0.971	0.006	0.9577	0.9714	0.9817
cagreer	0.9424	0.009	0.9224	0.9429	0.9597
kappar	0.4916	0.088	0.3145	0.4937	0.6579
agreel	0.968	0.006	0.9548	0.969	0.979
cagreel	0.946	0.009	0.927	0.947	0.962
kappal	0.4059	0.092	0.2287	0.4056	0.5852
Difference	0.0856	0.1052	−0.1246	0.086	0.29

TABLE 4.5

Bayesian Analog of the Oden and Schouten Indices of Agreement

Parameter	Mean	SD	2½	Median	97½
κ_{po} (Oden)	0.4513	0.0738	0.305	0.4518	0.5932
κ_w (Schouten)	0.4513	0.0738	0.305	0.4518	0.5932
P_o	0.9697	0.0052	0.9584	0.9702	0.9789
P_e	0.9446	0.0081	0.9275	0.945	0.959

(0.0738) of κ_w with the Shouten estimate of 0.50 (0.110), as reported by Shoukri[1]. Shoukri also reports estimates of P_o and P_e of 0.98 and 0.96, respectively, compared to posterior means of 0.97 and 0.94, respectively, from Table 4.5.

How should one generalize the above measures of agreement to any number of nominal scores and to more than two raters?

4.4 A Generalized Correlation Model

In order to compare two correlated kappas, Shoukri[1], Donner et al.[4], and Shoukri and Donner[8] present a generalization of the common correlation model, which is introduced in Section 2.8. The setting is two scenarios or situations and two raters, and kappa is estimated for each scenario and the two kappas compared. This was the case for the eye study example of Table 4.1, where two raters each examine both eyes of 840 patients. Kappa is estimated for the left and right eyes separately, and an overall kappa is constructed by a weighted average of the two kappas. Two approaches are taken, one by Oden, Equation 4.1, and the other by Schouten, Equation 4.2.

Based on Tables 4.1 through 4.3, and BUGS CODE 4.1, a Bayesian viewpoint was taken and the posterior distribution of an overall type kappa determined, and the results reported in Table 4.4. Also reported is the posterior analysis for the separate kappas for the right and left eyes. In addition, Bayesian analogs of the Oden and Schouten indices were reported in Table 4.5. The Bayesian analog of the Donner et al.[4] index of agreement will be taken in this section, but first the generalized correlation model is defined and explained.

There are two situations and for each situation there are two raters who assign binary scores to the same set of N patients. A kappa that measures the agreement between the two raters is estimated for each situation, and one would expect the two kappas to be correlated. Of interest to Donner et al. is the comparison between the two kappas. What is different about the Shoukri[1] and Donner et al.[4] approaches compared to that of Oden or Schouten is that the kappa of Donner et al. is based on the generalized correlation model.

Let $Y_{ijk}=1$ if the k-th rater assigns a 1 (success) to the i-th subject for the j-th situation, where $i=1, 2,..., N$, and $j, k=1, 2$. The generalized correlation model is nested and defined as follows: Suppose π is the marginal probability that an observation Y_{ijk} is a success across all N patients, and suppose the conditional distribution of Q_i given π is beta (a,b), where Q_i is the probability that subject i is assigned a 1, averaged over both situations. Furthermore, let Q_{ij} be the probability that subject i is assigned a 1 under scenario j, then the conditional distribution of Q_{ij} given Q_i is assigned a beta $(\lambda_j Q_i, \lambda_j(1-Q_i))$.

As important components of the model are two types of correlation, namely

$$\kappa_{c_j} = \text{corr}(X_{ij1}, X_{ij2}/Q_i) \tag{4.3}$$

and

$$\kappa_b = \text{corr}(X_{ijk}, X_{ij'k}) = (1+a+b)^{-1} \tag{4.4}$$

where $j \neq j'$ and $k \neq k'$. The first correlation is conditional on Q_i and is referred to as the within situation correlation, while the latter is called the between situation correlation.

Assuming the above and referring to Donner et al.[4], one may show that the joint distribution of (X_{i11}, X_{i12}) for the first scenario is

$$P_1(\kappa_{c_1}) = P_r(X_{i11}=0, X_{i12}=0)$$

$$= (1-\pi)^2(1-\kappa_b) + \kappa_{c_1}(1-\kappa_b)\pi(1-\pi) + \kappa_b(1-\pi)$$

$$P_2(\kappa_{c_1}) = \Pr(X_{i11} = 1, X_{i12} = 0 \text{ or } X_{i11} = 0, X_{i12} = 1)$$

$$= 2(1 - \kappa_{c_1})(1 - \kappa_b)\pi(1 - \pi) \tag{4.5}$$

and

$$P_3(\kappa_{c_1}) = \Pr(X_{i11} = 1, X_{i12} = 1)$$

$$= \pi^2(1 - \kappa_b) + \kappa_{c_1}(1 - \kappa_b)\pi(1 - \pi) + \kappa_b\pi$$

Equation 4.5 is valid for the first situation and an obvious second set of equations refer to the second situation.

It can also be shown that

$$\kappa_j = \text{corr}(X_{ij1}, X_{ij2}) = \kappa_b + \kappa_{c_j}(1 - \kappa_b) \tag{4.6}$$

for $j = 1, 2$ therefore Equation 4.5 can be written in terms of κ_j as

$$P_1(\kappa_1) = (1 - \pi)^2 + \kappa_1\pi(1 - \pi)$$

$$P_2(\kappa_1) = 2(1 - \kappa_1) + \pi(1 - \pi) \tag{4.7}$$

And

$$P_3(\kappa_1) = \pi^2 + \kappa_1\pi(1 - \pi)$$

A similar set of equations are valid for situation 2, by replacing κ_1 by κ_2.

According to Donner and Shoukri[4], the joint distribution of the frequencies V_{ij} and corresponding probabilities θ_{ij} is shown in Table 4.6.

The table is interpreted as follows: The numbers in the cells are from the eye experiment of Table 4.1, and θ_{11} is the probability that raters 1 and 2 assign a zero to a patient in both scenarios, therefore V_{11} is the number of patients who received a score of 0 by both raters for both eyes. For example, for the eye example of Table 4.1, θ_{11} is the probability that both examiners assign a 0 to both eyes, etc. In a similar way, θ_{32} is the probability rater 1 assigns a 1 and rater 2 assigns a 0 to a patient in situation 2 (or vice versa), while both assign a 1 to a patient in situation 1. Note the marginal probabilities (Equation 4.7) are the marginal totals given in the above table, thus, for example,

$$P_1(\kappa_2) = \theta_{11} + \theta_{21} + \theta_{31} = \theta_{.1}$$

Donner et al. employ maximum likelihood to estimate the two kappa coefficients and show that these estimators are given by:

TABLE 4.6

Joint Distribution for Generalized Correlation Model.

Category	Category				
	Rater 2				
Rater 1	(0,0)	(1,0) or (0,1)	(1,1)	Total	Marginal Situation 1
(0,0)	(V_{11},θ_{11})	(V_{12},θ_{12})	(V_{13},θ_{13})	$(V_{1.},\theta_{1.})$	$P_1(\kappa_1)$
	800	14	2	816	
(1,0) or (0,1)	(V_{21},θ_{21})	(V_{22},θ_{22})	(V_{23},θ_{23})	$(V_{2.},\theta_{2.})$	$P_2(\kappa_1)$
	12	3	0	15	
(1,1)	(V_{31},θ_{31})	(V_{32},θ_{32})	(V_{33},θ_{33})	$(V_{3.},\theta_{3.})$	$P_3(\kappa_1)$
	5	0	4	9	
Total	$(V_{.1},\theta_{.1})$	$(V_{.2},\theta_{.2})$	$(V_{.3},\theta_{.3})$	$(V_{..},1)$	
	817	17	6	840	
Marginal Situation 2	$P_1(\kappa_2)$	$P_2(\kappa_2)$	$P_3(\kappa_2)$		

$$\tilde{\pi} = (1/4N)[V_{2.} + V_{.2} + 2(V_{.3} + V_{3.})]$$

$$\tilde{\kappa}_1 = 1 - V_{.2}/(2N\tilde{\pi}_1(1-\tilde{\pi}_1))$$

$$\tilde{\kappa}_2 = 1 - V_{2.}/(2N\tilde{\pi}_2(1-\tilde{\pi}_2)) \tag{4.8}$$

$$\tilde{\pi}_1 = (2V_{.3} + V_{.2})/2N$$

$$\tilde{\pi}_2 = (2V_{3.} + V_{2.})/2N$$

and

$$\tilde{\kappa}_b = [(V_{22} + 2V_{23} + 2V_{32} + 4V_{33}) - 4N\tilde{\pi}^2]/4N\tilde{\pi}(1-\tilde{\pi})$$

With these estimators, Donner et al. estimate κ_1 (for the left eye) as 0.4034, κ_2 (for the right eye) as 0.54, and κ_b as 0.29, that is, the correlation between the two raters for paired scores between the right and left eye is estimated as 0.29.

The Bayesian analog (see Equation 4.8) will be based on the following parameters:

TABLE 4.7

Bayesian Posterior Analysis Based on Generalized Correlation Model

Parameter	Mean	SD	2½	Median	97½
Difference	−0.0997	0.1133	−0.3223	−0.1002	0.1243
π	0.0235	0.0038	0.0167	0.0232	0.0316
π_1	0.0223	0.0043	0.0147	0.0220	0.0316
π_2	0.0247	0.0046	0.0165	0.0244	0.0347
κ_1	0.4556	0.0995	0.2605	0.456	0.648
κ_2	0.5533	0.0914	0.3679	0.559	0.723
κ_b	0.303	0.0828	0.1535	0.2988	0.4743

$$\pi = (1/4)[\theta_{2.} + \theta_{.2} + 2(\theta_{.3} + \theta_{3.})]$$

$$\kappa_1 = 1 - \theta_{.2}/(2\pi_1(1 - \pi_1))$$

$$\kappa_2 = 1 - \theta_{.2}/(2\pi_2(1 - \pi_2)) \qquad (4.9)$$

$$\pi_1 = (2\theta_{.3} + \theta_{.2})/2$$

$$\pi_2 = (2\theta_{3.} + \theta_{2.})/2$$

and

$$\kappa_b = [(\theta_{22} + 2\theta_{23} + 4\theta_{33}) - 4\pi^2]/4\pi(1 - \pi)$$

The Bayesian analysis is based on Table 4.6 for the eye experiment and on the parameters in Equation 4.9, with a uniform prior for the nine multinomial parameters of the table. The kappa parameters are κ_1, for the left eye, and κ_2, for the right. The analysis is executed using 30,000 observations generated from the joint posterior distribution, a burn in of 5000, and a refresh of 1000. See BUGS CODE 4.2, where the results are reported as in Table 4.7.

The Bayesian analog agrees quite well with the Donner et al. analysis. For example, the posterior mean of κ_b is 0.303, while the MLE is 0.29, and one would most likely conclude there is no statistical difference between κ_1 and κ_2, the kappas for the left and right eyes, respectively. A 95% credible interval for the difference is ($-0.32, 0.12$) implying little evidence of a difference, which agrees with a non significant test of the hypothesis that the two kappas are the same (see Donner et al.[4]).

BUGS CODE 4.2

model;

is for the analysis of the eye data from Table 4.6

```
{
g[1,1]~dgamma(801,2)
g[1,2]~dgamma(15,2)
g[1,3]~dgamma(3,2)
g[2,1]~dgamma(13,2)
g[2,2]~dgamma(4,2)
g[2,3]~dgamma(1,2)
g[3,1]~dgamma(6,2)
g[3,2]~dgamma(1,2)
g[3,3]~dgamma(5,2)
h <- sum(g[,])
for ( i in 1 :3) { for ( j in 1 :3) {th[i,j] <- g[i,j]/h}}
```

the g[i,j] are the multinomial parameters for the 3×3 Table 4.6
the th[i,j] have a Dirichlet distribution

```
th1. <- sum(th[1,])
th2. <-sum(th[2,])
th3. <-sum(th[3,])
th.1 <-sum(th[,1])
th.2 <-sum(th[,2])
th.3 <-sum(th[,3])

pi <-(th2. + th.2 + 2*(th3. + th.3))/4

kappa1 <-1-th.2/(2*p1*(1-p1))

kappa2 <-1-th2./(2*p2*(1-p2))

p1 <-(2*th.3 + th.2)/2

p2 <-(2*th3. + th2.)/2

kappab <- ((th[2,2] + 2*th[2,3] + 4*th[3,3])-4*pi*pi)/(4*pi*(1-pi))

}
list( g= structure(.Data =c(1,1,1,1,1,1,1,1,1), .Dim=c(3,3)))
```

4.5 The G Coefficient and Other Indices of Agreement

The G coefficient for $c+1$ scores and four raters is defined as

$$G = P - (1 - P)$$

where P is the probability of unanimous agreement (all four agree) and

$$P = \sum_{i=0}^{i=c} \theta_{iiii} \tag{4.10}$$

Of course, the G coefficient is easily extended to five or more raters and, the main advantage of this index is that it refers to the probability of the main event of interest and has a convenient interpretation as follows: (a) if $P=1$, $G=1$, there is perfect agreement among the four raters for all subjects; (b) if $P=0$, $G=-1$, there is no unanimous agreement among the four for all subjects; and (c) if $P=1/2$, $G=0$, there is a split between the probability of agreement and probability of disagreement.

What is the posterior analysis for the G coefficient for the example of Table 4.1? The G coefficient for the right eye, and the G coefficient for the left eye are shown in Table 4.8. Note that P_r and P_l are the probabilities of agreement for the right and left eyes, respectively. As with using the kappa analysis, reported in Table 4.4, the agreement is higher for the right eye compared to the left eye, however, for the G coefficient, the difference is almost negligible. In addition, the probability of agreement for the right eye is only slightly higher than that for the left. See BUGS CODE 4.1 for the way these indices are calculated.

4.6 Homogeneity with Dependent Data

The ideas of homogeneity and agreement are linked, because if the raters do not agree one would expect lack of homogeneity between them as well. Of course, if raters do not agree there may be many additional reasons to why they do not. An examination of homogeneity is in order when one studies the agreement between raters. To begin our discussion of homogeneity,

TABLE 4.8

Posterior Analysis of G Coefficient of the Eye Data

Parameter	Mean	SD	2½	Median	97½
G	0.8999	0.0154	0.8677	0.9008	0.9279
P	0.9498	0.0077	0.9334	0.9502	0.9639
G_r	0.9419	0.0122	0.9153	0.9428	0.9632
P_r	0.9708	0.0062	0.9574	0.9712	0.9817
G_l	0.9371	0.0127	0.9097	0.9381	0.9595
P_l	0.9685	0.0063	0.9547	0.969	0.979

TABLE 4.9

Posterior Analysis for Homogeneity of Eye Data

Parameter	Mean	SD	2½	Median	97½
$\theta_{1.}$	0.9646	0.0063	0.9511	0.965	0.9761
$\theta_{.1}$	0.9658	0.0062	0.9524	0.9662	0.9769
$\theta_{2.}$	0.02119	0.0049	0.0125	0.0208	0.0320
$\theta_{.2}$	0.0235	0.0052	0.0143	0.0232	0.0348
$\theta_{3.}$	0.0141	0.0040	0.0073	0.0138	0.0231
$\theta_{.3}$	0.0106	0.0035	0.0048	0.0102	0.0185
δ_1	−0.0011	0.0071	−0.0152	−0.0011	0.0129
δ_2	−0.0023	0.0064	−0.0152	−0.0023	0.0101
δ_3	0.0035	0.0039	−0.0039	0.0034	0.0116

consider the eye study of Table 4.6. The Bayesian analysis is based on the generalized correlation model and reveals a posterior mean (SD) for kappa of the left and right eyes as 0.4556 (0.0995) and 0.5533 (0.0914), respectively. What does homogeneity tell us about the differences in agreement between the two examiners? In Table 4.9, $\theta_{1.}$ is the probability that rater 1 scores (0,0) for the left and right eyes, while $\theta_{.1}$ is the probability that rater 2 scores (0,0) for the right and left eyes. Also, $\theta_{2.}$ is the probability that rater 1 scores either a (1,0) or a (0,1) for the right and left eye and $\theta_{.2}$ is the corresponding probability that rater 2 scores (1,0) or (0,1) for the right and left eyes, while $\theta_{.3}$ is the probability that rater 2 score (1,1) for the right and left eyes. The other parameters are: $\delta_1 = \theta_{1.} - \theta_{.1}$, $\delta_2 = \theta_{2.} - \theta_{.2}$, and $\delta_3 = \theta_{3.} - \delta_3$.

The posterior distributions of the paired theta parameters are almost identical! For example the posterior means of $\theta_{1.}$ and $\theta_{.1}$ are 0.9646 and 0.9658, respectively, while the corresponding difference δ_1 has a 95% credible interval of (−0.0152, 0.0129), implying very little difference in the two probabilities of the two raters of the event (0,0) for scores assigned to the right and left eyes of a patient. The same inference applies to the other two events {(0,1) or (1,0)} and {(1,1)}.

The analysis is based on Table 4.6 and BUG CODE 4.3, which is executed with a burn in of 5000 observations with 30,000 observations generated from the joint posterior distribution of the parameters and a refresh of 100.

BUGS CODE 4.3

model;

is for the analysis of the eye data from Table 4.6

```
{
g[1,1]~dgamma(801,2)
g[1,2]~dgamma(15,2)
```

g[1,3]~dgamma(3,2)

g[2,1]~dgamma(13,2)

g[2,2]~dgamma(4,2)

g[2,3]~dgamma(1,2)

g[3,1]~dgamma(6,2)

g[3,2]~dgamma(1,2)

g[3,3]~dgamma(5,2)

h <- sum(g[,])

for (i in 1 :3) { for (j in 1 :3) {th[i,j] <- g[i,j]/h}}

the g[i,j] are the multinomial parameters for the 3×3 Table 4.6

the th[i,j] have a Dirichlet distribution

th1. <- sum(th[1,])

th2. <-sum(th[2,])

th3. <-sum(th[3,])

th.1 <-sum(th[,1])

th.2 <-sum(th[,2])

th.3 <-sum(th[,3])

delta1 <- th1.-th.1

delta2 <- th2.-th.2

delta3 <- th3.-th.3

list(g= structure(.Data =c(1,1,1,1,1,1,1,1,1), .Dim=c(3,3)))

As another example, consider Table 4.10 where two radiologists are examining CT images for the presence of prostate cancer, where 0 indicates no evidence and 1 signifies a definite presence of the disease. Also, the two radiologists return to each image and assign a score independently of what was assigned on the first image. The study recruited 1000 patients, and $R_{10}R_{20}$ denotes that a score of 0 was assigned on the first replication and 0 on the second replication, thus, for this event the two raters score 951 images, where both assigned a 0 (no evidence of disease) on both replications. On the other hand, for the event $R_{10}R_{21}$ (a 0 assigned for the first rep and a 1 for the second) for the first rater and $R_{11}R_{20}$ (a 1 is assigned for the first replication and a 0 to the second) for the second, there were no patients receiving these scores. Are the two raters homogenous?

A Bayesian analysis is reported below in Table 4.11 and shows the homogeneity of the two raters. For example, the posterior mean of rater 1 for the probability of the event $R_{10}R_{20}$ (a 0 is assigned to both replicates of the same image) is 0.9617 for rater 1 and 0.9519 for rater 2, and the posterior mean of the difference of the two probabilities is 0.0098 with a 95% credible interval of (− 0.0039, 0.0234). A similar result is obtained for the other three events. For

TABLE 4.10

CT Information for Two Radiologists and Diagnosis of Prostate Cancer

Rater 1	Rater 2				
	$R_{10}R_{20}$	$R_{10}R_{21}$	$R_{11}R_{20}$	$R_{11}R_{21}$	Total
$R_{10}R_{20}$	$n_{00}=951$	$n_{01}=11$	$n_{02}=10$	$n_{03}=3$	$n_{0.}=975$
	δ_{00}	δ_{01}	δ_{02}	δ_{03}	$\delta_{0.}$
	$w_{00}=1$	$w_{01}=.5$	$w_{02}=.5$	$w_{03}=0$	
$R_{10}R_{21}$	$n_{10}=5$	$n_{11}=2$	$n_{12}=0$	$n_{13}=1$	$n_{1.}=8$
	δ_{10}	δ_{11}	δ_{12}	δ_{13}	$\delta_{1.}$
	$w_{10}=.5$	$w_{11}=1$	$w_{12}=0$	$w_{13}=.5$	
$R_{11}R_{20}$	$n_{20}=4$	$n_{21}=0$	$n_{22}=3$	$n_{23}=2$	$n_{2.}=9$
	δ_{20}	δ_{21}	δ_{22}	δ_{23}	$\delta_{2.}$
	$w_{20}=.5$	$w_{21}=0$	$w_{22}=1$	$w_{23}=.5$	
$R_{11}R_{21}$	$n_{30}=2$	$n_{31}=0$	$n_{32}=0$	$n_{33}=6$	$n_{3.}=8$
	δ_{30}	δ_{31}	δ_{32}	δ_{33}	$\delta_{3.}$
	$w_{30}=0$	$w_{31}=.5$	$w_{32}=.5$	$w_{33}=1$	
Total	$n_{.0}=962$	$n_{.1}=13$	$n_{.2}=13$	$n_{.3}=12$	$N=1000$
	$\delta_{.0}$	$\delta_{.1}$	$\delta_{.2}$	$\delta_{.3}$	

TABLE 4.11

Posterior Analysis for Homogeneity of the CT Information

Parameter	Mean	SD	2½	Median	97½
$\delta_{0.}$	0.9617	0.0064	0.9481	0.9622	0.9733
$\delta_{.0}$	0.9519	0.0067	0.9378	0.9522	0.964
$\delta_{1.}$	0.0118	0.0033	0.0061	0.0115	0.0193
$\delta_{.1}$	0.0166	0.0039	0.0097	0.0163	0.0253
$\delta_{2.}$	0.0117	0.0033	0.0061	0.0114	0.0192
$\delta_{.2}$	0.0157	0.0039	0.0089	0.0153	0.0242
$\delta_{3.}$	0.0146	0.0044	0.0074	0.0142	0.0245
$\delta_{.3}$	0.0157	0.0039	0.9909	0.0154	0.0242
dif0	0.0098	0.0069	−0.0039	0.0099	0.0234
dif1	−0.0048	0.0046	−0.0143	−0.0048	0.0042
dif2	−0.0039	0.0046	−0.0132	−0.0038	0.0050
dif3	−0.0010	0.0046	−0.0098	−0.0012	0.0087

example, the probability of rater 1 for the event $R_{11}R_{21}$ (a 1 is assigned to both reps) is 0.0146, while for rater 2 it is 0.0157, with a 95% credible interval for the difference of (− 0.0098, 0.0087)! An overall conclusion of homogeneity is not a risky implication. The analysis was not based on the generalized correlation model, but instead it was assumed the 1000 patients were selected at random from some well-defined population so that the cell frequencies of Table 4.11 follow a multinomial distribution with 16 cells. The assumption of a uniform density for the cell probabilities implies that the posterior distribution of them is Dirichlet. Since kappa is a function of the probability of agreement and the chance probability of agreement, homogeneity can convey only part of the story of overall agreement.

A Bayesian analysis was executed with BUGS CODE 4.4 using 35,000 observations generated from the posterior distribution of the 12 parameters of Table 4.11, with a refresh of 100, and a burn in of 1000.

BUGS CODE 4.4

model

is for the analysis of the CT data from Table 4.10

```
{
g[1]~dgamma(952,2)
g[2]~dgamma(12,2)
g[3]~dgamma(11,2)
g[4]~dgamma(4,2)
g[5]~dgamma(6,2)
g[6]~dgamma(3,2)
g[7]~dgamma(1,2)
g[8]~dgamma(2,2)
g[9]~dgamma(5,2)
g[10]~dgamma(1,2)
g[11]~dgamma(3,2)
g[12]~dgamma(3,2)
g[13]~dgamma(3,1)
g[14]~dgamma(1,2)
g[15]~dgamma(1,2)
g[16]~dgamma(7,2)
h <- sum(g[])
for ( i in 1 :16) { a[i] <- g[i]/h}
# the a[i] are the multinomial parameters for the 4×4 Table 4.10.
# the a[i] have a Dirichlet distribution
delta0. <-a[1] + a[2] + a[3] + a[4]
delta.0 <-a[1] + a[5] + a[9] + a[13]
delta1. <-a[5] + a[6] + a[7] + a[8]
```

```
delta.1 <-a[2]+a[6]+a[10]+a[14]
delta2. <-a[9]+a[10]+a[11]+a[12]
delta.2 <-a[3]+a[7]+a[11]+a[15]
delta3. <-a[13]+a[14]+a[15]+a[16]
delta.3 <-a[4]+a[8]+a[12]+a[16]

dif0 <-delta0.-delta.0
dif1 <-delta1.-delta.1
dif2 <-delta2.-delta.2
dif3 <-delta3.-delta.3
}
list( g=c(1,1,1,1,1,1,1,1,1,1,1,1,1,1,1,1))
```

4.7 Logistic Regression and Agreement

It is often the case that the scores assigned by raters depend on certain subject and study covariates. For example, suppose that two radiologists are assigning binary scores to cancer patients undergoing an MRI, where a score of 0 indicates no evidence of a lung lesion, but 1 signifies evidence of a lung lesion. One would expect a patient's scores to depend on such covariates as age, gender, stage of disease, and previous and current treatments. There may be other information such as size of lesion and family history, and design characteristics (study site, history and experience of the health care workers, including the raters), etc. that could affect the scores assigned by raters and hence affect measures of agreement like kappa or the G coefficient. In logistic regression one could label the dependent variable as 1 if the raters agree (with a 0 or 1) on a patient's score, otherwise, label the dependent variable with a 0, signifying no agreement. Thus, the information for each patient would be the value of the dependent variable, and the corresponding values of the patient's covariates, and the main purpose of the regression would be to asses the effect of the raters on agreement, as well as the effects of the other covariates. The dependent variable in a logistic regression model targets the cell entries of the agreement table.

This section will present two examples: (a) One taken from Shoukri[1] which involves two populations and two tests for the diagnosis of tuberculosis; and (b) a similar example that involves the diagnosis of lung cancer with two radiologists using CT, but where there is a true replication of the study. Thus, the two raters see each images of all patients and provide a binary score for the diagnosis. After three weeks, the two radiologists return to score each image a second time. Thus there is a dependence between the two scores on the same image.

The Shoukri example is taken from an analysis of Shoukri and Mian[8], where the two skin tests for the diagnosis of tuberculosis each assign a binary score to each patient, with 0 indicating no evidence of disease and 1 denotes evidence of tuberculosis. The procedure was repeated with a second population conducted at the Minnesota State Sanitarium, where the first study was carried out at a southern school district. Since the populations are completely different in regard to the type of patients and geographic location, one would not expect a dependence or correlation between the two. Consider Table 5.8 of Shoukri, which is presented as Table 4.12.

Thus, there are a total of 1401 subjects, of which 79 are from a southern school district and 1322 form a Minnesota Sanitarium. There are 52 who tested negative from the first population, and 367 tested negative from the second. Shoukri employed a logistic regression where the dependent variable is 0 if the patient tested negative and is a 1 if positive, and where the independent variables are indicators of the raters and of the populations.

The analysis presented here is Bayesian and based on the model

$$\text{Logit}(\theta_{ij}) = \beta_1 + \beta_2 X_1(i) + \beta_3 X_2(j) \tag{4.11}$$

for the first population, where $i, j = 0, 1$, and θ_{ij} is the probability that the two tests agree when the first test gave a score of i and test 2 a score of j. A similar model for the second population is given by

$$\text{Logit}(\phi_{ij}) = \delta_1 + \delta_2 X_1(i) + \delta_3 X_2(j) \tag{4.12}$$

where ϕ_{ij} is the probability the two tests for tuberculosis agree, when test 1 assigns a score of i and rater 2 a score of j. Thus, the two tests agree when a (0,0) is assigned by both or when (1,1) is assigned by both. On the other hand, the two tests disagree when the assigned scores are (0,1) or (1,0) assigned by test 1 and test 2, respectively. Thus two separate analyses were executed, one for population 1 and the other for population 2.

A uniform prior was put on the 2×2 table for the first population and also independently for the 2×2 table for the parameters of the second population, resulting in Dirichlet posterior distributions for the two multinomial

TABLE 4.12

Test Results for Skin Tests

Test 1 Response	Test 2 Response	Population 1	Population 2	Total
0	0	52	367	419
0	1	9	37	46
1	0	4	31	35
1	1	14	887	901
Total		79	1322	1401

Source: Shoukri, M. M., and I. U. M. Mian. 1996. Maximum likelihood estimation of the kappa coefficient from bivariate logistic regression. *Stat. Med.* 15, 1409.

populations. The analysis is presented in two parts, where the first consists of estimating the agreement between two tests with the kappa and the *G* coefficients. The second part consists of performing a logistic regression for both populations and estimating the effects of each test on the logit of the probability that the two tests agree, etc. (Table 4.13)

Based on the 95% credible intervals, there appears to be some difference in the kappa and *G* coefficients of the two populations. This is also confirmed by the cagree1 and cagree2 posterior means, where the probability of agreement between test 1 and test 2 for the first populations is 0.6054 and is 0.5772 for the second.

The second part of the analysis is a presentation for the logistic regressions Equation 4.11 and Equation 4.12 for populations 1 and 2, respectively. Thus consider Table 4.14.

For the first population, the effect of test 1 on the logit is -1.726, as measured by the posterior mean, while for the second population the effect is 0.325, and for the second rater for the first population, the effect on the logit is -1.328 and is -2.604 for the second population. With regard to the first population, there appears to be no difference in the effect on the logit between tests 1 and 2, but on the other hand for the second population, there does appear to be some difference in the effects of the two tests. Concerning the difference in the effects on the logit of test 1 between the two populations, there does appear to be a difference, based on the 95% credible interval for beta[2] − delta[2]. Also, the credible interval of (0.5866, 1.959) for beta[3] − delta[3], implies a large difference in the effects of test 2 on the logit between the two populations. Of course, the results of the logistic regression complement those for the conventional approach to estimating agreement by kappa.

TABLE 4.13

Bayesian Posterior Analysis for Agreement of Two Tests for Diagnosis of Tuberculosis in Two Populations

Parameter	Mean	SD	2½	Median	97½
agree1	0.8191	0.0421	0.7292	0.8218	0.8941
agree2	0.9472	0.0061	0.9346	0.9474	0.9586
cagree1	0.6054	0.0395	0.5357	0.6032	0.6892
cagree2	0.5772	0.0095	0.5589	0.5769	0.5965
kappa1	0.54	0.1015	0.3312	0.5447	0.7235
kappa2	0.8751	0.0144	0.8457	0.8756	0.902
G1	0.638	0.0842	0.4596	0.6434	0.7857
G2	0.8944	0.0123	0.8689	0.8948	0.9172
K12	−0.3338	0.102	−0.5463	−0.329	−0.1481
G12	−0.2554	0.0843	−0.4346	−0.2505	−0.1049

TABLE 4.14

Logistic Regression for Population 1 and Population 2

Parameter	Mean	SD	2½	Median	97½	Odds Ratio
beta[1]	0.1428	0.2142	−0.275	0.1418	0.5647	1.153
beta[2]	−1.726	0.3145	−2.358	−1.72	−1.127	0.1779
beta[3]	−1.328	0.3013	−1.929	−1.325	−0.747	0.2650
delta[1]	−0.9566	0.0612	−1.078	−0.9563	−0.8375	
delta[2]	0.325	0.0736	0.1811	0.3248	0.4702	
delta[3]	−2.604	0.1786	−2.966	−2.6	−2.265	
beta[2]−beta[3]	−0.3974	0.4007	−1.189	−0.3948	0.3832	
delta[2]−delta[3]	2.929	0.1728	2.602	2.925	3.28	
beta[2]−delta[2]	−2.051	0.3234	−2.699	−2.046	−1.433	
beta[3]−delta[3]	1.276	0.3503	0.5866	1.277	1.959	

BUGS CODE 4.5

model;

{

g1~dgamma(53,2)

g2~dgamma(10,2)

g3~dgamma(5,2)

g4~dgamma(15,2)

g5~dgamma(368,2)

g6~dgamma(38,2)

g7~dgamma(32,2)

g8~dgamma(888,2)

h1 <- g1 + g2 + g3 + g4

h2 <- g5 + g6 + g7 + g8

th00 <- g1/h1

th01 <-g2/h1

th10 <-g3/h1

th11 <-g4/h1

ph00 <- g5/h2

ph01 <- g6/h2

ph10 <- g7/h2

ph11 <- g8/h2

```
G1 < - agree1- (1-agree1)
G2 < - agree2- (1-agree2)

th0. < - th00 + th01
th.0 < - th00 + th10
th1. < - th10 + th11
th.1 < - th01 + th11

ph0. < - ph00 + ph01
ph.0 < - ph00 + ph10
ph1. < - ph10 + ph11
ph.1 < - ph01 + ph11

agree1 < - th00 + th11
agree2 < - ph00 + ph11

cagree1 < - th0.*th.0 + th1.*th.1
cagree2 < - ph0.*ph.0 + ph1.*ph.1

kappa1 < - (agree1-cagree1)/(1-cagree1)
kappa2 < - (agree2-cagree2)/(1-cagree2)
k12 < -kappa1-kappa2

# logistic regression for Shoukri page 77.

y11~dbin(theta[1,1],79)
y12~dbin(theta[1,2],79)
y21~dbin(theta[2,1],79)
y22~dbin(theta[2,2],79)

z11~dbin(phi[1,1],1322)
z12~dbin(phi[1,2],1322)
z21~dbin(phi[2,1],1322)
z22~dbin(phi[2,1],1322)

# this is the logistic for population 1
for(i in 1 :2){for (j in 1 :2){logit(theta[i,j]) < -beta[1] + beta[2]*x1[i] + beta[3]*x2[j]}}
# the logistic for population 2
for(i in 1 :2){for (j in 1 :2){logit(phi[i,j]) < -delta[1] + delta[2]*x1[i] + delta[3]*x2[j]}}
dbd1 < -beta[1]-delta[1]
dbd2 < -beta[2]-delta[2]
dbd3 < -beta[3]-delta[3]
for (i in 1 :3){ beta[i]~dnorm(0.000,.00001)}
```

for (i in 1 :3){ delta[i]~dnorm(0.000,.00001)}

for(i in 1 :2){for(j in 1 :2){ p[i,j] <-
(exp(beta[1]+beta[2]*x1[i]+beta[3]*x2[j]))/(1+exp(beta[1]+beta[2]*x1[i]
+beta[3]*x2[j])))}}

for(i in 1:2){for(j in 1:2){ q[i,j] <-
(exp(delta[1]+delta[2]*x1[i]+delta[3]*x2[j]))/(1+exp(delta[1]+delta[2]*x1[i]
+delta[3]*x2[j])))}}

db23 <- beta[2]-beta[3]

dd23 <-delta[2]-delta[3]

kaplog <-(agreel-cagreel)/(1-cagreel)

agreel <- theta[1,1]+theta[2,2]

cagreel <-(theta[1,1]+theta[1,2])*(theta[1,1]+theta[2,1])+(theta[2,1]
+theta[2,2])*(theta[1,2]+theta[2,2])

}

list(x1=c(0,1), x2=c(0,1), y11=52, y12=9, y21= 4,y22=14, z11= 367,
z12=37,z21= 31,z22= 887))

list(g1=1,g2=1,g3=1,g4=1,g5=1,g6=1,g7=1,g8=1,
beta=c(0,0,0),delta=c(0,0,0))

Suppose that two radiologists are conducting a study for the diagnosis of lung cancer. They examine one image for each patient with a score of 1 if there is no evidence of the disease and a 2 if there is evidence of the disease. After about three weeks, they return and grade each image again, using the same scoring system as in the first 'look' at a subject's image. Thus the study has two scores per subject and one would expect a correlation between the two scores of a subject, and one would want to focus on the agreement between the two radiologists at each replication and to see the effect of the raters on the scores of the image , at each replication, and overall. In Table 4.14, $R_{11}R_{21}$ designates the event where a score of 1 (no evidence of the disease), is assigned on both replications of an image. There where 665 subjects who were scored a 1 on both reps by both raters. On the other hand, there were only four subjects who are scored a 2 (definite evidence of disease) by both raters on each replication. The probability that both raters score a 2 on both reps is $\theta[2,2,2,2]$. Thus in general $\theta[i,j,k,l]$ is the probability that rater 1 scores a j on rep i and that rater 2 scores a l on rep k, where $i, j, k, l = 1, 2$. Also, the study is designed so that the frequencies of the 16 categories follow a multinomial distribution with $N=697$ subjects and parameters $\theta[i,j,k,l]$, thus the marginal cell frequencies follow a binomial distribution. For example, the first cell has a binomial distribution with parameters $\theta[1,1,1,1]$ and $N=697$ (Table 4.15).

BUGS CODE 4.6 is based on the following logistic model:

$$\text{logit}(\theta[i,j,k,l]) = \beta_1 + \beta_2 X_{11} + \beta_3 X_{12} + \beta_4 X_{21} + \beta_5 X_{22} \tag{4.13}$$

TABLE 4.15

CT Information for Two Radiologists and Diagnosis of Lung Cancer

	Rater 2				
Rater 1	$R_{11}R_{21}$	$R_{11}R_{22}$	$R_{12}R_{21}$	$R_{12}R_{22}$	Total
$R_{11}R_{21}$	665	7	8	1	681
	$\theta[1,1,1,1]$	$\theta[1,1,1,2]$	$\theta[1,1,2,1]$	$\theta[1,1,2,2]$	
$R_{11}R_{22}$	3	1	0	2	6
	$\theta[1,2,1,1]$	$\theta[1,2,1,2]$	$\theta[1,2,2,1]$	$\theta[1,2,2,2]$	
$R_{12}R_{21}$	2	0	2	1	5
	$\theta[2,1,1,1]$	$\theta[2,1,1,2]$	$\theta[2,1,2,1]$	$\theta[2,1,2,2]$	
$R_{12}R_{22}$	1	0	0	4	5
	$\theta[2,2,1,1]$	$\theta[2,2,1,2]$	$\theta[2,2,2,1]$	$\theta[2,2,2,2]$	
Total	671	8	10	8	$N=697$

where β_1 is a constant, β_2 is a constant measuring the effect of rater 1 on the logit at the first replication, β_3 measures the effect on the logit of rater 1 at the second replication, β_4 is the effect of the second rater at the first rep, and lastly β_5 is the effect of the logit of the second rater at the second look of an image. The X_{ij} are the independent variables, where i refers to the rater and j to the rep, with $i, j = 1, 2$, thus X_{21} has possible values 1 and 2, where if $X_{21} = 1$, the second rater at the first rep assigns a score of 1, otherwise $X_{21} = 2$, that is, at the first rep, the second rater assigns a score of 2, indicating the image provides evidence of lung cancer. With regard to the code below, there are two list statements, where the first provides the cell frequencies with the four dimensional vector y, and the values of the independent variables X_{ij} are also given, while the second list statement lists the initial values of the regression coefficients, which are labeled as $b[i]$, $i = 1, 2, \ldots, 5$. These are given non informative prior distributions that are normal with means zero and precisions 0.00001 (variance = 100,000).

BUGS CODE 4.6

model

```
{
y[1,1,1,1]~dbin(theta[1,1,1,1],697)
y[1,1,1,2]~dbin(theta[1,1,1,2],697)
y[1,1,2,1]~dbin(theta[1,1,2,1],697)
y[1,1,2,2]~dbin(theta[1,1,2,2],697)
y[1,2,1,1]~dbin(theta[1,2,1,1],697)
y[1,2,1,2]~dbin(theta[1,2,1,2],697)
y[1,2,2,1]~dbin(theta[1,2,2,1],697)
```

y[1,2,2,2]~dbin(theta[1,2,2,2],697)
y[2,1,1,1]~dbin(theta[2,1,1,1],697)
y[2,1,1,2]~dbin(theta[2,1,1,2],697)
y[2,1,2,1]~dbin(theta[2,1,2,1],697)
y[2,1,2,2]~dbin(theta[2,1,2,2],697)
y[2,2,1,1]~dbin(theta[2,2,1,1],697)
y[2,2,1,2]~dbin(theta[2,2,1,2],697)
y[2,2,2,1]~dbin(theta[2,2,2,1],697)
y[2,2,2,2]~dbin(theta[2,2,2,2],697)

for(i in 1 :2){for(j in 1 :2){for(k in 1 :2){for(l in 1 :2){logit(theta[i,j,k,l]) <-
b[1]+b[2]*X11[i]+b[3]*X12[j]+b[4]*X21[k]+b[5]*X22[l]}}}}

for(i in 1 :5) { b[i]~dnorm(0.0,.00001)}

for(i in 1 :2){for(j in 1 :2){ for(k in 1 :2){ for(l in 1 :2){q[i,j,k,l] <-
exp(b[1]+b[2]*X11[i]+b[3]*X12[j]+b[4]*X21[k]+b[5]*X22[l])/(1+exp(b[1]+
b[2]*X11[i]+b[3]*X12[j]+b[4]*X21[k]+b[5]*X22[l]))}}}}

}

list(y = structure(.Data = c(665,7,8,1,3,1,0,2,2,0,2,1,1,0,0,4),.Dim = c(2,2,2,2)),
X11 = c(1,2),X12 = c(1,2), X21 = c(1,2), X22 = c(1,2)))

list(b = c(0,0,0,0,0))

In order to perform the analysis, 75,000 observations were generated from the joint posterior distribution, with a burn in of 5000 and a refresh of 100. Table 4.16 gives the results of the analysis, where the main focus is on the coefficients of the logistic regression. Their values provide evidence of good agreement between the two radiologists.

Based on Table 4.16, one may conclude that the effects on the logit of each raters' score is approximately the same between replications. For example, the effect of rater 1 on the logit at rep 1 is −6.763 and is −6.655 for rep 2, essentially the same effect. A similar result holds for the second rater. Also it appears that the effect on the logit of rater 1 at rep 1 is the same as the effect on the logit at rep 1 for rater 2. The D parameter is the sum of squared deviations of the observed cell frequencies from their expected values, based on the logistic model. The odds for $b[2]$ is interpreted as follows: The possible values of the corresponding independent variable are 1 and 2, thus the odds of the event that this value is 2 is the value given. Note that the odds of scoring a 1 is much more likely than scoring a 2. Overall, this implies good agreement between the two radiologists at each rep and overall.

This section serves as an introduction to modeling of agreement studies, and will be studied in depth in Chapter 5, where a log linear model approach is taken to the problem of assessing agreement between observers.

TABLE 4.16

Posterior Analysis for Logistic Regression of Two Radiologists Diagnosing for Lung Cancer

Parameter	Mean	SD	2½	Median	97½	Odds
D	950.6	374	391.1	895.5	1814	
$b[1]$	28.29	0.9202	26.58	28.26	30.07	1.932×10^{12}
$b[2]$	−6.763	0.3585	−7.515	−6.747	−6.103	0.001155757
$b[3]$	−6.655	0.3541	−7.376	−6.638	−6.021	0.001287568
$b[4]$	−6.136	0.2827	−6.72	−6.127	−5.61	?
$b[5]$	−6.259	0.2953	−6.86	−6.25	−5.706	?

Exercises

1. Prove or disprove that the Shouten parameter (Equation 4.2) and Oden parameter (Equation 4.1) are identical if the Oden weights are chosen as follows: $w_{00} = w_{11} = 1$ and the off-diagonal weights are given a value of zero. The Schouten weights are provided by Table 4.1.

2. Consider Table 4.17 that lists the outcome of the scoring of two pathologists for two replicates of the study with 27 patients. A 0 indicates absence of dysplasia and 1 denotes evidence of the disorder. The results are from a study of Baker[9]. Using a uniform prior, perform a Bayesian analysis and estimate the kappa index for each replicate, for the difference in the two kappas, and the between replicate correlation κ_b. Refer to BUGS CODE 4.2 and Equation 4.9. Also, plot the posterior densities of these parameters and explain if there is a difference in the two kappas.

3. Consider again Table 4.17, but suppose a previous related experiment with 18 patients gave the outcomes in Table 4.18. (a) Using a Bayesian approach and employing the table below as prior information, perform a Bayesian analysis based on the generalized correlation model with 25,000 observations generated from the posterior distribution of the parameters, with a burn in of 1000, and a refresh of 100. (b) Also perform the classical analysis of Donner et al. based on Equations 4.8. (c) How would the Donner approach utilize the prior information in the table? (d) Compare the classical and Bayesian methods. How well do they agree?

4. Based on Table 4.1, verify Table 4.4.

5. Verify Table 4.5.

6. Verify Table 4.7.

TABLE 4.17

Joint Distribution for Generalized Correlation Model

Category					
Category	Pathologist 1				
Pathologist 2	(0,0)	(1,0) or (0,1)	(1,1)	Total	Marginal replication 1
(0,0)	(V_{11}, θ_{11})	(V_{12}, θ_{12})	(V_{13}, θ_{13})	$(V_{1.}, \theta_{1.})$	$P_1(\kappa_1)$
	9	5	6	20	
(1,0) or (0,1)	(V_{21}, θ_{21})	(V_{22}, θ_{22})	(V_{23}, θ_{23})	$(V_{2.}, \theta_{2.})$	$P_2(\kappa_1)$
	0	1	0	1	
(1,1)	(V_{31}, θ_{31})	(V_{32}, θ_{32})	(V_{33}, θ_{33})	$(V_{3.}, \theta_{3.})$	$P_3(\kappa_1)$
	1	1	4	6	
Total	$(V_{.1}, \theta_{.1})$	$(V_{.2}, \theta_{.2})$	$(V_{.3}, \theta_{.3})$	$(V_{..}, 1)$	
	10	7	10	27	
Marginal replication 2	$P_1(\kappa_2)$	$P_2(\kappa_2)$	$P_3(\kappa_2)$		

TABLE 4.18

Joint Distribution for Generalized Correlation Model

Category					
Category	Pathologist 1				
Pathologist 2	(0,0)	(1,0) or (0,1)	(1,1)	Total	Marginal replication 1
(0, 0)	(V_{11}, θ_{11})	(V_{12}, θ_{12})	(V_{13}, θ_{13})	$(V_{1.}, \theta_{1.})$	$P_1(\kappa_1)$
	7	3	2	12	
(1,0) or (0,1)	(V_{21}, θ_{21})	(V_{22}, θ_{22})	(V_{23}, θ_{23})	$(V_{2.}, \theta_{2.})$	$P_2(\kappa_1)$
	0	1	0	1	
(1,1)	(V_{31}, θ_{31})	(V_{32}, θ_{32})	(V_{33}, θ_{33})	$(V_{3.}, \theta_{3.})$	$P_3(\kappa_1)$
	2	1	2	5	
Total	$(V_{.1}, \theta_{.1})$	$(V_{.2}, \theta_{.2})$	$(V_{.3}, \theta_{.3})$	$(V_{..}, 1)$	
	9	5	4	18	
Marginal replication 2	$P_1(\kappa_2)$	$P_2(\kappa_2)$	$P_3(\kappa_2)$		

7. Consider the following generalization of Table 4.1 for the eye example, where each rater scores each eye with a 0, 1, or 2. That is for each rater the possible outcomes are R_iL_j, where $i, j = 0, 1, 2$, that is the right eye receives a score of i and the left a score of j, thus there are nine possible paired ways to score the right and left eyes per rater, a total of 81 possibilities for two raters. (a) Describe how to analyze such an experiment with a Bayesian approach. (b) Describe a table for the outcomes of the right eye. (c) Describe a table for the outcomes of the left eye. (d) Suppose the scores are interpreted as follows: 0 signifies definitely no evidence of atrophy, a 1 implies uncertain evidence of atrophy, and a 2 means there is definitely evidence of atrophy, then assign frequencies to the 81 cells of the table for 840 patients and perform a posterior analysis for the overall kappa, a kappa for the left eye only, a kappa for the right eye only, then compare the two kappas. With 55,000 observations generated from the joint posterior distribution, a burn in of 5000, and a refresh of 100, write the code similar to that of BUGS CODE 4.1 in order to execute the analysis. Plot the posterior densities of the three kappas (overall, for the right, and for the left eye).

8. Compare the Bayesian analysis of the eye data given in Table 4.5 to the Bayesian analog of the generalized correlation model given in Table 4.7. (a) Is there an advantage to the analysis of Table 4.5 to that of Table 4.7? (b) If there is an advantage or disadvantage to the generalized correlation model, explain the difference in the two approaches.

9. Under what conditions does the generalized correlation model reduce to the common correlation model described in Section 2.8? Verify your conclusions.

10. How would you generalize the generalized correlation model to three scores per eye, in the eye study of Table 4.6? You may have to search the literature.

11. Using BUGS CODE 4.1, verify Figures 4.1–4.3 with 55,000 observations generated from the joint posterior distribution, with a refresh of 100 and a burn in of 5000.

12. Three non Bayesian approaches were used for the eye data of Table 4.1. The two weighting schemes due to Oden and Schouten and one based on the generalized correlation model of Donner et al.[4] and reported by Shoukri[1]. What are the advantages and disadvantages of each compared to the other? Does the Bayesian analog of these approaches add anything to one's understanding of agreement for dependent observations?

13. Verify Table 4.9. The two raters appear to be homogenous in the sense the two have the same probability of the same event, however the two kappas for the right and left eyes are different with posterior means 0.4556 and 0.5333. Explain how the two kappas can be different, but the raters appear to be homogeneous.

TABLE 4.19

Posterior Analysis for the Shouten and Oden Indices of Agreement

Parameter	Mean	SD	2½	Median	97½
κ	0.379	0.0598	0.2651	0.3778	0.4985
κ_{pol} (Oden)	0.4357	0.0685	0.3018	0.4358	0.568
κ_w (Schouten)	0.4357	0.0685	0.3018	0.4358	0.568

TABLE 4.20

Posterior Distribution Based on Table 4.10

Parameter	Mean	SD	2½	Median	97½
G	0.896	0.0147	0.8651	0.8968	0.9226
kappa	0.379	0.0597	0.265	0.378	0.4977
Agree	0.948	0.0073	0.9325	0.9484	0.9613
cagree	0.9161	0.0105	0.8941	0.9166	0.9354

14. (a) Based on Table 4.10, verify Table 4.11. (b) Also, show that the posterior analysis for the overall κ, the Oden kappa, κ_{pol}, and the Schouten kappa, κ_w, is as illustrated in Table 4.19. Why are the Oden and Schouten posterior means the same?

15. See Table 4.10 and show that the posterior analysis for agreement is given by Table 4.20. The kappa parameter is the overall measure of agreement, made up of the two components, agree and cagree, where the former is the probability of agreement (the probability of the diagonal cells of the table) and cagree is the probability of agreement assuming independence of the two radiologists. The other measure of agreement is the G coefficient with a posterior mean of 0.896, indicating very high agreement. (a) Write your own program and use 30,000 observations for the joint posterior distribution of the above four parameters, with a burn in of 5000 observations and a refresh of 100. (b) How do the results for agreement impinge on the results for homogeneity given by Table 4.11? Do these results help explain the overall conclusions about homogeneity?

16. Based on Table 4.12 and BUGS CODE 4.5, verify Table 4.13.

17. Based on Table 4.12 and BUGS CODE 4.5, verify Table 4.14.

18. Complete the column for the odds of the various effects of Table 4.14 and provide an interpretation.

19. Based on BUGS CODE 4.6, verify Table 4.16. Use a burn in of 5000 observations and generate 75,000 observations from the joint posterior distribution with a refresh of 100. Complete the column for the odds.

TABLE 4.21

CT Information for Two Radiologists and Diagnosis of Lung Cancer: A Preliminary
Study

Rater 1	Rater 2				Total
	$R_{11}R_{21}$	$R_{11}R_{22}$	$R_{12}R_{21}$	$R_{12}R_{22}$	
$R_{11}R_{21}$	8	2	2	0	12
	$\theta[1,1,1,1]$	$\theta[1,1,1,2]$	$\theta[1,1,2,1]$	$\theta[1,1,2,2]$	
$R_{11}R_{22}$	3	3	1	2	9
	$\theta[1,2,1,1]$	$\theta[1,2,1,2]$	$\theta[1,2,2,1]$	$\theta[1,2,2,2]$	
$R_{12}R_{21}$	2	1	2	3	8
	$\theta[2,1,1,1]$	$\theta[2,1,1,2]$	$\theta[2,1,2,1]$	$\theta[2,1,2,2]$	
$R_{12}R_{22}$	9	2	4	0	15
	$\theta[2,2,1,1]$	$\theta[2,2,1,2]$	$\theta[2,2,2,1]$	$\theta[2,2,2,2]$	
Total	22	8	9	5	$M=44$

20. Based on Table 4.15 and a revision of BUGS CODE 4.6, execute a
    Bayesian analysis for the agreement between the two radiologists.
    Find the posterior mean and standard deviation of kappa for the first
    rep, kappa for the second rep, and the overall kappa. Compare the
    two kappas for the two reps.
21. Table 4.21 presents the results of a preliminary study with 44 patients,
    using CT for the diagnosis of lung cancer. As with Table 4.15, 1
    denotes a negative finding and 2 denotes evidence of the disease. (a)
    Consider the results of Table 4.21 as prior information and update
    the posterior results of Table 4.16. In order to execute the analysis,
    refer to BUGS CODE 4.6 and generate 65,000 observations from the
    posterior distribution, with a burn in of 10,000 with a refresh of 100.
    Perform the analysis for the coefficients of the logistic regression
    (Equation 4.13) by computing the posterior mean, standard devia-
    tion, median, and a 95% credible interval for each coefficient. (b)
    Plot the posterior density of the posterior distribution of the squared
    deviations (observed versus expected cell frequencies) D. (c) Provide
    a scatter plot of the observed versus expected cell frequencies. (d)
    Does this plot imply a good fit of the logistic model to the data?

References

1. Shoukri, M. M. 2004. *Measures of interobserver agreement*. Boca Raton: Chapman
   & Hall/CRC.
2. Oden, N. 1991. Estimating Kappa from binocular data. *Stat. Med.* 10, 1301.

3. Schouten, H. J. A. 1993. Estimating kappa from binocular data and comparing marginal probabilities. *Stat. Med.* 12, 2207.

4. Donner, A., M. M. Shoukri, N. Klar, and N. Bartfay. 2002. Testing the equality of two dependent kappa statistics. *Stat. Med.* 11, 373.

5. Agresti, A. 1992. Modeling pattern of agreement and disagreement. *Stat. Meth. Med. Res.* 1, 201.

6. Tanner, M., and M. A. Young. 1985. Modeling agreement among raters. *J. Am. Stat. Assoc.* 80, 175.

7. Von Eye, A., and E. Y. Mun. 2005. *Analyzing rater agreement.* London: Lawrence Erlbaum Associates.

8. Shoukri, M. M., and I. U. M. Mian. 1996. Maximum likelihood estimation of the kappa coefficient from bivariate logistic regression. *Stat. Med.* 15, 1409.

9. Baker, S. G., L. S. Friedman, and M. K. B. Parmer. 1961. Using replicate observations in observer agreement studies with binary observations. *Biometrics* 47, 1327.

5

Modeling Patterns of Agreement

5.1 Introduction

Using models for agreement was introduced in Section 4.7 of Chapter 4, where the logistic model was employed to measure the effect of two tests on the scores used to diagnose tuberculosis with a skin test. Two populations were involved and the objective is to measure the effect of the two tests on agreement and to compare the agreement between the two populations. A similar approach is taken here, where the logistic linear model is again employed, however, the approach is extended in order to examine more subtle effects of agreement between two or more raters and to incorporate factors that could have a possible influence on the scores of the raters and hence on the various measures that estimate agreement.

There are two important books that introduce the interested reader to the subject of agreement. Shoukri[1] primarily emphasizes overall measures of agreement, such as kappa and the intraclass correlation. On the other hand, the book by Von Eye and Mun[2] primarily emphasizes the use of the log linear model to explore agreement between raters, and their method will be followed to some extent. Both books introduce agreement topics for continuous scores, but their primary interest is in other topics. The log linear model of Von Eye and Mun is similar to Tanner and Young[3], who also used the model, however, the Von Eye and Mun method uses an interesting and novel strategy for model selection and diagnostics. The Bayesian approach taken here is based on the logistic linear model, which is quite similar to the log linear, and the strategy to select a model, using a sequence of hierarchical nested larger models, is similar to Von Eye and Mun.

The model diagnostics for the Bayesian method is somewhat different than the strategy of Von Eye and Mun, and is based on comparing the observed and expected (expectations computed according to a tentative model) cell frequencies and estimating the Euclidean distance between the observed and expected frequencies. The Euclidean distance has a posterior distribution which can be compared to the posterior distribution of the residuals of competing models.

Von Eye and Mun employ a model selection technique that begins with a base model that includes the main effects of the raters and is appropriate if the raters are assigning scores independently, then at the next stage the model is enlarged to include so called weight vectors (equal and unequal weights) that put more emphasis on the diagonal cells, then at the next stage, if appropriate and subject covariates are relevant, they are also included in the model. At each stage of this strategy, the Euclidean distance goodness of fit statistic is computed to gauge the fit of the model. The optimal model is selected based on the one with the smallest chi square value. At each stage, the significance of the added variables can be judged by conventional tests of hypotheses.

Chapter 5 is divided into two broad parts: one for nominal scores, and the other part for ordinal scores. With regard to ordinal scores, a linear by linear association vector is introduced to take into account the ordinal nature of the responses. For ordinal as with nominal, the model selection strategy is the same and the optimal one selected on the basis of comparing the posterior distribution of the Euclidean distance between the observed and expected cell frequencies of the agreement table. Often interesting hypotheses can be built into the design matrix of the model. For example, rater-specific trends (e.g. are there more frequencies above the main diagonal of the agreement table than below) that compare the two raters are examined with the logistic model.

An interesting feature of Von Eye and Mun[2] approach is to employ a design matrix that models the main effects of the raters as contrasts between the scores of a rater and some reference score. For example, if a rater uses three nominal scores, say 1, 2, and 3, the design matrix uses two columns for the effects of a rater, where one column expresses the contrast between scores 1 and 3, and the other expresses the contrast between scores 2 and 3. This imposes a constraint on the three levels of the rater, which of course is needed in order to uniquely identify the parameters of the model.

The chapter is concluded by examining exploratory methods that identify the type of agreement or disagreement at the end of model selection. Once a model is selected, the way the raters agree or disagree should be investigated. This is done by employing the so-called techniques referred to as configural frequency analysis by Von Eye and Mun, which amounts to estimating the direction (are they positive or negative?) in the way the observed and expected frequencies differ. A complete Bayesian analysis is taken where by the posterior distribution of the differences (observed minus expected cell frequencies) will provide the investigator with a conclusion about the pattern of agreement between the two raters.

This chapter is essentially on regression, where agreement studies with continuous observations are presented. If categorical scores are appropriate the logistic model measures the rater main effects on the logit of the probability of the cell frequencies. If additional factors such as certain weight vectors and/or covariates are included, the model measures the effect of the levels of those factors on the logit of the cell probabilities, adjusted for the other factors in the model. For a particular model, the fit will produce residuals, based on the observed and expected cell frequencies. The Euclidean distance between the observed and expected (expected with respect to the tentative model at hand) is adopted as the appropriate measure of how well the model fits the data. It is important to remember that the residuals have a posterior distribution with a posterior mean and variance and that these measures of fit can be compared between competing models. The residuals have a posterior distribution, because they are functions of the expected frequencies, which in turn depend on the posterior distribution of the cell probabilities of the agreement table.

Some of the examples will include those that were studied in earlier chapters, plus some of those analyzed by Von Eye and Mun, and in the latter scenario, the Bayesian analysis can be compared to the conventional one made up of chi square goodness of fit tests and likelihood ratio statistics. In other examples, the one important asset of the Bayesian method is illustrated with information from previous experiments which can be treated as prior information and combined with the present experiment to arrive at the posterior distribution for the final result.

Modeling patterns of agreement via the logistic model allows one to determine the effects each rater has in assigning scores to the subjects and allows one to compare the effects between raters. For example, trends of one rater can be compared to the other, in that one rater may be assigning higher (or lower) scores than the other raters. One of main strengths of the modeling approach is that subject covariates can be included in the analysis and their effects measured on the manner the raters are assigning scores. Thus, modeling increase one's understanding of agreement and adds greatly to the usual measures of agreement such as the G coefficient and kappa.

5.2 Nominal Responses

Suppose the 3×3 table for agreement is considered as in Table 5.1, where ϕ_1 is the probability that raters A and B assign a 1 to a subject, and ϕ_9 is the probability that rater A assigns a 3 and rater B assigns a 3, etc. The corresponding

TABLE 5.1

Two Raters and Three Nominal Scores

Rater A	Rater B			Total
	1	2	3	
1	ϕ_1	ϕ_2	ϕ_3	
2	ϕ_4	ϕ_5	ϕ_6	
3	ϕ_7	ϕ_8	ϕ_9	
Total				

cell frequencies are denoted by n_i, where $n = \sum_{i=1}^{i=9} n_i$. The logistic linear model is defined as

$$\text{logit}(\phi_i) = \sum_{j=1}^{j=p} b[j]X_j[i] \qquad (5.1)$$

where,

$$\text{logit}(\phi_i) = \log[\phi_i/(1 - \phi_i)]$$

and $i = 1, 2, \ldots, 9$. Note that ϕ_i is the cell probability corresponding to the i-th cell of Table 5.1, where i is read from top to bottom and from left to right. Thus, the model consists of p factors which might have some effect on the logit of the cell probabilities. It is obvious how to extend the model to more than two raters and scores. Also, the $b[j]$ are considered as scalars constants to be estimated, and the X_j are known independent variables, the effect of which are on the logits as measured by the corresponding $b[j]$. By convention, the column X_1 is a 9×1 vector of ones, while the other vectors of the design matrix are specified by the user, where $X_j[i]$ is the i-th row of X_j.

There are alternative ways to model the agreement responses of Table 5.1, and one of the most popular is the log linear model, which was first used by Tanner and Young and is the model of choice by Von Eye and Mun.

In the log linear case the logit transformation of Equation 5.1 is replaced by the log function, thus the log linear model is specified by

$$\log(\phi_i) = \sum_{j=1}^{j=p} b[j]X_j[i] \qquad (5.2)$$

and the effect of the j-th independent variable, as measured by $b[j]$, is on the log scale of ϕ_i, instead of on the logit scale. I found very little difference in both approaches, however it can happen that the natural constraints ($0 \le \phi_i \le 1$ and $\sum_{i=1}^{i=9} \phi_i = 1$) on the cell probabilities can be violated with the log linear model.

Note that if Equation 5.1 is true, then it follows that the cell probabilities are given by

$$\phi_i = \frac{\exp\left(\sum_{j=1}^{j=p} b[j]X_j[i]\right)}{1 + \exp\left(\sum_{j=1}^{j=p} b[j]X_j[i]\right)} \qquad (5.3)$$

thus, it is seen how the various experimental factors affect the cell probabilities. If the agreement study is well-designed, where the n subjects are selected at random from some well-defined population, then the cell frequencies will have a joint multinomial distribution and marginally the individual cell frequencies will have a binomial distribution. Consequently, if a Bayesian analysis uses a Dirichlet prior distribution, the cell probabilities will jointly follow a Dirichlet distribution, and the marginal distribution of a single ϕ_i will follow a beta posterior distribution. Three types of prior information will be employed in this book: (a) a uniform prior distribution; (b) an improper type prior distribution; and (c) previous experimental results that are relevant to the experiment at hand.

There are several strategies for selecting an appropriate model for the agreement table, and that will very much depend on the information available for the study. In addition to the rater scores of the table, there may be information on the raters and information about the subjects. For example, there may be information on sex, gender and additional information on medical history including previous and present treatments. Certain scores of the raters may have implications about medial treatment and one would want to take that into account. Also, additional information about design of the study may be relevant. For example, is the study carried on at many sites, where one would expect the location of a site to affect the agreement scores of the rater, and so on.

The strategy taken here is the one adopted by Von Eye and Mun[2] in their book *Analyzing Rater Agreement*, where they proceed through various stages by beginning with a simple model (called the base model) that incorporates only the effects of the raters on agreement scores. Depending on the information available and on the objectives of the study, they consider models containing weight vectors that assign either equal or unequal weights to the diagonal cells of the agreement (Table 5.1). The purpose here is to weigh the cell vectors according to the importance the investigator places on that particular score. Thus, if one believes that the raters agree equally likely on each possible score, equal weights are appropriate, but on the other hand if one believes that the agreement among raters for a particular score is more frequent than for other scores, then unequal weights are in order. This type of consideration is a particular pattern of agreement that can be explored by means of the logistic model. Still another way the model can be augmented is with covariate information on the subjects or study design. The covariates can be either discrete or continuous. If such information affects the agreement pattern, then it should be included in the model. At each stage of the model building process, the observed and expected cell frequencies can be compared, via the Euclidean distance, or by Bayesian analogs of the chi square or the likelihood-ratio chi-square test for goodness of fit.

Another feature of the Von Eye and Mun model adopted in this chapter is to use the effects approach for the design matrix of the model. To see this, consider the 3×3 table above with two raters, and suppose the columns of model (Equation 5.1) are chosen as:

$$X_1 = (1,1,1,1,1,1,1,1,1)'$$
$$X_2 = (1,1,1,0,0,0,-1,-1,-1)'$$

$$X_3 = (0,0,0,1,1,1,-1,-1,-1)' \tag{5.4}$$
$$X_4 = (1,0,-1,1,0,-1,1,0,-1)'$$

and

$$X_5 = (0,1,-1,0,1,-1,0,1,-1)'$$

This choice of design matrix is interpreted as follows: The first column is constant for the overall mean of the logits of the cell probabilities, while the second column is the contrast between score 1 and score 3 of rater A. In a similar fashion, the third column is the contrast between score 2 and score 3 for rater A, while the analogous contrasts for rater B, are specified by columns four and five. Note that the design matrix is non singular and the main effect parameters $b[j], j = 1, 2, \ldots, p$ are identifiable.

Our model building strategy is initiated with an example form Von Eye and Schuster[4], where two psychiatrists score a subject as either not depressed, scored as a 1, mildly depressed, with a score of 2, and clinically depressed, with a score of 3. Note these are considered nominal scores, and the information from the study is given in Table 5.2, where the cell frequencies and the corresponding probabilities are displayed. The majority ($82/129 = 0.635$) of patients are clinically depressed, and a small percentage, $3/129 = 0.023$, are mildly depressed, while $11/129 = 0.085$ were classified by both raters as not depressed. Overall, the probability of agreement is 0.744 with a G coefficient of 0.488, which indicates fairly good agreement. According to Von Eye and Mun[2], kappa $= 0.375$ (0.079), with a chance probability of agreement of 0.591, which again implies fair agreement in the diagnosis of depression between the two psychiatrists.

What does the logistic model tell us about the pattern of agreement or disagreement between the two? In order to implement the Bayesian analysis, the assumption of a multinomial experiment for Table 5.2 is made, which implies the cell frequencies follow a multinomial distribution, and individually each cell frequency follows a binomial distribution. The Bayesian analysis is executed using 45,000 observations, where the first 5000 are for the burn in using a refresh of 100. See BUGS CODE 5.1 for the details,

TABLE 5.2

Two Psychiatrists and Three Depression Scores

	Psychiatrist B			
Psychiatrist A	1	2	3	Total
1	ϕ_1, 11	ϕ_2, 2	ϕ_3, 19	32
2	ϕ_4, 1	ϕ_5, 3	ϕ_6, 3	7
3	ϕ_7, 0	ϕ_8, 8	ϕ_9, 82	90
Total	12	13	104	$N = 129$

TABLE 5.3

Posterior Analysis for Logistic Base Model of Depression

Parameter	Mean	SD	2½	Median	97½
$b[1]$	−3.219	0.1907	−3.607	−3.213	−2.861
$b[2]$	0.0745	0.1898	−0.296	0.0726	0.4514
$b[3]$	−1.662	0.2722	−2.232	−1.65	−1.162
$b[4]$	−0.9545	0.2309	−1.428	−0.9487	−.5233
$b[5]$	−0.8656	0.2259	−1.319	−0.8613	−.4368
d24	1.029	0.2937	0.4694	1.024	1.62
d35	−0.796	0.3467	−1.494	−0.7908	−0.1324
ob2	1.097	0.2113	0.7438	1.075	1.571
ob3	0.1968	0.0526	0.1921	0.3129	0.5926
ob4	0.3953	0.0906	0.2399	0.3873	0.5926
ob5	0.4316	0.0972	0.2673	0.4226	0.6461
LR	54.67	17.75	21.14	54.3	90.62
χ^2	80.73	25.28	48.79	75.2	144.8
Dis	21.09	3.154	15.65	20.84	27.89

where the first list statement specifies the columns (Equation 5.4) as the design matrix, and where the second list statement determines the starting values of the $b[]$ vector. The component of the $b[]$ vector are the coefficients of the regression model and they measure the effect of the corresponding column, this $b[2]$ measures the effect of the contrast score 1 minus score 3 of rater A. In a similar fashion, $b[4]$, measures the effect of the same contrast for rater B. The difference in these two is denoted by d24. The corresponding odds are denoted by ob2, ob3, etc. where ob2 is the odds corresponding to the effect $b[2]$, while ob4 is the odds of the effect $b[4]$. Also computed are the expected cell frequencies and the deviations of the observed from the expected cell frequencies. See Table 5.3 for the Bayesian analysis of the logistic model.

BUGS CODE 5.1

Model;

```
# von eye and mun example page 33
{
for ( i in 1:N){y[i]~dbin(p[i],129)}

for(i in 1:N) {logit(p[i])<- b[1]*X1[i]+b[2]*X2[i]+b[3]*X3[i]+b[4]*X4[i]+b[5]*X5[i]}

# b[2] is the effect of the contrast (1–3) for rater A
# b[3] is the effect of the contrast (2–3) for rater A
```

```
# b[4] is the effect of the contrast (1–3) for rater B
# b[5] is the effect of the contrast (2–3) for rater B
ob2<- exp(b[2])
# obs is the odds for the event measured by b[2]
ob3<- exp(b[3])
ob4<- exp(b[4])
ob5<- exp(b[5])

for( i in 1:N){

# E[] the expected cell frequencies
E[i]<-p[i]*129
# dev[] the deviations
dev[i]<-y[i]−E[i]

sdev[i]<- dev[i]*dev[i]
sdevn[i]<-sdev[i]/E[i]
lr[i]<- y[i]*log(y[i]/E[i])

}

LR<- 2*sum(lr[])
chisq<- sum(sdevn[])

dis<-sqrt(sum(sdev[]))
# prior distribution for the coefficients
 for(i in 1:5){ b[i]~dnorm(0.0,.000001)}
d24<- b[2]-b[4]
od24<- exp(d24)
d35<- b[3]-b[5]
od35<- exp(d35)
}
list(N=9,y=c(11,2,19,1,3,3,0,8,82),
X1=c(1,1,1,1,1,1,1,1,1),X2=c(1,1,1,0,0,0,−1,−1,−1),
X3=c(0,0,0,1,1,1,−1,−1,−1), X4=c(1,0,−1,1,0,−1,1,0,−1),
X5 =c(0,1,−1,0,1,−1,0,1,−1))

# X1 is the vector for the constant
# X2 is the contrast between the score 1 and 3 for rater A
# X3 is the contrast between the score 2 and 3 for rater A
```

\# X4 is the contrast between the score 1 and 3 for rater B

\# X5 is the contrast between the score 2 and 3 for rater B

list(b = c(0,0,0,0,0))

What are the rater effects? The effect of the contrast, score 1 minus score 3, of rater A, as measured by the posterior mean of $b[2]$, is 0.0745, with a 95% credible interval of (−0.296, 0.4514) on the logit scale, while the corresponding odds of that contrast is 1.097 with a 95% credible interval of (0.7438, 1.571). For rater B, the effect $b[4]$ of the same contrast (score 1 minus score 3) on the logit scale is −0.9545 with a 95% credible interval of (−1.428, −0.5233) and a corresponding odds of 0.3953, with a 95% credible interval of (0.2399,0.5926). Do the raters have the same effect on the contrast, score 1 minus score 3? The 95% credible interval for the difference in the effect of rater A (on the contrast) minus the effect of rater B, is given by the posterior mean of d24, which is 1.029 (0.2937) and a 95% credible interval of (0.4694, 1.62), suggesting that most likely there is a difference in the two effects, thus in this case, the two raters are not behaving in the same way when assigning scores 1 and 3! Graphical evidence of the above posterior analysis for the contrast score 1 minus score 3 is presented in Figure 5.1 and Figure 5.2.

How well does the logistic model fit the data? The scatter plot (Figure 5.3) shows the expected versus the observed cell frequencies and the linear regression line with slope 0.841 (0.649, 1.033) and intercept 2.277 (−3.188, 7.743), indicating only a fair fit to the cell frequencies. The intervals are the 95% confidence intervals as computed with SPSS, version 11.

It is difficult to judge the fit of the cell frequencies to the logistic model, because there are no competing models by which the goodness of fit can be compared. The observed and expected frequencies are given in Table 5.4.

One may show the sum of squared deviations is 379.08 and that the distance between the observed and expected observations is 19.47. From a Bayesian point of view, the posterior mean of the Euclidean distance is 21.11, a median of 20.84, and a 95% credible interval of (15.7, 28.11).

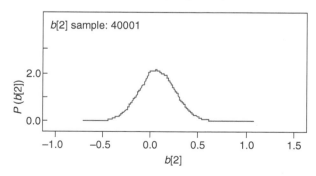

FIGURE 5.1

Posterior density logit of score 1 minus score 3 for rater A.

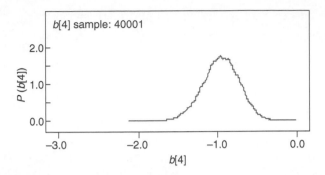

FIGURE 5.2
Posterior density logit of score 1 minus score 3 for rater B.

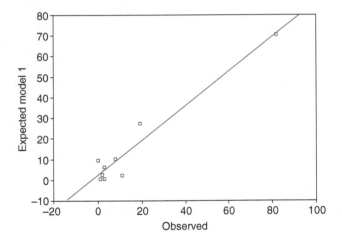

FIGURE 5.3
Expected versus observed depression information.

TABLE 5.4

Observed and Expected Observations Logistic
Model Depression Information

Cell	Expected $E[]$	Observed $y[]$
1	2.237	11
2	2.433	2
3	27.34	19
4	0.4228	1
5	0.461	3
6	6.15	3
7	9.37	0
8	10.14	8
9	70.49	82

Bayesian analogs of the chi-square and likelihood ratio goodness of fit tests are

$$\chi^2 = \sum_{i=1}^{i=n} (y[i] - E[i])^2 / E[i] \tag{5.5}$$

and

$$LR = 2 \sum_{i=1}^{i=n} y[i] \log(y[i]/E[i]) \tag{5.6}$$

respectively, where $E[i]$ is the expected cell frequency for cell i, assuming the expectation is taken with respect to the tentative model being considered at the moment. Of course, the measures of goodness of fit are related to the Euclidean distance, which is

$$\text{Dis} = \sqrt{\sum_{i=1}^{i=n} (y[i] - E[i])^2} \tag{5.7}$$

and the three will be presented at each stage of model development. Note for a perfect fit, the three have the value zero! For the present case of the base logistic model for the depression data of Table 5.2, the posterior analysis for the three measures of goodness of fit is presented in Table 5.3.

See BUGS CODE 5.1 for the statements of computing the three measures of fit. Two stages of model development have been accomplished, and now an additional stage that incorporates information about the patient's depression is taken into account and the model augmented to include equal weights for the diagonal cells of the agreement Table 5.2. If the investigator believes that the raters are giving the same importance to each score, then the equal weight vector

$$X_6 = (1,0,0,0,1,0,0,0,1)'$$

can be added to the design matrix of Equation 5.1. Of course, the addition of this vector might not add any predictive power to the model, indicating that the raters are not giving equal importance to the assigned scores. Does the equal weight vector give a better goodness of fit to the logistic base model? For the second stage model, the posterior analysis gives Table 5.5.

The posterior analysis implies a much better goodness of fit, because the Euclidean distance for the base model is 21.09, as measured by the posterior mean, and now for the second stage model has a posterior mean of 8.628. There was also a dramatic decrease in the chi-square goodness of fit from a posterior mean of 80.73 for the base model, and now a value of 15.92, for the second stage. Von Eye and Mun[2] expand the model building stage to

TABLE 5.5

Posterior Analysis for Depression Information with Equal Weight Vector

Parameter	Mean	SD	2½	Median	97½
$b[1]$	−3.882	0.2704	−4.452	−3.868	−3.391
$b[2]$	0.5677	0.239	0.119	0.5599	1.058
$b[3]$	−1.49	0.2983	−2.115	−1.478	−0.9947
$b[4]$	−1.27	0.2827	−1.855	−1.26	−0.7463
$b[5]$	−0.24	0.264	−0.7623	−0.2375	0.2733
$b[6]$	2.028	0.3187	1.424	2.018	2.676
χ^2	15.92	8.834	7.546	13.49	38.68
Dis	8.628	2.404	5.117	8.224	14.45
LR	13.33	17.33	−19.43	12.96	48.55

inclusion of additional information given by a subject covariate, namely a paranoia rating expressed as an average for each cell,

$$X_7 = (17,27,3,16,45,14,1,3,3)'$$

and when this is included in the logistic model, the posterior analysis is shown in Table 5.6. Measures of goodness of fit such as LR, χ^2, and Dis (Euclidean distance) have decreased from their model 2 values of 13.33, 15.92, and 8.628, respectively, to 9.23, 16.85, and 7.977, respectively, for their model 3 values, which indicates a very slight decrease, thus, the addition of the subject covariates have added very little to the goodness of fit. A look at the 95% credible interval of (−0.3449, −0.0393) for $b[7]$ implies that the added variable does contribute to the overall predictive value of the model, over and above that given by model 2, which did not include the covariates. Note that $b[7]$ measures the effect of the covariate for paranoia on the logit scale. Thus although the goodness of fit has slightly decreased, the addition of the covariate should be retained as part of the description of the effect on the cell probability logits.

Our analysis of the depression data ends here, however the reader is referred to the exercises at the end of the chapter to gain more experience with this Bayesian type of approach to model selection.

The depression data was aggregated over each cell, however, there are cases where the individual information per subject should not be summarized or aggregated, but instead should be analyzed on a per subject basis. For example, consider the following example of a small lung cancer pathology study with two pathologists who are grading biopsy specimens. The specimens are graded as follows: 1 designates no evidence of disease; 2 denotes some evidence of the disease; and 3 means that there is definite evidence of lung cancer. The patients had indications of earlier respiratory problems detected by an exercise stress test, by an experimental CT imaging program and other tests, and had been referred for a lung biopsy by their personal physician. They had not been previously treated for lung cancer. This is not a well-defined population in the sense of being a randomized study, but instead is more representative of

TABLE 5.6

Posterior Analysis Depression Data Model 3

Parameter	Mean	SD	2½	Median	97½
$b[1]$	−2.434	0.6574	−23.68	−2.45	−1.093
$b[2]$	0.7368	0.3032	−3.684	0.7165	1.382
$b[3]$	0.6643	1.032	0.1962	0.5677	2.892
$b[4]$	−2.102	0.5773	−1.125	−2.035	−1.15
$b[5]$	1.621	0.9164	−3.371	1.529	3.593
$b[6]$	4.421	1.22	2.392	4.28	7.111
$b[7]$	−0.1771	0.0800	−0.3449	−0.1689	−0.0393
LR	9.23	17.42	−23.45	8.705	44.67
χ^2	16.85	22.25	4.648	11.71	52.26
Dis	7.977	2.857	3.516	7.59	14.61

TABLE 5.7

Lung Cancer Study with Two Pathologists

Pathologist A	Pathologists B			Total
	1	2	3	
1	32	20	15	67
2	9	15	9	33
3	12	3	8	23
Total	53	38	32	123

the community population of subjects with above average risk for lung cancer. The pathology results are summarized in Table 5.7.

In addition to the pathology results, there is additional information on certain patient characteristics: (a) age; (b) gender; and (c) results of exercise stress test, as positive or negative. The descriptive statistics for the covariates for the subjects are portrayed in Table 5.8. Of course, the pertinent questions to be investigated are: (a) what is the agreement between pathologists? and (b) what is the effect of the patient covariates on the assignment of scores to the subjects? The first question is easily answered by the G coefficient or by kappa, while the later will be approached with a logistic model that includes covariates. Another course of action is to stratify the patients on the basis of age groups, however, this is usually less efficient than using the age directly, because grouping often induces a loss of information about age variability.

Our approach to the analysis will consist of two fundamental phases: (a) the first is where model selection begins initially with the base model, then at the second stage an equal or unequal weight vector is added, then at the final stage, the average age for each cell is added. At each stage the effects of each factor are computed and the three measures (chi-square, likelihood ratio, and Euclidean distance) of goodness of fit estimated; (b) The second phase takes

TABLE 5.8

Lung Cancer Study: Demographics and Patient Covariates

Pathologist A Score	Pathologist B Score	Age (mean)	Age (SD)	Male (Fraction)	Stress Test Positive (Fraction)	N
1	1	50.94	10.94	0.5	0.3437	32
1	2	52.25	11.58	0.55	0.4500	20
1	3	61.47	8.57	0.4667	0.4667	15
2	1	50.56	8.233	0.6667	0.5556	9
2	2	61.07	6.649	0.5333	0.6667	15
2	3	72.78	3.993	0.6667	0.6667	9
3	1	57.58	5.648	0.500	0.7500	12
3	2	70.67	5.508	1.000	1.000	3
3	3	73.63	4.897	0.75	0.875	8

into account gender as a binary covariate, and proceeds with model selection as follows: the base model which includes the two strata, and the main effects (the first contrast where score 1 is compared to score 3 for rater A, and in a similar manner for the corresponding contrast of the second pathologist; the second contrast where score 2 is compared to score 3 for pathologist A, and in a similar way for the corresponding contrast of the second pathologist). After the base model, the four interactions between gender and the two main effects of the two pathologists. Lastly, the average age of the 18 cells (nine cells per gender) is added as a covariate. At each stage of model selection, the three measures of goodness of fit are computed as well as the effects of the raters and their interactions.

The base model for the lung cancer information of Table 5.7 and Table 5.8 is

$$\text{logit}(\varphi_i) = \sum_{j=1}^{j=p} b[j]X_j[i]$$

where $p=5$, $i=1, 2, \ldots , 9$, and the 9×1 columns of the design matrix are given by Equation 5.4.

The column X_1 is the vector of ones, X_2 is the contrast of score 1 (no evidence of disease) minus score 3 (definite evidence of disease) of pathologist A, while X_3 is the contrast of score 2 (ambiguous evidence of disease) minus score 3 for the same pathologist. The 1 minus 3 contrast for pathologist B is given by X_4, while X_5 gives the 2 minus 3 contrast. Thus, the effect of the contrast 1 minus 3 for pathologist A is measured by $b[2]$, etc.

This is the base model and is the reference in the model building strategy, whereby the other models will be judged.

For the lung cancer data of Table 5.7, what are the estimates of the coefficients in the model? Consider BUGS CODE 5.2, which was executed with 70,000 observations, using 5000 as a burn in, with a refresh of 500.

BUGS CODE 5.2

model;

lung cancer chapter 5 base model

{

for (i in 1:N){y[i]~dbin(p[i],123)}

for(i in 1:N) {logit(p[i])<- b[1]*X1[i] + b[2]*X2[i] + b[3]*X3[i] + b[4]*X4[i] + b[5]*X5[i]}

for(i in 1:N){

# E[] the expected cell frequencies
E[i]<-p[i]*123

# dev[] the deviations
dev[i]<-y[i]-E[i]
sdev[i]<- dev[i]*dev[i]
sdevn[i]<-sdev[i]/E[i]
lr[i]<- y[i]*log(y[i]/E[i])
LR<- 2*sum(lr[])

chisq<- sum(sdevn[])

dis<-sqrt(sum(sdev[]))

# prior distribution for the coefficients
 for(i in 1:5){ b[i]~dnorm(0.0,.000001)}

d24<- b[2]-b[4]
d35<- b[3]-b[5]

}

list(N=9, y=c(32,20,15,9,15,9,12,3,8),

# these are the columns of the design matrix
X1=c(1,1,1,1,1,1,1,1,1),X2=c(1,1,1,0,0,0, −1, −1, −1),
X3=c(0,0,0,1,1,1, −1, −1, −1), X4=c(1,0, −1,1,0, −1,1,0, −1),
X5 =c(0,1, −1,0,1, −1,0,1, −1))

list(b=c(0,0,0,0,0))

A posterior analysis for the base model estimates the effects of the five columns in the model, and the inference of primary interest is the difference between columns $b[2]$ and $b[4]$, which is the difference d24 of the 1 minus 3 contrast of the two pathologists. Of equal interest is the difference d35 in the 2 minus 3 scores of the two pathologists. Table 5.9 provides the Bayesian characteristics of the parameters of the model, as well as the three measures

TABLE 5.9

Posterior Analysis for Lung Cancer with the Logistic Base Model

Parameter	Mean	SD	2½	Median	97½
$b[1]$	−2.224	0.1072	−2.439	−2.222	−2.02
$b[2]$	0.6874	0.1316	0.4308	0.6868	0.9474
$b[3]$	−0.1437	0.1497	−0.4405	−0.143	0.146
$b[4]$	0.3283	0.1322	0.0696	0.3281	0.5854
$b[5]$	−0.0648	0.1416	−0.3462	−0.0632	0.2082
d24	0.3591	0.1847	−0.0023	0.3589	0.7211
d35	−0.07886	0.2064	−0.4815	−0.0788	0.3247
LR	13.71	20.77	−25.72	13.32	55.5
χ^2	13.74	3.748	9.229	12.83	23.33
Dis	12.57	1.698	10.18	12.27	16.63

of the goodness of fit, the chi-square, likelihood ratio, and the Euclidean distance.

The main purpose of this analysis is to detect the pattern of agreement between the two pathologists, and it is seen that the effect of the contrast 1 minus 3 of pathologist A on the logit is 0.6874 (as measured by the posterior mean) with a 95% credible interval of (0.4308, 0.9474) and that the effect on the logit of the 2 minus 3 contrast of pathologist B is 0.3283 with a 95% credible interval of (0.0696, 0.5854). The posterior distribution of the difference between the two pathologist has mean 0.3591 with credible interval of (−0.0023, 0.7211) indicating somewhat of a difference, that is, the effect on the contrast 1 minus 3 is different for the two pathologists. Note the inferences are on the logit scale, however, one can compute the odds of each effect. For example, the odds corresponding to $b[2]$ is approximately $\exp(0.6874) = 1.988$, and that for the $b[4]$ is $\exp(0.3283) = 1.388$, thus the odds ratio of pathologist A relative to pathologist B is $1.988/1.388 = 1.432$. What is the interpretation of the odds ratio? The difference between the two of the 1 minus 3 contrast is 0.3591 on the logit scale, but on the odds scale, the appropriate way to compare the two pathologists is by the ratio. See the exercises for additional information about odds and odds ratio.

Suppose the base model is expanded to include the weight vector which assigns equal weights to the diagonal cells of the agreement table. If equal weights are assigned, one implies the same importance is being attached by both raters to the three possible scores 1, 2, and 3. That is, score 3 is as important as score 2 and score 1 as far as agreement between the two is concerned, and this is tested by adding the weight vector to the mode. Of course, equal weights are not always viable because there are situations where unequal weights are appropriate, and this is developed further in the exercise section.

The addition of the weight vector changes estimates of the coefficients of the models as well as the goodness of fit statistics. For example, the Euclidean distance is reduced from 12.54 for the base model, to 10.57 for model with the weight vector. Also, the effect to the weight vector on the logit is estimated as −0.0465 with a 95% credible interval of (−0.3325, 0.2317) implying some effect, but not a substantial one, thus, on the one hand, the addition of the vector appears to reduce the goodness of fit, but on the other hand, by itself, has little effect on the logit.

It appears that the addition of the weight vector does not affect the difference in the contrast score 1 minus score 3 between the two pathologists, that is the two appear to differ in that contrast, and also it appears the addition of the weight vector does not affect the 2 minus 3 contrast between the two, that is, there appears to be no difference between the two. To see this compare Table 5.9 and Table 5.10.

Augmenting the model with the age cell means of Table 5.7 is now considered. Suppose X_7 denotes this vector of covariate values, then the posterior analysis is portrayed by Table 5.11.

With the addition of the age covariate, the Euclidean distance of 10.83 does not change from its previous value of 10.57, however, the effect of age on the logit is 0.5096 with a 95% credible interval of (0.0608, 0.9475) implying that its addition is a valuable predictor of the cell probabilities and corresponding logits. The addition does have a slight effect on the rater main effects (the two contrasts: 1 minus 3; and 2 minus 3), but not much. As measured by the posterior mean of d24, there is a slight difference between the two raters with respect to the contrast 1 minus 3, but with respect to contrast 2 minus 3, there appears to be very little difference. The observed cell frequencies are plotted for the three previous logistic models and they are labeled 1, 2, and 3, where 1 is the base model, 2 denotes the base model plus the equal weight vector, and model 3 is model 2 augmented by the average cell ages. Upon inspection of Figure 5.4, the largest vertical variation is displayed by the base model, but models 2 and 3 appear to have about the same dispersion, thus, based on the graphical evidence, models 2 and 3 appear to give equivalent inferences

TABLE 5.10

Posterior Analysis for Lung Cancer: Logistic Base Model Plus Equal Weights

Parameter	Mean	SD	2½	Median	97½
$b[1]$	−2.388	0.134	−2.657	−2.386	2.132
$b[2]$	0.6623	0.1324	0.404	0.6613	0.9225
$b[3]$	−0.1397	0.1513	−0.4387	−0.1384	0.1522
$b[4]$	0.2491	0.139	−0.0242	0.2499	0.5192
$b[5]$	−0.0465	0.1431	−0.3325	−0.0452	0.2317
d24	0.4132	0.1933	0.0296	0.4115	0.8077
d35	−0.0931	0.2202	−0.5241	−0.0932	0.3397
LR	10.15	20.82	−29.5	9.727	52.04
χ^2	10.49	3.877	5.652	9.642	20.33
Dis	10.57	2.134	7.307	10.27	15.55

TABLE 5.11

Posterior Analysis for Lung Cancer: Logistic Base Model Plus Equal
Weights and Age

Parameter	Mean	SD	2½	Median	97½
$b[1]$	−1.265	2.348	−5.956	−1.235	3.56
$b[2]$	0.5607	0.2538	0.0654	0.5575	1.067
$b[3]$	−0.1453	0.1518	−0.4486	−0.1447	0.1469
$b[4]$	0.1139	0.3189	−0.5254	0.1128	0.7457
$b[5]$	−0.0545	0.1441	−0.3415	−0.0532	0.2252
$b[6]$	−0.0187	0.0389	−0.0989	−0.0190	0.0589
$b[7]$	0.5096	0.2263	0.0608	0.5107	0.9475
d24	0.4468	0.2102	0.0365	0.4553	0.863
d35	−0.0908	0.2219	−0.5292	−0.0894	0.342
LR	10.59	20.73	−28.84	10.16	52.18
χ^2	11.08	4.079	5.868	10.2	21.33
Dis	10.83	2.17	7.464	10.54	15.82

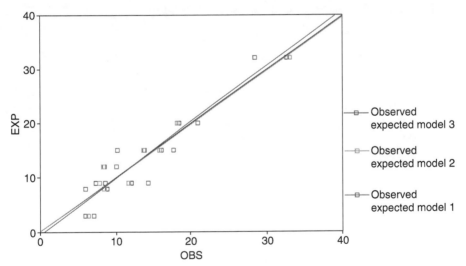

FIGURE 5.4
Observed versus expected models 1, 2, and 3 (lung cancer information).

about the pattern of rater agreement, while models 2 and 3 appear to be an
improvement over model 1.

Often a trend is obvious in the pattern of agreement, and occurs in the
training of novice radiologists, who tend to give either lower or higher scores
compared to a trained professional. Consider the lung cancer data of Table
5.7, where it seems that there are higher scores above the diagonal. How
should this trend be investigated? Let

$$X_6 = (0,1,1,-1,0,1,-1,-1,0)'$$

be the sixth and last column of the design matrix, where the elements above the diagonal are given a weight of 1, while those below are given a weight of −1, and the diagonal cells are weighted with 0. The other five columns refer to the base model for the lung cancer data, where the first column is a vector of ones, the second models the contrast 1 minus 3 for rater A etc. By revising BUGS CODE 5.1 with the addition of the trend vector X_6 the posterior analysis produces 0.6674 (0.3865) as the estimate (sd) of the effect for the trend vector and an estimated Euclidean distance (sd) of 12.08 (1.944), all based on 55,000 observations generated from the joint posterior distribution with a burn in of 5000 and a refresh of 500. It can be shown that the 95% credible intervals for $b[6]$, the effect of the trend vector, and the Euclidean distance are (−0.0763, 1.437) and (9.308, 16.61) respectively. The value for the posterior mean of $b[6]$ and its associated credible interval, imply that X_6 is a useful contribution to the design matrix and implies that it is very likely that the cell frequencies above the diagonal are greater than those below, also implying the rater B is the one scoring higher than rater A. It can also be shown that there is a small difference in the effect of the 1 minus 3 contrast for the two raters.

Phase two of the analysis for the lung cancer data is begun with stratifying on gender, which results in the following table, and proceeds by adding additional information such as gender by pathologist's interactions, equal and unequal weight vectors, and finally the cell age averages as a covariate (Table 5.12).

TABLE 5.12A

Lung Cancer Information for Females

Pathologist A	Pathologist B			
	1	2	3	Total
1	16	9	8	33
2	3	7	3	13
3	6	0	2	8
Total	25	16	13	54

TABLE 5.12B

Lung Cancer Information for Males

Pathologist A	Pathologist B			
	1	2	3	Total
1	16	11	7	34
2	6	8	6	20
3	6	3	6	15
Total	28	22	19	69

The posterior analysis is based on BUGS CODE 5.3 and is executed with 70,000 observations generated from the joint posterior distribution of all the parameters, with a burn in of 5000, and a refresh of 500. As before the model coefficients and the goodness of fit parameters will be estimated. Major emphasis will be placed on the pattern of agreement and on the goodness of fit of successive models. The logistic model is

$$\text{logit}(\varphi_i) = \sum_{j=1}^{j=p} b[j]X_j[i] \tag{5.1}$$

where $p = 6$, $i = 1, 2, \ldots, 18$, and the columns of the design matrix are specified in the first list statement of BUGS CODE 5.3. Note that the 18×1 column vector X_2 is the gender column, column X_3 is the contrast 1 minus 3 for pathologist A which is repeated for both strata.

BUGS CODE 5.3

```
model;
# lung cancer chapter 5 base model with gender
{
for ( i in 1:N){y[i]~dbin(p[i],123)}
for(i in 1:N) {logit(p[i])<- b[1]*X1[i]+b[2]*X2[i]+b[3]*X3[i]+b[4]*X4[i]
+b[5]*X5[i]+b[6]*X6[i]}

for( i in 1:N){
# E[] the expected cell frequencies
E[i]<-p[i]*123
# dev[] the deviations
dev[i]<-y[i]-E[i]
sdev[i]<- dev[i]*dev[i]
sdevn[i]<-sdev[i]/E[i]
lr[i]<- y[i]*log(y[i]/E[i])
}
LR<- 2*sum(lr[])
chisq<- sum(sdevn[])
dis<-sqrt(sum(sdev[]))
# prior distribution for the coefficients
 for(i in 1:6){ b[i]~dnorm(0.0,.000001)}
d35<- b[3]-b[5]
d46<- b[4]-b[6]
```

```
}
list(N = 18,y = c(16,9,8,3,7,3,6,0,2,16,11,7,6,8,6,6,3,6),
X1 = c(1,1,1,1,1,1,1,1,1,1,1,1,1,1,1,1,1,1),
X2 = c(1,1,1,1,1,1,1,1,1, −1, −1, −1, −1, −1, −1, −1, −1, −1),
X3 = c(1,1,1,0,0,0, −1, −1, −1,1,1,1,0,0,0, −1, −1, −1),
X4 = c(0,0,0,1,1,1, −1, −1, −1,0,0,0,1,1,1, −1, −1, −1),
X5 = c(1,0, −1,1,0, −1,1,0, −1,1,0, −1,1,0, −1,1,0, −1),
X6 = c(0,1, −1,0,1, −1,0,1, −1,0,1, −1,0,1, −1,0,1, −1))
list(b = c(0,0,0,0,0,0))
```

Table 5.13 presents the posterior analysis for the lung cancer data stratified by gender, and its effect is −0.1333 on the logit scale with a 95% credible interval of (−0.3212, 0.0512) indicating some effect. The coefficient $b[3]$ measures the effect of the contrast 1 minus 3 of rater A on the logit, while $b[5]$ is the corresponding effect, averaged over strata, for pathologist B.

Goodness of fit parameters, such as the Euclidean distance, signify a good fit, if compared to its values when gender was not included. See Table 5.11 as a reference. Notice the relatively "large" standard deviation for the likelihood ratio parameter, which is defined as the log of a ratio between observed and expected cell frequencies. The next stage of the model selection process is to add the interactions between gender and raters, of which there are four: (a) gender by contrast 1 minus 3 for rater A; (b) gender by contrast 2 minus 3 for rater A; (c) gender by contrast 1 minus 3 for rater B; and (d) gender by contrast 2 minus 3 for rater B. This will add four columns to the base model, and they are denoted by $X_7 - X_{10}$, where X_7 is formed by the product of columns $X_2 \times X_3$, etc. and added to the first list statement of BUGS

TABLE 5.13

Posterior Analysis for Lung Cancer: Logistic Base Plus Gender

Parameter	Mean	SD	2½	Median	97½
$b[1]$	−2.992	0.1056	−3.203	−2.989	−2.789
$b[2]$	−0.1333	0.0946	−0.3212	−0.1329	0.0512
$b[3]$	0.6413	0.1268	0.3937	0.6406	0.8907
$b[4]$	−0.1293	0.1459	−0.4184	−0.1278	0.1517
$b[5]$	0.3042	0.1267	0.0558	0.3042	0.5512
$b[6]$	−0.0593	0.1367	−0.3334	−0.0581	0.2056
d35	0.337	0.1776	−0.0108	0.336	0.686
d46	−0.07	0.1993	−0.4606	−0.0692	0.3233
LR	21.45	21.68	−20.24	21.03	64.76
χ^2	19.66	4.691	13.73	18.63	31.59
Dis	10.48	1.321	8.559	10.27	13.62

CODE 5.3. As above for the base model without interactions, the program is executed with 70,000 observations with a burn in of 5000, and a refresh of 100 to give Table 5.14. The effects of the four gender by rater interactions are estimated by the posterior means of $b[7]$, $b[8]$, $b[9]$, and $b[10]$. Inspecting the above graphs for the effect of the interactions, reveal that the effects of the gender by rater B interactions are quite small, but on the other hand, as for the gender by rater A interactions, they appear to have some effect on the logit. Of course, the addition of the four interactions also affect the effects of the other parameters. Consider the effect of the contrast 1 minus 3 of rater A versus the effect of the same contrast for rater B, which is measured by d36 and has a posterior mean of 0.3555 with a 95% credible interval of (0.000642, 0.7167). This implies that the two pathologists are behaving somewhat differently when assigning scores 1 minus 3. Looking at d46, it seems as if the two have the same effect on the contrast 2 minus 3. Note there is a slight increase in the posterior mean of the goodness of fit parameters such as the Euclidean distance, however the increase is negligible. At the last stage of model building, the age covariate is added as column X_{11} with the following posterior analysis (Table 5.15).

What does this tell us about the pattern of observer agreement? The three indices of goodness of fit have not changed with respect to the previous model, thus, it appears that the addition of age to the model does not improve it. What does this say about the four gender by rater interactions (adjusted)? It appears that they are negligible with their effect on the probit, which implies the pattern of agreement is the same for both genders!

TABLE 5.14

Posterior Analysis for Lung Cancer: Logistic Base Plus Gender Plus Gender by Rater Interactions

Parameter	Mean	SD	2½	Median	97½
$b[1]$	−3.032	0.1096	−3.225	−3.03	−2.825
$b[2]$	−0.2116	0.1097	−0.4284	−0.2105	0.0005186
$b[3]$	0.6745	0.1303	0.4212	0.6734	0.9335
$b[4]$	−0.1279	0.1515	−0.4301	−0.1265	0.1638
$b[5]$	0.3189	0.1287	0.0667	0.3189	0.5704
$b[6]$	−0.0627	0.1394	−0.3406	−0.9616	0.205
$b[7]$	0.18	0.1304	−0.07335	0.1789	0.4381
$b[8]$	0.08775	0.1291	−0.1661	0.0875	0.3407
$b[9]$	−0.0356	0.1508	−0.332	−0.03538	0.2596
$b[10]$	−0.0270	0.1397	−0.3036	−0.0266	0.2474
d35	0.3555	0.1819	0.0000642	0.3546	0.7167
d46	−0.0651	0.2062	−0.4707	−0.0637	0.3771
LR	23.08	21.74	−18.44	22.68	66.94
χ^2	23.23	6.772	14.31	21.86	39.91
Dis	11.13	1.55	8.862	10.94	14.69

TABLE 5.15

Posterior Analysis for Lung Cancer: Logistic Base Plus Gender Plus Gender by Rater Interactions and Age

Parameter	Mean	SD	2½	Median	97½
$b[1]$	−5.111	2.096	−9.292	−5.095	−0.9713
$b[2]$	−0.1595	0.1214	−0.4014	−0.1578	0.0748
$b[3]$	0.8605	0.2289	0.4166	0.8586	1.315
$b[4]$	−0.132	0.152	−0.4353	−0.1304	0.1615
$b[5]$	0.5569	0.2733	0.0227	0.556	1.097
$b[6]$	−0.0234	0.1449	−0.3093	−0.0219	0.2568
$b[7]$	0.2327	0.1407	−0.0391	0.2315	0.511
$b[8]$	0.0865	0.13	−0.1666	0.0861	0.3419
$b[9]$	−0.0745	0.1563	−0.3842	−0.0736	0.2297
$b[10]$	0.0271	0.1502	−0.2703	0.0273	0.3188
$b[11]$	0.0340	0.0343	−0.0339	0.0339	0.1023
d35	0.3037	0.1969	−0.0852	0.3048	0.6891
d46	−0.1085	0.2091	−0.5222	−0.1088	0.2996
LR	23.14	21.77	−18.45	22.73	66.96
χ^2	23.23	7.031	13.89	21.87	40.85
Dis	10.92	1.66	8.292	10.71	14.74

5.3 Ordinal Responses

Von Eye and Mun[2] introduce rater agreement with ordinal scores with a study of Bortz, Lienert, and Boehnke[5]. The study had two journalists rate 100 TV programs by assigning percentages to each show, where the percentage was the percent of the TV audience that will view that particular program. The percentages are grouped into six categories: (1) 0–10%; (2) 11–20%; (3) 21–30%; (4) 31–40%; (5) 41–50%; and (6) at least 51%, with the following results. Von Eye and Mun report a kappa = 0.151 (0.054), which indicates poor agreement, the probability of agreement estimated as only 0.29, and the chance probability of agreement as 0.163. Also, the G coefficient is estimated as −0.42, again confirming rather poor agreement between TV journalists. What is the pattern of agreement and how can the ordinal nature of the rating be emphasized to the advantage of the analysts?

The usual kappa measure and the logistic model considered up to now do not take advantage of the ordinal nature of the information, therefore Agresti[6,7] proposed the following model:

$$\text{logit}(\varphi_i) = \sum_{j=1}^{j=p} b[j]X_j[i] \tag{5.8}$$

where $p = 12$ and the independent variables include the constant vector, 10 vectors for the main effects of the two raters, and a vector that defines a linear by linear association

$$X_{12} = (u_1u_1, u_1u_2, \ldots, u_1u_6, u_2u_1, u_2u_2, \ldots, u_2u_6; \ldots; u_6u_1, u_6u_2, \ldots, u_6u_6).$$

Therefore, $b[12]$ is the effect of the linear by linear interaction. Although the model applies to Table 5.16, it is obvious how to extend it to any number of possible scores. The choice for the values of u_i are usually given by either: (a) the midpoint of the ordinal interval; or (b) the mean of the continuous scores from which the groups are determined; or (c) as an integer. For example, if one uses $u_i = i$, the distance between the categories is determined, thus taking advantage of the ordinal nature of the scores. See Goodman[8], Clogg and Shihadeh[9], Agresti[6,7] for history of the use of the linear by linear association, and for choices for the u_i values. Of course, Equation 5.8 can be expanded to include weight (equal or unequal) vectors and other covariates.

In order to initiate the model building strategy, the base model for the TV data will be analyzed. The base model does not include linear by linear association model, but does include the constant vector and the main effects of the two journalists. The main effect vectors are the following contrasts: 1 minus 6, 2 minus 6, 3 minus 6, 4 minus 6, and 5 minus 6. Thus the design matrix is 11×11 and of full rank, and the corresponding $b[]$ vector of Equation 5.8 measures the corresponding effects of those design columns, therefore $b[1]$ measures the effect of the overall mean on the logit scale, while $b[2]$ measures the effect of the contrast 1 minus 6 for journalist A. The Bayesian approach is executed with 50,000 observations generated from the joint posterior distribution, with a burn in of 5000 and a refresh of 100. Table 5.17 presents the posterior analysis, and a normal non informative prior was put on the 11 coefficients of the logit model.

This analysis provides the reference for successive models, and one sees that the posterior mean of $b[2]$ is 0.3547, which estimates the effect of the

TABLE 5.16

Two Journalists Rating 100 TV Shows

	Journalist B						
Journalist A	1	2	3	4	5	6	Total
1	5	8	1	2	4	2	22
2	3	5	3	5	5	0	21
3	1	2	6	11	2	1	23
4	0	1	5	4	3	3	16
5	0	0	1	2	5	2	10
6	0	0	1	2	1	4	8
Total	9	16	17	26	20	12	100

TABLE 5.17

Posterior Distribution for Scores of Two Journalists

Parameter	Mean	SD	2½	Median	97½
$b[1]$	−3.729	0.1173	−3.967	−3.726	−3.507
$b[2]$	0.3547	0.2115	−0.0696	0.358	0.7579
$b[3]$	0.308	0.2146	−0.1273	0.3118	0.7164
$b[4]$	0.4041	0.2062	−0.0111	0.4077	0.7988
$b[5]$	0.0168	0.2377	−0.4656	0.0234	0.4622
$b[6]$	−0.4876	0.2885	−1.085	−0.476	0.0453
$b[7]$	−0.6066	0.3031	−1.241	−0.5919	−0.0505
$b[8]$	0.07176	0.2314	−0.3983	0.0766	0.512
$b[9]$	0.0704	0.2314	−0.3993	0.0755	0.5075
$b[10]$	0.5223	0.1983	0.1218	0.5268	0.9021
$b[11]$	0.2418	0.2166	−0.1941	0.2462	0.6515
d27	0.9613	0.3705	0.256	0.9531	1.71
d38	0.2362	0.316	−0.3801	0.2339	0.8559
d49	0.3337	0.3087	−0.2682	0.3329	0.9421
d510	−0.5064	0.3092	−1.118	−0.5035	0.0917
d611	−0.7294	0.36	−1.446	−0.7231	−0.0366
LR	62.46	20.29	23.77	62.12	103.3
χ^2	69.63	13.15	51.6	67.23	102.1
Dis	13.15	.7067	12.1	13.04	14.86

contrast 1 minus 6 for journalist A, while the estimated effect $b[7]$ of the same contrast for journalist B is −0.6066. Overall, the difference in these two effects is 0.9613 with a 95% credible interval of (0.256, 1.71) indicating a difference in the effects of the two raters for contrast 1 minus 6, therefore the two are not behaving in the same manner when assigning scores to TV shows. On the other hand, it appears the two are behaving in a similar way when estimating the other four contrasts. Measures of goodness of fit for the base model are estimated by the posterior medians as 62.12, 67.23, and 13.04 for the likelihood ratio, chi-square, and the Euclidean distance, respectively. See BUGS CODE 5.4, where only the first 11 columns were used for the design matrix.

BUGS CODE 5.4

model

Von Eye and Mun example page 56 with unequal weight vector and linear by linear association

```
{
for ( i in 1:N){y[i]~dbin(p[i],100)}
```

```
for(i in 1:N) {logit(p[i])<-
b[1]*X1[i] + b[2]*X2[i] + b[3]*X3[i] + b[4]*X4[i] + b[5]*X5[i] +
b[6]*X6[i] + b[7]*X7[i] + b[8]*X8[i] + b[9]*X9[i] + b[10]*X10[i] + b[11]*X
11[i] + b[12]*X12[i] + b[13]*X13[i]

}
```

```
# b[2] is the effect of the contrast (1-6) for rater 1
# b[3] is the effect of the contrast (2-6) for rater 1
# b[4] is the effect of the contrast (3-6) for rater 1
# b[5] is the effect of the contrast (4-6) for rater 1
# b[6] is the effect of the contrast (5-6) for rater 1

# b[7] is the effect of the contrast (1-6) for rater 2
# b[8] is the effect of the contrast (2-6) for rater 2
# b[9] is the effect of the contrast (3-6) for rater 2
# b[10] is the effect of the contrast (4-6) for rater 2
# b[11] is the effect of the contrast (5-6) for rater 2
# b[12] is the L by L association
# b[13] is the unequal weight vector

d27<-b[2]-b[7]
d38<-b[3]-b[8]
d49<-b[4]-b[9]
d510<-b[5]-b[10]
d611<-b[6]-b[11]
for( i in 1:N){
# E[] the expected cell frequencies
E[i]<-p[i]*100
# dev[] the deviations
dev[i]<-y[i]-E[i]
sdev[i]<-dev[i]*dev[i]
sdevn[i]<-sdev[i]/E[i]
lr[i]<-y[i]*log(y[i]/E[i])
}
LR<- 2*sum(lr[])
chisq<- sum(sdevn[])

dis<-sqrt(sum(sdev[]))
# prior distribution for the coefficients
for(i in 1:13){ b[i]~dnorm(0.0,.00001)}

}
```

list(N = 36, y = c(5,8,1,2,4,2, 3,5,3,5,5,0,

1,2,6,11,2,1, 0,1,5,4,3,3,

0,0,1,2,5,2, 0,1,1,2,1,4),

X1 = c(1,1,1,1,1,1,1,1,1,1,1,1,1,1,1,1,1,1,1,1,1,1,1,1,1,1,1,1,1,1,1,1,1,1,1,1),

X2 = c(1,1,1,1,1,1, 0,0,0,0,0,0, 0,0,0,0,0,0,0,0,0,0,0,0,0,0,0,0,0,0, −1, −1, −1, −1, −1, −1),

X3 = c(0,0,0,0,0,0, 1,1,1,1,1,1, 0,0,0,0,0,0, 0,0,0,0,0,0,0,0,0,0, −1, −1, −1, −1, −1, −1),

X4 = c(0,0,0,0,0,0,0,0,0,0,0,0,1,1,1,1,1,1,

0,0,0,0,0,0,0,0,0,0,0,0, −1, −1, −1, −1, −1, −1),

X5 = c(0,0,0,0,0,0, 0,0,0,0,0,0, 0,0,0,0,0,0,

1,1,1,1,1,1,0,0,0,0,0,0, −1, −1, −1, −1, −1, −1),

X6 = c(0,0,0,0,0,0, 0,0,0,0,0,0, 0,0,0,0,0,0,

0,0,0,0,0,0,1,1,1,1,1,1, −1, −1, −1, −1, −1, −1),

X7 = c(1,0,0,0,0, −1,1,0,0,0,0, −1, 1,0,0,0,0, −1,

1,0,0,0,0, −1,1,0,0,0,0, −1,1,0,0,0,0, −1),

X8 = c(0,1,0,0,0, −1, 0,1,0,0,0, −1, 0,1,0,0,0, −1,

0,1,0,0,0, −1,0,1,0,0,0, −1, 0,1,0,0,0, −1),

X9 = c(0,0,1,0,0, −1,0,0,1,0,0, −1,0,0,1,0,0, −1,

0,0,1,0,0, −1, 0,0,1,0,0, −1, 0,0,1,0,0, −1),

X10 = c(0,0,0,1,0, −1,0,0,0,1,0, −1,0,0,0,1,0, −1,

0,0,0,1,0, −1, 0,0,0,1,0, −1, 0,0,0,1,0, −1),

X11 = c(0,0,0,0,1, −1, 0,0,0,0,1, −1, 0,0,0,0,1, −1,

0,0,0,0,1, −1, 0,0,0,0,1, −1, 0,0,0,0,1, −1),

X12 = c(1,2,3,4,5,6, 2,4,6,8,10,12,3,6,9,12,15,18,

4,8,12,16,20,24, 5,10,15,20,25,30, 6,12,18,24,30,36),

X13 = c(1,0,0,0,0,0,0,2,0,0,0,0,0,0,3,0,0,0,0,0,0,4,0,0,0,0,0,0,5,0,0,0,0,0,0,6))

list(b = c(0,0,0,0,0,0,0,0,0,0,0,0,0))

A logistic model incorporating the linear by linear association as the next candidate for the TV show ratings. Thus, BUGS CODE 5.4 is executed excluding the weight vector X_{13} with 50,000 observations and a burn in of 5,000 with a refresh of 100, and the table below provides the posterior analysis (Table 5.18).

What has been the effect of adding the linear by linear association vector to the base model? Firstly, the Euclidean distance, which measures the distance between the observed and expected cell frequencies, is reduced from a posterior mean of 13.15 to 11.36. Also, the addition of the linear by linear interaction as measured by $b[12]$ has a 95% credible interval of (0.1335, 0.3609), implying that the addition was useful. The addition does not appear to alter the pattern of agreement between the two journalists. For example, the

TABLE 5.18

Posterior Distribution for Scores of Two Journalists, Base Model Plus Linear by Linear Interaction

Parameter	Mean	SD	2½	Median	97½
$b[1]$	−6.862	0.8212	−8.362	−6.8	−5.435
$b[2]$	2.49	0.5714	1.463	2.456	3.686
$b[3]$	1.725	0.4299	0.9409	1.703	2.632
$b[4]$	1.003	0.2729	0.4784	0.9991	1.552
$b[5]$	−0.3165	0.2496	−0.8282	−0.3093	0.1533
$b[6]$	−1.839	0.4527	−2.79	−1.818	−1.013
$b[7]$	0.9234	0.4634	0.0321	0.9171	1.851
$b[8]$	1.125	0.3466	0.4655	1.117	1.825
$b[9]$	0.553	0.2715	0.0153	0.5526	1.083
$b[10]$	0.3365	0.2005	−0.067	0.3407	0.7194
$b[11]$	−0.7368	0.3157	−1.387	−0.7266	−0.1449
$b[12]$	.2365	0.0582	0.1335	0.2323	0.3609
d27	1.566	.4639	.7055	1.551	2.51
d38	.6005	.3611	−.0943	.5934	1.325
d49	.4501	.3219	−.1735	.4475	1.091
d510	−.653	.3249	−1.301	−.6503	−.0322
d611	−1.102	.401	−1.909	−1.092	−.335
LR	43.07	20.27	4.381	42.73	83.57
χ^2	48.42	13.34	33.01	45.63	79.69
Dis	11.36	.901	9.934	11.25	13.45

difference in the effect of the contrast 1 minus 6 between the two raters is measured by d27, which has a credible interval of (0.7055, 2.51) implying the two differ in this regard, but this was also the case for the base model, when the linear by linear interaction was not a predictor.

Lastly, an unequal weight vector is added as a predictor labeled by X_{13} and defined in BUGS CODE 5.4, and the posterior analysis is presented in Table 5.19.

The change in the Euclidean distance is negligible, from a value of 11.36 for the second model to 11.48 for the present and the value of $b[13]$ which measures the effect of the weight vector has a 95% credible interval of (−0.0358, 0.2465) implying that the weight vector has made somewhat of a contribution to the overall fit, and as with the previous model the pattern in agreement between the two is about the same, and has been little affected by the addition of the unequal weight vector. The overall pattern of agreement is measured by d27, d38, d49, d510, and d611. For example, d611 is the difference $(b[6] − b[11])$ in the effect of contrast 5 minus 6 between the two journalists and has a credible interval of (−1.983, −0.3349), implying some difference. The model development is based on the three previous models

TABLE 5.19

Posterior Distribution for Scores of Two Journalists, Linear by Linear Interaction Plus Unequal Weight Vector

Parameter	Mean	SD	2½	Median	97½
$b[1]$	−6.546	0.8452	−8.257	−6.528	−4.978
$b[2]$	2.304	0.5887	1.194	2.292	3.478
$b[3]$	1.592	0.4418	0.7597	1.582	2.479
$b[4]$	0.9591	0.276	0.4258	0.9574	1.506
$b[5]$	−0.3047	0.2501	−0.8098	−0.2996	0.1685
$b[6]$	−1.73	0.4631	−2.666	−1.718	−0.8589
$b[7]$	0.7392	0.4788	−0.2092	0.7414	1.671
$b[8]$	0.9854	0.359	0.2861	0.9838	1.691
$b[9]$	0.4717	0.2768	−0.0773	0.4729	1.007
$b[10]$	0.3595	0.2008	−0.0406	0.3624	0.7465
$b[11]$	−0.5895	0.3323	−1.257	−0.5854	0.0498
$b[12]$	0.2047	0.0623	0.0872	0.2042	0.3292
$b[13]$	0.106	0.0717	−0.0358	0.106	0.2465
d27	1.565	0.4633	0.6923	1.551	2.508
d38	0.6069	0.365	−0.0916	0.6028	1.333
d49	0.4874	0.3369	−0.1665	0.4849	1.157
d510	−0.6642	0.3333	−1.327	−0.6601	−0.0188
d611	−1.14	0.4197	−1.983	−1.133	−0.3349
LR	42.08	20.31	3.412	41.75	82.84
χ^2	47.17	11.55	31.95	44.84	75.79
Dis	11.4	0.9788	9.86	11.28	13.69

labeled 1, 2, and 3, and one may plot the observed versus expected cell frequencies of the three.

5.4 More than Two Raters

Suppose three pathologists are involved in the lung cancer study of Table 5.7, where three histological grades are assigned to each of the 123 slides of lesion specimens (Table 5.20).

The histological grades are interpreted as follows: 1 specifies a well differentiated specimen; 2 identifies fair differentiation, and 3 indicates a poorly differentiated cell, where poorly differentiated cells are more indicative of a malignant lesion. Of course, more raters imply a more complex agreement pattern, but the logistic model allows one to analyze a variety of agreement/disagreement patterns.

TABLE 5.20

Lung Cancer with Three Pathologists

Pathologist B	Pathologist C 1	2	3	Total
A=1				
1	8	5	4	17
2	2	4	2	8
3	3	1	2	6
	13	10	8	31
A=2				
1	13	8	6	27
2	4	6	3	13
3	5	1	3	9
	22	15	12	49
A=3				
1	11	7	5	23
2	3	5	3	11
3	5	1	3	9
	19	13	11	43

Our approach of model building is to begin with the base model for the three pathologists, where only the main effects of the three are included with the following columns of the design matrix.

$$X_1 = (1,1,1,1,1,1,1,1,1,1,1,1,1,1,1,1,1,1,1,1,1,1,1,1,1,1,1)'$$

$$X_2 = (1,1,1,1,1,1,1,1,1,0,0,0,0,0,0,0,0,0,-1,-1,-1,-1,-1,-1,-1,-1,-1)'$$

$$X_3 = (0,0,0,0,0,0,0,0,0,1,1,1,1,1,1,1,1,1,-1,-1,-1,-1,-1,-1,-1,-1,-1)' \quad (5.9)$$

$$X_4 = (1,1,1,0,0,0,-1,-1,-1,1,1,1,0,0,0,-1,-1,-1,1,1,1,0,0,0,-1,-1,-1)'$$

$$X_5 = (0,0,0,1,1,1,-1,-1,-1,0,0,0,1,1,1,-1,-1,-1,0,0,0,1,1,1,-1,-1,-1)'$$

$$X_6 = (1,0,-1,1,0,-1,1,0,-1,1,0,-1,1,0,-1,1,0,-1,1,0,-1,1,0,-1,1,0,-1)'$$

and

$$X_7 = (0,1,-1,0,1,-1,0,1,-1,0,1,-1,0,1,-1,0,1,-1,0,1,-1,0,1,-1,0,1,-1)'$$

It is seen that X_1 is the column of ones, and the main effects of the three pathologists are represented by the remaining six columns. For example, X_2 is the contrast score 1 minus score 3 for pathologist A, while X_7 is the contrast score 2 minus score 3 for pathologist C, thus the 27×7 design matrix is of full rank 7. This parameterization is sufficient to explore the agreement/disagreement pattern among the three.

The corresponding logistic model is

$$\log it(\varphi_i) = \sum_{j=1}^{j=p} b[j]X_j[i]$$

(5.10)

where $p=7$, and $i=1, 2, ..., 27$. Note that i varies over the cells of Table 5.20, from left to right and from top to bottom, thus $i=1$ corresponds to the 11 cells of the first part of Table 5.20, while $i=27$, corresponds to the 33 cells of the third part of the table. The coefficients of Equation 5.10 corresponding to the columns are the effects of that column on the logit, therefore, for example, $b[2]$ is the effect of the contrast score 1 minus score 3 of pathologist A. In order to explore the pattern of agreement/disagreement, the following effects are defined: $d24=b[2]-b[4]$ is the difference in the contrast score 1 minus score 3 between pathologist A and B, while $d57=b[5]-b[7]$ is the difference in the contrast score 2 minus score 3 between pathologists B and C (Table 5.21).

With regard to the goodness of fit, the posterior mean of the Euclidean distance is 8.193, with a 95% credible interval of (6.33, 11.02), and the analysis is executed with 75,000 observations generated from the joint posterior distribution, with a burn in of 5000 and a refresh of 100.

At the second stage of model building, the design matrix is expanded with the columns that place equal weights on the diagonal cells of pairs of pathologists.

$$X_8 = (1,1,1,0,0,0,0,0,0,\ 0,0,0,1,1,1,0,0,0,\ 0,0,0,0,0,0,1,1,1)'$$

$$X_9 = (1,0,0,1,0,0,1,0,0,1,0,0,1,0,0,1,0,0,0,1,0,0,1,0,0,1)'$$

TABLE 5.21

Posterior Analysis for Three Pathologists of Lung Cancer Study

Parameter	Mean	SD	2½	Median	97½
$b[1]$	−3.432	0.1062	−3.646	−3.43	−3.228
$b[2]$	−0.2777	0.142	−0.5655	−0.2746	−0.00649
$b[3]$	0.2048	0.1273	−0.04391	0.2093	0.4566
$b[4]$	0.6232	0.1253	0.3786	0.6234	0.8706
$b[5]$	−0.1609	0.1461	−0.452	−0.1594	0.1213
$b[6]$	0.322	0.1252	0.0733	0.3225	0.5659
$b[7]$	−0.05352	0.1354	−0.3217	−0.0526	0.2096
d24	−0.9031	0.1899	−1.278	−0.8991	−0.534
d46	0.3016	0.1764	−0.0413	0.3009	0.6478
d35	0.3693	0.1936	−0.00729	0.3679	0.7499
d57	−0.1073	0.1986	−0.4969	−0.1065	0.2812
LR	16.56	22.04	−25.82	16.3	60.52
χ^2	16.87	4.798	10.62	15.85	28.81
Dis	8.193	1.225	6.33	8.021	11.02

and

$$X_{10} = (1,0,0,0,1,0,0,0,1,1,0,0,0,1,0,0,0,1,1,0,0,0,1,0,0,0,1)' \qquad (5.11)$$

where X_8 is the equal weight vector for pathologists A and B, X_9 for pathologists A and C, and X_{10} for B and C. In a similar way to the base model, the posterior analysis for the expanded model is executed with 75,000 observations, a burn in of 5000, and a refresh of 100 (Table 5.22).

The addition of the weight vectors indicate that the contributions of equal weights for A and B and for A and C are negligible, however equal weights for B and C do indeed add something to the overall fit of the logistic model. This is also confirmed by the corresponding 95% credible intervals (-0.3883, 0.4564), (-0.4487, 0.3588), and (0.0028, 0.7808), respectively. The reduction is the Euclidean distance from 8.193 to 8.059 is minimal, indicating the addition of the three equal weight vectors added very little to the overall fit to the data via the logistic model. One could consider adding an equal weight vector X_{11} for all three pathologists A, B, and C simultaneously, where

$$X_{11} = (1,0,0,0,0,0,0,0,0,0,0,0,0,1,0,0,0,0,0,0,0,0,0,0,0,0,1)' \qquad (5.12)$$

Two scatter plots will show the observed versus expected cell frequencies for the lung cancer study with three pathologists and reveal that the addition

TABLE 5.22

Posterior Analysis for Three Pathologists of Lung Cancer Study

Parameter	Mean	SD	2½	Median	97½
$b[1]$	-3.582	0.1626	-3.911	-3.579	-3.271
$b[2]$	-0.2857	0.1508	-0.5888	-0.3832	0.0033
$b[3]$	0.2133	0.1292	-0.0418	0.2127	0.4631
$b[4]$	0.6023	0.1278	0.353	0.6014	0.8539
$b[5]$	-0.1618	0.1478	-0.4552	-0.1601	0.1236
$b[6]$	0.2423	0.1338	-0.0193	0.2432	0.5034
$b[7]$	-0.0251	0.1378	-0.2995	-0.0237	0.2399
$b[8]$	0.0373	0.2143	-0.3883	0.0377	0.4564
$b[9]$	-0.0382	0.2054	-0.4487	-0.0363	0.3588
$b[10]$	0.3951	0.1982	0.0028	0.3964	0.7808
d24	-0.8879	0.2029	-1.298	-0.8842	-0.5001
d46	0.36	0.1934	-0.0136	0.3586	0.7437
d35	0.375	0.1987	-0.0095	0.3735	0.7678
d57	-0.1367	0.2146	-0.5571	-0.1363	0.2816
LR	15.45	22.08	-26.98	15.17	59.48
χ^2	16.43	5.63	8.644	15.42	30.14
Dis	8.059	1.543	5.632	7.857	11.64

of the equal weight vectors for the three pairs of raters does not add to the overall fit of the model to the data.

5.5 Other Methods for Patterns of Agreement

Suppose one has examined agreement data, say between two raters, and computed kappa and other indices of agreement, which gives one some idea of the overall agreement, but more information is required and the logistic model is employed to determine the patterns of agreement between the two. For example, are there any trends, that is to say, is one rater more prone to assign higher scores than the other, or do the raters assign equal importance to each score, or are some scores more important for agreement than others? The logistic model is also useful in estimating the main effects of raters and deciding if the main effects of the two are more or less equivalent. These questions and others can be answered with a modeling approach. For example, subject covariates can be included since they might have some effect on the assignment of scores by the raters, also covariates make it possible to compare certain groups of subjects (male versus female, older versus younger, etc.). Are their other methods that will help one better understand the agreement between two raters?

Von Eye and Mun[2] introduce exploratory techniques that indeed elucidate the nature of agreement, and such methods they call configured frequency analysis (CFA), and such techniques focus on a cell by cell comparison of the observed versus the corresponding expected cell frequency. See Lienert and Krauth[10] and Von Eye[11] who describe the early methods of configured frequency analysis. In order to introduce the subject, consider Table 5.23, which provides the scores of two raters who are assigning diagnostic scores to 464 patients at high risk for prostate cancer. Each man had their PSA level read and were administered an MRI in order to determine their risk of the disease. Each rater was a member of a committee of medical oncologists, radiologists, and pathologists, thus what is reported is actually the consensus of the group, and the scores should be interpreted as: 1 implies the committee believes there is no evidence of disease, while a 2 indicates some evidence of prostate cancer, and finally a 3 denotes definite evidence of the disease (Table 5.23).

A quick look shows that there is high agreement with a kappa of 0.665 and a G coefficient of 0.675, while the probability of agreement is 0.8362, with a probability of chance agreement 0.511. It also appears that rater B is assigning higher scores than A. A Bayesian approach using an improper prior

$$\prod_{i=1}^{i=9} \theta_i^{-1}$$

$$(5.13)$$

TABLE 5.23

Cell Frequencies and Probabilities for Prostate Cancer Study

Rater A	Rater B 1	2	3	Total
1	283, θ[1]	23, θ[2]	18, θ[3]	324
2	10, θ[4]	72, θ[5]	11, θ[6]	93
3	8, θ[7]	6, θ[8]	33, θ[9]	47
Total	301	101	62	464

for the cell probabilities determines the posterior mean (SD) of kappa as 0.6651 (0.0334), the posterior mean (SD) of G as 0.6725 (0.0342), and the probability of agreement of 0.8363 (0.0171), while the corresponding 95% credible intervals are (0.5973, 0.7282), (0.6021, 0.737), and (0.8011, 0.8685). A BUGS program is executed with 50,000 observations with a burn in of 5000, and a refresh of 100.

An examination for patterns of agreement is based on a logistic model that includes five column vectors for the design matrix namely where the first column is a 9×1 vector of ones, and the remaining columns are the 1 minus 3 and 2 minus 3 contrasts for the main effects of the two raters. Of course, this has been used in many previous examples, and is also found in the exercises. I found that the main effects of the two raters were the same, in that the effects of the contrast 1 minus 3 is the same for both, as is it is for the contrast 2 minus 3.

However, the fit of the model, as measured by the Euclidean distance, to the data is quite poor with a posterior mean (SD) of 132 (5) and an associated credible interval of (122, 142). This is the largest value for this goodness of fit index I have seen, considering all the examples so far. Of course, this is the usual analysis for studying agreement, first with the kappa statistic followed by detection of agreement patterns via the logistic model. At this point, the modeling building strategy would be to enlarge the model by including equal weight vectors and various patient covariates, however, it is at this juncture, with the base model, that Von Eye and Mun[2] introduce Configural Frequency Analysis or CFA.

Their technique depends on comparing the observed versus the expected cell frequencies, where the expected are computed with respect to the base model, and where with the Bayesian approach, the posterior mean of expected cell frequencies are used in the comparison. For the base model the observed versus expected cell frequencies are shown in Table 5.24, T and A designate the direction of the difference. If the observed is larger than the expected, the group designation is *type* (T), while if the observed is smaller than the expected, the label is *anti type* (A). If the posterior probability of the difference (taking into account the direction) is large (i.e., greater than say 0.95), the designation is considered significant (denoted with an *), otherwise insignificant. For example, for cell 1, the posterior probability that the observed is

TABLE 5.24

Observed and Expected Cell Frequencies, Prostate Cancer Study

Observed	283	23	18	10	72	11	8	6	33
Expected	203.2	74.54	46.42	64.66	17.78	10.5	33.21	8.642	5.06
Cell	1	2	3	4	5	6	7	8	9
Group	T*	A*	A*	A*	T*	T	A*	A	T

greater than expected is essentially 1, however, for cell 6, the probability that the observed is less than the observed is approximately 0.5, thus the label T is not considered significant. Note that the word significant is not used in the usual sense, such as a significant *p*-value, but instead depends only on the posterior probability of the event.

The posterior probabilities of these events are induced by the posterior distribution of the expected cell frequencies and are computed via WinBUGS. What do the *type/antitype* designations tell us about agreement? Since the goodness of fit is quite bad, one would expect more significant *types* and *antitypes* compared to a better fitting model. Consider cell 1, where there is a significant type, then the two raters agree much more often than expected by chance, but on the other hand for cell 3, rater B is assigning a higher score than rater A 18 times, significantly lower than expected number of observations (46), predicted by the model. Thus a cell by cell comparison of observed versus, provides additional information beyond that given by kappa (or G) and the estimated main effects of logistic model. If there is good agreement, one would expect *type* designations on the main diagonal and *antitype* labels for the off-diagonal cells, which is certainly the case for the prostate cancer data.

I recommend that the CFA be performed in conjunction with kappa estimation and fitting the information with the log linear model, beginning with the base model, which includes only the main effects of the raters. The CFA should be performed at each stage of the model fitting process, where if the fit does indeed improve, one expects fewer significant designations of *type/antitype*. To that end, suppose the scores are considered ordinal and a linear by linear interaction term is included so that the logistic model is

$$\mathrm{logit}(\varphi_i) = \sum_{j=1}^{j=p} b[j]X_j[i] \tag{5.10}$$

where $p = 6$, and $X_6 = (1,2,3,2,4,6,3,6,9)$ is the linear by linear interaction term. A posterior analysis reveals that the interaction term does contribute to the goodness of fit, with a reduction of 48% from a posterior mean $(SD) = 132$ (5) for the former model to a value of 68 (4.172) for the present one. A complete account of the posterior analysis is given by Table 5.25.

What is striking about the analysis is the addition of the linear by linear interaction term is quite meaningful, because the 95% credible interval is (1.689, 2.373), which goes hand in hand with the dramatic reduction in the

TABLE 5.25

Posterior Analysis for Prostate Cancer, Linear by Linear Interaction

Parameter	Mean	SD	2½	Median	97½
$b[1]$	−11.03	0.7748	−12.6	−11.01	−9.555
$b[2]$	4.916	0.3506	4.247	4.908	5.624
$b[3]$	0.2579	0.1221	0.0226	0.2562	0.5009
$b[4]$	4.335	0.3009	3.764	4.328	4.941
$b[5]$	0.1225	0.1103	−0.0885	0.125	0.3418
$b[6]$	2.021	0.1743	1.689	2.018	2.373
d24	0.5803	0.1796	0.232	0.5785	0.937
d35	0.1324	0.1435	−0.147	0.1317	0.4164
Dis	68.09	4.172	60.43	67.87	76.91

Euclidean distance! Note also there is strong evidence that there is no difference in the effects of the 1 minus 3 contrast for the two raters, however, the same claim cannot be made for the 2 minus 3 contrast.

What does this imply about the configural frequency analysis? The expected cell frequencies are

$$E_2 = (260.3, 58.41, 5.321, 39.21, 33.61, 20.22, 1.535, 8.996, 36.48) \qquad (5.14)$$

and when compared to the observed cell frequencies

$$Y = (283, 23, 18, 10, 72, 11, 8, 6, 33) \qquad (5.15)$$

determine the following *type/antitype* designations

$$(T, A^*, T^*, A^*, T^*, A, T^*, A, A) \qquad (5.16)$$

Note that cell 5 is a significant *type*, that is, the raters agree on some evidence of disease much more than expected and it appears that the posterior probability that the expected cell 5 frequency is less than the observed value of 72 appears to be close to 1, and cell 5 is designated as a significant *type*. The Bayesian analysis was performed with 60,000 generations, with a burn in of 5000, and a refresh of 100. On the other hand, cell 9 is designated as an *antitype*, but is not significant because the posterior probability that $E[9]$ is greater than 33 is approximately 0.4. Because of the better fit, there are fewer significant cell deviations, form a total of six with the base model to five with the model that included the linear by linear interaction term.

How should the analysis of the prostate data be continued? Is their additional information that might improve the fit of the model? Fortunately, the medical history of the subjects include information about previous cancers, which are given as Table 5.26.

The majority of subjects did not have a previous history of disease, and both groups appear to have about the same level of agreement as the large

TABLE 5.26A

Prostate Cancer Study with Previous History of Cancer

	Rater B			
Rater A	1	2	3	Total
1	57	5	4	66
2	2	24	3	19
3	1	2	8	11
Total	60	21	15	96

TABLE 5.26B

Prostate Cancer Study without Previous History of Cancer

	Rater B			
Rater A	1	2	3	Total
1	226	18	14	258
2	8	58	8	74
3	7	4	25	36
Total	241	80	47	368

aggregated group. The splitting of the group into two gives one valuable additional information about how the two agree, and the main interest is: do the two agree in the same manner for both groups?

It is shown that the group effect (those with a history of cancer versus those without a history) $b[2]$ makes an important contribution to the model with a 95% credible interval of $(-0.9353, -0.6907)$, which is accompanied by a large reduction in the Euclidean distance from the previous model which had a value (SD) of 68.01 (4.1720) to the present value (SD) = 58.37 (4.191). The effect of the 1 minus 3 contrast of rater A is measured by $b[3]$ and has a posterior mean (SD) of 4.498 (0.3391) compared to the posterior mean of $b[5]$ of 3.943 for rater B for the same contrast. The two raters appear not to have the same effect on the 1 minus 3 contrast, since the 95% credible interval for the difference in the two effects is (0.2151, 0.9028). Note that $b[7]$ measures the effect of the linear by linear interaction, has 95% credible interval of (1.549, 2.211) and appears to make an important contribution to the goodness of fit (Table 5.27). What remains is to perform a CFA based on the deviations of the observed and expected cell frequencies, which are shown in Table 5.28.

Consider cell 8 which has a *type* designation, but is not considered significant, because the posterior probability of $E[8]$ being less than 2 (the observed cell frequency) is approximately 0.4. In order for the cell 8 deviation to be labeled significant, the probability would have had to exceed 0.95.

Of course, additional information can be added to the model in order to improve the goodness of fit, but this will be referred to the exercises.

TABLE 5.27

Posterior Analysis for Prostate Cancer Study with Two Groups of Patients and Linear by Linear Interaction

Parameter	Mean	SD	2½	Median	97½
$b[1]$	−11.38	0.7564	−12.97	−11.34	−10.01
$b[2]$	−0.8115	0.0623	−0.9352	−0.8108	−0.6907
$b[3]$	4.498	0.3391	3.877	4.482	5.205
$b[4]$	0.2616	0.1189	0.0316	0.2598	0.5
$b[5]$	3.943	0.289	3.414	3.931	4.545
$b[6]$	0.1525	0.107	−0.0534	0.1515	0.3659
$b[7]$	1.857	0.1693	1.549	1.849	2.211
d35	0.5548	0.1755	0.2151	0.5529	0.9028
d46	0.109	0.1396	−0.163	0.1087	0.3828
Dis	58.37	4.191	50.92	50.05	67.46

TABLE 5.28

Observed Versus Expected Cell Frequencies and *Types/Antitypes* for Prostate Cancer

Observed	Expected	Type/Antitype	Cell
57	61	A	1
5	9.9	A	2
4	0.94	T*	3
2	6.413	A*	4
14	6.047	T*	5
3	3.555	A	6
1	0.2853	T*	7
2	1.643	T	8
8	6.121	T	9
226	201.3	T?	10
18	46.23	A*	11
14	4.714	T*	12
8	30.64	A*	13
58	28.96	T*	14
8	17.39	A*	15
7	1.434	T*	16
4	8.168	A*	17
25	29.31	A	18

5.6 Summary of Modeling and Agreement

When conducting a model building strategy with the logistic model, one begins with estimating kappa, which gives an overall measure of agreement, then at the second stage, one investigates the main effects of the two (or more) raters with the base model. At this stage, the posterior analysis consists of estimating the main effects and comparing the two raters. An essential element is to determine the posterior distribution of the Euclidean distance, or some other index of goodness of fit, which is to be used as a reference value for future values of the index. In addition, at each stage, a CFA (configural frequency analysis) is performed by designating the cell frequencies as either a *type* or an *antitype*, and determining the significance of the designation. This provides more information on the pattern of agreement, relative to the present tentative model. Upon adding additional information to the model with columns to the design matrix, one repeats the posterior analysis by estimating and comparing the main effects of the raters, evaluating the goodness of fit, and exercising a CFA. The process is terminated once one is satisfied that one has a parsimonious model and that a further reduction in the goodness of fit is unlikely.

Another Bayesian approach to modeling agreement between raters is presented by Johnson and Albert[12], and is based on the latent variable method, where by the range of an underlying continuous variable is partitioned in such a way that uniquely determines the values of an ordinal variable that is used by the raters to assign scores to subjects.

Exercises

1. Two mammographers are screening 766 patients for breast cancer using a five point nominal scoring system, where 1 indicates definitely no evidence of disease, 2 denotes probably no evidence of disease, 3 refers to evidence that is inconclusive, 4 implies probable evidence of a lesion, and 5 means there is definite evidence of breast cancer. Refer to Table 5.29:

TABLE 5.29

Two Mammographers Screening for Breast Cancer (Frequencies and Average Age)

| Radiologist A | Radiologist B | | | | | |
	1	2	3	4	5	Total
1	402, 38.7	36, 44.1	20, 48.9	12, 55.7	6, 69.4	476
2	11, 45.3	56, 40.2	29, 57.1	18, 65.2	4, 71.1	118
3	7, 51.2	13, 44.1	39,48.6	20, 55.5	17, 62.7	96
4	5, 55.8	4, 59.7	8, 60.6	15, 50.2	14, 69.9	46
5	4, 66.1	7, 70.3	6, 69.9	4, 65.7	9, 67.2	30
Total	429	116	102	69	50	766

(a) Perform a Bayesian posterior analysis for kappa and the G coefficient, using 45,000 observations generated from the posterior distribution of the parameters, with a burn in of 5000, and a refresh of 200 observations. Assume a uniform prior distribution for the 25 cell probabilities and assume the individual cell frequencies follow a binomial distribution, where $n = 766$.

(b) Build the design matrix for the logistic base model using the following columns, each of length 25:

$$X_1 = (1,1,1,1,1,1,1,1,1,1,1,1,1,1,1,1,1,1,1,1,1,1,1,1,1)'$$

$$X_2 = (1,1,1,1,1,0,0,0,0,0,0,0,0,0,0,0,0,0,0,0, -1, -1, -1, -1, -1)'$$

$$X_3 = (0,0,0,0,0,1,1,1,1,1,0,0,0,0,0,0,0,0,0,0, -1, -1, -1, -1, -1)'$$

$$X_4 = (0,0,0,0,0,0,0,0,0,0,1,1,1,1,1,0,0,0,0,0, -1, -1, -1, -1, -1)'$$

$$X_5 = (0,0,0,0,0,0,0,0,0,0,0,0,0,0,0,1,1,1,1,1, -1, -1, -1, -1, -1)'$$

$$X_6 = (1,0,0,0, -1,1,0,0,0, -1,1,0,0,0, -1,1,0,0,0, -1,1.0.0.0. -1)'$$

$$X_7 = (0,1,0,0, -1,0,1,0,0, -1,0,1,0,0, -1,0,1,0,0, -1,0,1,0,0, -1)'$$

$$X_8 = (0,0,1,0, -1,0,0,1,0, -1,0,0,1,0, -1,0,0,1,0, -1,0,0,1,0 -1)'$$

$$X_9 = (0,0,0,1, -1,0,0,0,1, -1,0,0,0,1, -1,0,0,0,1, -1,0,0,0,1, -1)'$$

This form of the design matrix is called the effects form, because the second column is the contrast of score 1 minus score 5 for mammographer A, and the 6-th column is the corresponding contrast for mammographer B. Also, the third column is the contrast of score 2 minus score 5 for mammographer A, etc. A non singular design matrix of rank 9 is the result, and the corresponding logistic model is

$$\text{logit}(\varphi_i) = \sum_{j=1}^{j=p} b[j]X_j[i]$$

where $p = 9$, $i = 1, 2, \ldots, 25$.

(c) With a burn in of 10,000 and a refresh of 100, generate 100,000 observations from the joint posterior distribution of the nine parameters of the model. Use a non informative prior for the b[] coefficients, namely b[i]~dnorm(0.0,.00001) for the analysis, and compute the mean, SD, median, 2½ and 97½ percentiles of each posterior distribution.

(d) Based on (c), perform a posterior analysis for the LR, chi-square, and Euclidean distance goodness of fit statistics and plot the posterior density of the Euclidean distance.

(e) Based on (c), what is the effect of the contrast 1 minus 5 for mammographers A and B? Are the effects the same? Why or why not?

(f) Add an additional column X_{10} to the design matrix that puts equal weights on the diagonal cells. What is the posterior distribution of $b[10]$, the effect of the equal weight vector? Does this addition, improve the goodness of fit? Find the posterior distribution of $b[10]$, LR, chi-square and Euclidian distance goodness of fits measures.

(g) Suppose the following average age of the subjects are given cell-wise in the above table. Use X_{11} for this covariate. The average age is the second number after the comma in each cell. Does age reduce the goodness of fit statistics? What is the posterior distribution of the goodness of fit measures LR, chi square and Euclidean distance?

(h) Plot the expected cell frequencies versus the observed cell frequencies for the model that includes the age covariate and the equal weight vector.

(i) What is your choice for the 'best' model for this data set ? Plot the expected versus observed for the best fitting model.

(j) Add an additional column to the design matrix that models the following pattern of agreement. Radiologist B assigns higher scores than Radiologist A. See Von Eye and Mun[2]. Test the hypothesis that Rater B assigns higher scores than Rater A.

2. Refer to Table 5.7 and Table 5.8 for the lung cancer data and stratify on the results of the Karnofsky score, where there are 56 subjects who tested negative and the remaining 67 tested positive. The Karnofsky score is considered positive if the score is below 70 (a score of 90 indicates very little evidence of disease, while a score of 20 indicates patient is very sick) and is considered negative otherwise.

(a) Using this information (from Table 5.30), build a logistic model stratified on Karnofsky Score. For the base model use six columns for the design matrix, where the first is a constant 9×1 vector of ones, the second column identifies the stratum (negative or positive stress Karnofsky score), column three contrasts score 1 minus score 3 for the first rater A etc. Based on BUGS CODE 5.3, execute the Bayesian analysis for the three measures of goodness of fit and the five coefficients of the model.

(b) To the base model add the four Karnofsky (positive vs negative) by rater interactions and conduct a Bayesian analysis similar to the part (a) with 55,000 observations generated from the posterior distribution, with a burn in of 5000 and a refresh of 200. Plot the posterior density of the three measures of goodness of fit. Does the addition of the interactions produce a better fitting model compared to the base model? See Table 5.13, Table 5.14, and Table 5.15, and use the same format to report the results of your answer.

(c) Add to the previous model, an equal weight vector that assigns a 1 to the diagonal cells of the agreement tables for each strata. Based on the Euclidean distance, does this improve the fit compared to the previous model?

(d) Does adding the average age cell values improve the fit of the model?

(e) What are the effects of the contrast 2 minus 3 of the two pathologists?

(f) What is your choice of the best overall fitting model? Justify your answer.

(g) What are the effects of the two raters on the contrast score 1 minus score 2 on the logit scale?

(h) Estimate kappa for the two strata and compare them.

TABLE 5.30A

Lung Cancer Pathology Stratified by Karnofsky Score: Negative Result

Pathologist A	Pathologist B			Total
	1	2	3	
1	21	11	8	40
2	4	5	3	12
3	3	0	1	4
Total	28	16	12	56

TABLE 5.30B

Lung Cancer Pathology Stratified by Karnofsky Score: Positive Result

Pathologist A	Pathologist B			Total
	1	2	3	
1	11	9	7	27
2	5	10	6	21
3	9	3	7	19
Total	25	22	20	67

3. Refer to Table 5.16 for the 100 TV shows rated by two journalists and verify Tables 5.17 through 5.19 of the posterior analysis of the three candidate models. Which model is the best among the three? Does the linear by linear interaction improve the goodness of fit above that of the base model as measured by chi-square?

4. Refer to Table 5.16 and execute a Bayesian analysis for the base model, plus a linear by linear interaction column, and a vector that tests the assertion that journalist B is providing higher scores than journalist A. You will have to modify BUGS CODE 5.4.

5. Refer to Table 5.16 and write a paragraph that explains why the linear by linear interaction term determines the distance between the scores assigned to the TV shows.

6. Use the following values for the linear by linear interaction term in the logistic model for the analysis of frequencies in Table 5.31, where the u_i values are the midpoints of the continuous measurements (% of audience who will watch the show) underlying the six groups. Perform a Bayesian analysis using the base model (constant term plus the rater main effects), plus the linear by linear interaction term. What is the effect of this interaction term? What is the Euclidean distance with this model and what is it with the base model? Is the goodness of fit improved?

TABLE 5.31

Expected Cell Frequencies

u_i	5	15	25	35	45	75
i	1	2	3	4	5	6

7. The following example is taken from Agresti[7] (see Table 5.32): Two pathologists assign the following ordinal scores to the 118 slides: (1) negative; (2) atypical squamous carcinoma; (3) carcinoma in situ; (4) squamous carcinoma in early stromal invasion; and (5) invasive carcinoma. Conduct a Bayesian analysis with 55,000 observations generated from the posterior distribution of the parameters, with a burn in of 5000 and a refresh of 100. Analyze the following models: (a) the base model with constant and the main effects of the two raters; (b) the base model plus a linear by linear interaction term; (c) the model in (b) plus an equal weight vector for the diagonal cells. Use a normal (0.00, 0.00001) distribution for the coefficients of the logistic model and use these coefficients for your analysis. Also compute the goodness of fit parameters (Euclidean distance, chi-square, and the likelihood ratio). Which model gives the best fit? Overlay the three scatter plots of observed versus expected cell frequencies corresponding to the three models. Incidentally, Agresti computes kappa as 0.498 and the probability of agreement as 0.636, with a chance probability of agreement of 0.273. Perform a Bayesian analysis and estimate kappa and the G coefficient. What prior distribution will you employ?

8. Based on Table 5.20, write the necessary BUGS code and verify the posterior analyses in Table 5.21 and Table 5.22. Use a normal (0, 0.00001) prior distribution for the coefficients of the logistic model (Equation 5.1) and generate 75,000 observations from the joint posterior distribution with a burn in of 5000, and a refresh of 100. Table 5.21 is the posterior analysis of the base model and Table 5.22 shows the analysis for the base model plus the three equal weight vectors for the three pairs of pathologists.

TABLE 5.32

Histology Scores of Two Pathologists

Pathologist A	Pathologist B					
	1	2	3	4	5	Total
1	22	2	2	0	0	26
2	5	7	14	0	0	26
3	0	2	36	0	0	38
4	0	1	14	7	0	22
5	0	0	3	0	3	6
Total	27	12	69	7	3	118

Source: Agresti, A. 2002. *Categorical data analysis.* 2nd ed. New York: John Wiley & Sons Inc.

9. Based on problem 8, add the vector X_{11} (the equal weight vector for the three raters simultaneously) to the logistic model with 10 columns for the design matrix. Execute a Bayesian analysis and determine if the additional vector adds to the overall goodness of fit. Justify your results. Plot the observed versus expected cell frequencies for the fitted model and fit a linear regression line. Use the posterior mean of the expected observations for the expected values!

10. Produce the two scatter plots of the observed versus expected cell frequencies of Figure 5.4. The graph has the simple linear regression lines for the two plots. What are the slopes and intercepts for the two regressions?

11. Consider Table 5.20 and investigate the trend of the raters. Do any pathologists tend to assign higher scores compared to the others? How would this question be investigated with a Bayesian approach? What vector must be added to the design matrix to test the assertion that pathologist A is assigning higher scores compared to the other two.

12. Von Eye and Mun[2] present an example of three raters, where three psychiatrists re-diagnose the depression status of 163 patients. Recall that 1 designates no depression, 2 means mild depression, and 3 indicates clinical depression. (See Table 5.33)
    (a) Using the base logistic model of seven column vectors (Equation 5.9), determine the pattern of agreement/disagreement among the three psychiatrists with a Bayesian analysis. Report your results in the format of Table 5.21.
    (b) Expand the model to ten columns of the design matrix by adding the equal weight vectors X_8, X_9, and X_{10} (Equation 5.11) and determine if the addition improves the overall fit of the model. Use a Bayesian approach and report your results in the format of Table 5.22.
    (c) As the third stage of the model building process, add the X_{11} equal weight vector (for all three raters simultaneously) and determine the goodness of fit.

(d) What is your choice for the best fitting model? Justify your choice.

(e) Plot the observed versus expected cell frequencies for the three models. Based on this information, what is the optimal model?

TABLE 5.33

Diagnosis of Depression with Three Psychiatrists

	Psychiatrist C			
Psychiatrist B	1	2	3	Total
A=1				
1	4	3	6	13
2	2	1	3	6
3	2	2	17	21
	8	6	26	40
A=2				
1	0	1	2	3
2	1	1	1	3
3	0	0	4	4
	1	2	7	10
A=3				
1	0	1	3	4
2	0	1	8	9
3	0	4	96	100
	0	6	107	113

13. Refer to Table 5.20 for the lung cancer data with three raters, and using an improper prior for the 27 cell frequencies, verify the posterior analysis for kappa and the G coefficient is shown in Table 5.34. Produce the BUGS CODE necessary to verify the above results. I used 20,000 observations generated from the posterior distribution with a burn in of 2000, and a refresh of 100. Note, both kappa and the G coefficient imply poor agreement between the three pathologists.

TABLE 5.34

Posterior Analysis of Agreement for Lung Cancer Data

Parameter	Mean	SD	2½	Median	97½
agree	0.138	0.0308	0.08364	0.1361	0.2039
cagree	0.1096	0.0075	0.0958	0.1091	0.1259
G	−0.7239	0.0617	−0.8327	−0.7279	−0.5921
kappa	0.0319	0.0335	−0.0821	0.0300	0.103

14. Write BUGS code and verify the posterior analysis of Table 5.25. (a) What is the effect of the contrast 1 minus 3 for raters A and B? (b) Verify the significant *type/antitype* designation of Equation 5.15 by referring to the plots of the posterior densities of the expected observations $E[i]$, $i = 1, 2, ..., 9$.

15. Continue with the prostate cancer study by adding the following age covariate cell averages for the 18 cells of the two groups of subjects. Expand Table 5.27 to include $b[8]$ which measures the effect of the age covariate. The age column vector is

$$A = (57,59,63,55,65,69,59,70,75,43,48,52,47,51,53,51,54,65)$$

Perform a posterior analysis by generating 66,000 observations from the joint posterior distribution, with a burn in of 6000 and a refresh of 200: (a) What are the estimated main effects? (b) Compare the 2 minus 3 contrast between the two raters; (c) Does the addition of the age covariate improve the goodness of fit; (d) What is the posterior mean of the Euclidean distance; (e) Perform a CFA by labeling the cell frequencies as either a *type* or *antitype*; (f) Which deviations are significant? Explain how significance is determined. Interpret the significant deviations and summarize the pattern of agreement between the two raters.

16. Perform a Bayesian analysis for the prostate cancer information of Tables 5.26 but add four more columns to the design matrix: (1) the group (with or without history of previous cancer, column X_2) by contrast 1 minus three for rater A (column X_3) interaction; (2) the group by contrast 2 minus 3 for rater A interaction; (3) the group by contrast 1 minus 3 for rater A (column X_5) interaction; and (4) the

TABLE 5.35

Posterior Analysis for Prostate Cancer, Logistic Model with Interactions

Parameter	Mean	SD	2½	Median	97½
$b[1]$	−11.51	0.7401	−13.97	−11.45	−10.17
$b[2]$	−0.7332	0.0789	−0.8906	−0.7318	−0.5834
$b[3]$	4.555	0.3438	3.927	4.53	5.287
$b[4]$	0.2894	0.1331	0.0309	0.2887	0.5556
$b[5]$	3.932	0.2862	3.41	3.917	4.536
$b[6]$	0.1796	0.1211	−0.05592	0.1782	0.4197
$b[7]$	1.895	0.1654	1.596	1.882	2.243
$b[8]$	−0.04376	0.1222	−0.2808	−0.0444	0.1998
$b[9]$	0.0291	0.1117	−0.1911	0.0294	0.2459
$b[10]$	−0.144	0.1107	−0.3599	−0.1445	0.0756
$b[11]$	0.0322	0.1056	−0.1773	0.0327	0.2376
Dis	55.76	3.746	49.04	55.5	64.01

group by contrast 2 minus 3 for rater B (column X_6) interaction. Use a multiplicative (multiply the appropriate columns) interaction and generate 55,000 observations from the joint posterior distribution with a burn in of 5000, and a refresh of 100. Verify the posterior analysis in Table 5.35.

(a) Interpret the above four interactions given by $b[8] - b[11]$; (b) Is there an interaction between the group and the 1 minus 3 contrast for rater A? (c) Did the addition of the four interactions improve the overall goodness of fit as measured by the Euclidean distance? (See Table 5.27); (d) Are some of the interactions unnecessary in the model? (e) What is the best fitting model?

17. Refer to Tables 5.16 through 5.18 and BUGS CODE 5.4 for the TV show information. Plot the observed versus expected cell frequencies for models 1, 2, and 3. From the plots, can one select the best model that fits the data?

References

1. Shoukri, M. M. 2004. *Measures of interobserver agreement*. Boca Raton, London, New York: Chapman & Hall/CRC.
2. Von Eye, A., and E. Y. Mun. 2005. *Analyzing rater agreement, manifest variable methods*. Mahwash, NJ, London: Lawrence Erlbaum Associates.
3. Tanner, M., and M. A. Young. 1985. Modeling agreement among raters. *J. Am. Stat. Assoc.* 80, 175.
4. Von Eye, A., and C. Schuster. 2000. Log linear models for rater agreement. *Milticiencia* 4, 38.
5. Bortz, J., G. A. Lienert, and K. Boehnke. 1990. *Vertielungsfreie methoden in der biostatistik* (Distribution Free Methods in Biostatistics). Berlin: Springer-Verlag.
6. Agresti, A. 1988. A model for agreement between ratings on an ordinal scale. *Biometrics* 44, 539.
7. Agresti, A. 2002. *Categorical data analysis*. 2nd ed. New York: John Wiley & Sons.
8. Goodman, L. A. 1979. Simple models for the analysis of association in cross classifications having ordered categories. *J. Am. Stat. Assoc.* 74, 537.
9. Clogg, C. C., and E. S. Shihadeh. 1994. *Statistical methods for ordinal variables*. Thousand Oaks, CA: Sage.
10. Lienert, G. A., and J. Krauth. 1975. Configural frequency analysis as a statistical tool for defining types. *Ed. Psych.Meas.* 35, 231.
11. Von Eye, A. 2002. *Configural frequency analysis, methods, models, and applications*. Mahwah, NJ: Lawerence Erlaum Associates.
12. Johnson, V. E., and J. H. Albert. 1999. *Ordinal data modeling*. New York: Springer-Verlag.

6

Agreement with Quantitative Scores

6.1 Introduction

With a small number of ordinal or nominal scores, two raters have more of a chance of agreeing compared to raters who are using continuous or quantitative scores. For example, if two raters are using say three scores to assign to subjects, they can disagree as follows: If rater A assigns a score of 1, then rater B can disagree by assigning a score of either 2 or 3. Thus for every assignment of a score by rater A, rater B can disagree in two ways. On the other hand, if raters are using a continuous score, in principal, they can disagree in an infinity of ways. Of course, on a practical level, even using continuous scores, there is only a finite number of possible score assignments that are actually used, but if that number is "large", say 100, there is a huge number of ways to disagree. In practice, since raters are judging the same subjects, their scores are correlated and the number of disagreements is limited, compared to randomly assigning scores. By quantitative or continuous scores is meant variables like weight, height, blood pressure, tumor size, SUV (standardized uptake values in radiology), blood glucose values. etc.

In addition, with continuous observations, the statistical approach is somewhat different than with discrete, and involves methods that are familiar to the statistician. First descriptive statistics are always required, including a listing of the sample mean, median, and standard deviation for each rater, and these are usually accompanied by graphical presentations of the data, including, histograms, box plots, scatter plots, etc. Graphical presentations are usually quite revealing and show at a glance the relevant inter and intra rater differences, and consequently are a valuable adjunct to other methods. The graphical techniques include goodness of fit type plots, including PP and QQ plots for normality.

There are several well-known methods that are employed to reveal the patterns of agreement between raters using continuous scores. The most obvious is simple linear regression, where the scores of one rater are regressed on the other and the slope and intercept estimated, along with their corresponding credible intervals. If two raters are in agreement, one would expect the true regression to go through the origin with a slope of 1. If the scores of two raters are the same, the fitted linear regression will go through the origin with a slope of 1.

Patterns that deviate from a simple linear regression should be explored, because they are evidence of rater trends and complex disagreements. Of course, along with regression, the correlations between the scores of pairs of raters are also estimated.

Another method that can be utilized in order to study agreement is to estimate the intraclass correlation coefficient, which is the common correlation between pairs of raters who are judging the same set of subjects. A Bayesian approach based on the normality of the observations will be performed with one-way and two-way models. We saw a similar approach for the intraclass correlation when analyzing agreement with discrete scores (see Chapter 3 and Chapter 4). The old standby, the analysis of variance, is also quite helpful when exploring agreement. A typical scenario is when there are several sources of variation: (a) between subjects; (b) between raters; (c) between replications within raters; and (d) the error, i.e., what is left over after accounting for the preceding sources of variation. Each source of variation has a component of variance, thus the fraction of variation attributable to each source is estimated and might reveal interesting patterns of agreement/disagreement.

Some of the examples with continuous scores appear in previous chapters. For example, the wine tasting example was introduced in Chapter 1 where Californian and French wines were judged by a panel of experts. The degree to which the raters agree was explored with regression and graphical methods.

The author worked for several years with the diagnostic radiology division of the University of Texas MD Anderson Cancer Center in Houston, Texas and assisted with the design and analysis of many clinical studies with a large variety of imaging devices, including X-ray, computed tomography (CT), magnetic resonance imaging (MRI), and nuclear medicine. Several examples involving those studies include: (a) assessing agreement between several radiologists who are measuring the size of lung tumors with CT; (b) imaging the liver with CT and measuring blood flow and volume, where four readers are involved; (c) using positron emission tomography (a nuclear medicine procedure) to estimate the standardized uptake value (SUV) in non-small-cell lung cancer patients with five readers; and (d) another case of estimating tumor size with several radiologists, but with MRI.

6.2 Regression and Correlation

Figure 6.1 shows box plots of the tumor size of lung lesions measured by CT and interpreted by five radiologists. Each reader measured the tumor size on 40 lesions and the sizes were reestimated using the same image. Thus, there is a true replication and the main purpose was to assess rater agreement in the context of a phase II clinical trial. A phase II clinical trial usually investigates the response to treatment by measuring the size of the tumor at baseline and at successive times during the course of the treatment. Depending on the change of size, the patient is classified into one of several categories (complete response, partial response, etc.) and the success of the trial is judged accordingly. See Broemeling[1] for an introduction to reader agreement and phase II clinical trials.

Tumor sizes (measured in cm) and other observations were gathered by Erasmus et al.[2] using 33 patients with a total of 40 lesions and two types of tumor sizes were measured, namely: unidimensional and bidimensional, where the former are based on the longest diameter of the lesion, and the latter are based on the largest diameter and the next largest diameter, which is perpendicular to the largest diameter of the lesion. What is reported here are tumor sizes based on the unidimensional diameter, and the descriptive statistics are reported in Table 6.1.

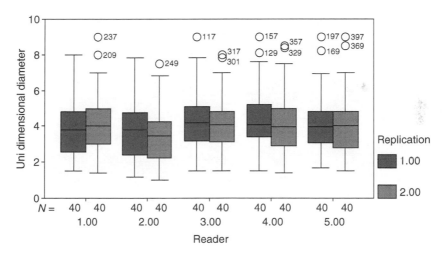

FIGURE 6.1
Unidimensional diameter tumor size.

TABLE 6.1A

Unidimensional Tumor Sizes of 40 Lesions (First Replication)

Reader	Mean	SD
1	3.92	1.61
2	3.70	1.51
3	4.42	1.55
4	4.36	1.61
5	4.14	1.55

TABLE 6.1B

Unidimensional Tumor Sizes of 40
Lesions (Second Replication)

Reader	Mean	SD
1	4.13	1.66
2	3.48	1.45
3	4.25	1.42
4	4.17	1.68
5	4.07	1.69

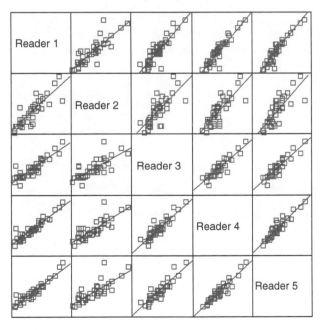

FIGURE 6.2
Tumor sizes for all pairs.

At first glance at Figure 6.1, the inter rater agreement appears pretty good, with the largest intra reader difference occurring with reader 2, who reports the smallest tumor size in both replications, which is also verified from the tables. The box plots also reveal large variations within each plot, which is portrayed in greater detail in the following scatter plots.

Figure 6.2 shows the scatter plots of the tumor sizes for all pairs of the five readers for the first replication and demonstrates a somewhat good overall agreement. Also presented is the fitted simple linear regression which regresses the 40 tumor sizes of each rater on the 40 of another. Better detail is shown in Figure 6.3 for reader 1 versus reader 2, where the 40 tumor sizes of reader 1 are paired with those of reader 2, and the fitter

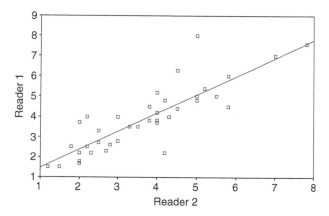

FIGURE 6.3
Reader 1 versus Reader 2.

linear regression shows a line that goes approximately through the origin. The greater detail of Figure 6.3 shows the vertical and horizontal variation of the 40 pairs and reveals the challenge in assessing agreement with continuous observation.

Note each plot has the fitted simple linear regression line super imposed. It appears to me that tumor sizes for reader 2 versus 3 have the most variation. The simple linear regression model is given by

$$y[i] = \beta_1 + \beta_2 x[i] + e[i] \tag{6.1}$$

where $y[i]$ is the tumor size of i-th patient measured by one reader, $x[i]$ is the tumor size of the i-th patient as measured by another rater (for the same replication), and the $e[i]$ are independent and distributed as $n(0, \text{tau})$, where tau is the precision of $e[i]$, $i = 1, 2, \ldots, 20$. The intercept and slope are the unknown parameters β_1 and β_2, respectively. With the Bayesian approach the parameters are given uninformative prior distributions, i.e. β_1 and β_2 are given independent $n(0, 0.00001)$ distributions, and tau is specified as a gamma(0.00001, 0.00001).

After examining the graphical evidence the regression of one reader scores on the others can be studied in greater detail by estimating the intercept and slope of the fitted lines, along with the 95% credible intervals. Table 6.2 shows the posterior analysis for several pairs of five readers, where the paired tumor sizes are regressed, one on the other. The computations are based on BUGS CODE 6.1 and the regression model, Equation 6.1.

In every case, the 95% credible interval for the intercept contains the value zero, suggesting the true line goes through the origin, however, the credible interval for the slope does not always include the value 1. For example, the regressions involving reader 2 scores as the dependent variable suggest a slight lack of agreement between reader 2 and readers 3, 4, and 5 and suggests that readers 3, 4, and 5 are assigning higher scores than rater 2.

TABLE 6.2

Bayesian Regression for Pairs of Raters Tumor Sizes

Pair	Intercept	95% CI	Slope	95% CI	Sigma	SD Sigma
1 vs 2	0.5561	(−0.175,1.292)	0.907	(0.725,1.09)	0.7592	0.1843
1 vs 3	−0.1454	(−0.898,0.602)	0.919	(0.758,1.08)	0.6036	0.1478
1 vs 4	−0.1314	(−0.689,0.421)	0.928	(0.810,1.04)	0.3783	0.0919
1 vs 5	−0.1017	(−0.637,0.430)	0.971	(0.850,1.09)	0.344	0.084
2 vs 3	0.41	(−0.563,1.37)	0.745	(0.542,0.954)	1.05	0.255
2 vs 4	0.3828	(−0.468,1.22)	0.761	(0.579,0.944)	0.863	0.209
2 vs 5	0.452	(−0.404,1.29)	0.785	(0.593,0.980)	0.886	0.215

Using SPSS® (version 11.0) produced almost the same outcomes for the regression analysis, because the Bayesian version uses an uninformative normal prior density of slope and intercept, and an uninformative gamma prior is placed on the precision (inverse of variance) about the regression line. With uninformative prior distributions, the Bayesian analysis will be similar to the 'classical' approach. It should be emphasized that the analysis assumes that the observations are normally distributed and this assumption may be checked with a QQ plot of the observations, using the normal as a references.

BUGS CODE 6.1

Model

```
# Erasmus example tumor size
# e.g. reader 2 vs 5
{
# Regression of y on x
for(i in 1:N) { y[i]~dnorm(mu[i],precy)
 mu[i] <- beta[1]+beta[2]*x[i]
}
# prior distribution of betas
for (i in 1:P){ beta[i] ~ dnorm(0.0,.000001)}
# prior distribution for the precision about regression line
precy ~ dgamma(.00001,.00001)
# standard deviation about regression line
sigma <- 1/precy}
# values of y and x
list( N=40, P=2,
y=c(3.30,3.80,2.00,1.20,4.00,3.30,4.00,5.20,7.80,3.00,5.50,2.70,4.50,1.80,4.00,2.00,
5.80,3.80,5.80,2.50,7.00,2.20,5.00,5.00,2.50,4.00,2.00,4.50,4.30,1.50,2.30,4.20,2.80,
2.00,2.20,5.00,5.00,3.00,3.50,4.20),
```

TABLE 6.3

Pearson Correlation Between Tumor Sizes of Five Readers (First and Second Replications)

Reader	1	2	3	4	5
1		0.855,0.830	0.887,0.853	0.930,0.919	0.937,0.911
2			0.762,0.837	0.809,0.854	0.804,0.795
3				0.890,0.905	0.904,0.906
4					0.939,0.934

x = c(4.00,3.90,2.10,2.00,4.20,3.90,4.20,5.00,8.20,3.10,5.50,2.90,4.20,3.20,5.10,3.80, 4.50,4.50,6.00,3.50,6.90,4.60,4.20,5.00,2.50,3.90,2.50,6.50,4.00,1.70,2.50,3.50,3.00, 3.00,3.60,4.00,9.00,3.30,3.10,5.00))

initial values of betas

list(beta = c(0,0))

Table 6.3 shows a strong correlation between the scores of the five readers and this evidence along with the regression analysis implies the inter rater agreement is strong, however, the intra reader agreement also needs to be investigated (see Exercises 1–3).

6.3 The Analysis of Variance

This next example is from the Diagnostic Imaging Division of MD Anderson Cancer Center and introduces the analysis of variance approach to the study of inter and intra reader agreement. As with the previous study, there are several raters assigning scores to non-small cell lung cancer patients, but instead of CT, the imaging is based on a nuclear medicine technique with positron emission tomography (PET) along with the radionuclide FDG (2-deoxy-2-[18]fluoro D-glucose). PET is a gamma camera that measures the amount of radioactivity emitted by FDG, and, in this case, measures the amount of uptake by the lung lesion. This procedure consists of administering FDG intravenously and detecting the resulting radiation with a gamma camera and estimates the metabolism of the lesion in standardized uptake value (SUV) units, and had been used in phase II clinical trials to measure the tumor response to therapy. The purpose of the Marom et al.[3] study is to assess the inter and intrarater variability of raters who are measuring the SUV activity with the help of a computed tomography PET image. The PET image is viewed on the screen by the radiologist, who outlines the tumor and from this outline, the computer calculates the SUV of the tumor. The SUV is a measure of the malignancy of the tumor, thus if the therapy is active, the SUV units decrease, but on the other hand, if the lesion grows, the SUV activity increases. See Broemeling[1] for a brief description of the imaging techniques that are used in the division of diagnostic imaging of any large health care center, including nuclear medicine and the associated gamma cameras such as PET and single photon emission computed tomography (SPECT).

Positron emission tomography-computed tomography (PET-CT) examinations of 20 consecutive patients referred for initial evaluation of newly diagnosed non-small-cell lung cancer were retrospectively reviewed by five experienced radiologists, who independently measured the maximal SUV/body weight of the primary tumors. Inter observer and intra observer variation is assessed by four statistical techniques: correlation, regression, Bland–Altman, and the analysis of variance.

Both inter observer and intra observer SUV measurements are highly reproducible, with Pearson correlation coefficients greater than 0.95 and 0.94, respectively. Good inter observer and intra observer agreement was shown with regression analysis and the overall conclusion was that there was excellent agreement in the measurement of SUV among the five radiologists. See Marom et al.[3] for more details about this important study. It was one of the first to study inter observer and intra observer variability with SUV. The presentation here, differs form the analyses described in Marom et al. Only three radiologists are employed (because only three had two replications each), and the analysis of variance is somewhat different from that of the original study. Instead of a so-called fixed analysis, a version using random effects is used, and is based on the model,

$$y[i, j, k] = \theta + a[i] + b[j] + d[k] + e[i, j, k] \tag{6.2}$$

where, $i = 1, 2, j = 1, 2, 3$, and $k = 1, 2, \ldots, 20$. θ is a constant, the $a[i]$~dnorm(0,tau$a$), the $b[j]$~dnorm(0,tau$b$), the $d[k]$~dnorm(0,tau$d$), and the $e[i,j,k]$~dnorm(0,tau$w$), and all random variables are independent. Prior information is assigned as follows: θ~dnorm(0,0.0001), tau$a$~dgamma(0.00001,0.00001), tau$b$~dnorm(0.00001,0.00001), tau$d$~dgamma(0.00001,0.00001), and tau$w$~dgamma(0.00001,0.00001)., thus non informative priors are assigned to the last stage of this hierarchical model. Note that θ is the overall mean of the SUV values and $y[i,j,k]$ is the SUV value for the i-th rep, the j-th radiologist, and the k-th lesion. The major parameters are:

$$\sigma_w^2 = 1 / \text{tauw}$$

$$\sigma_a^2 = 1 / \text{taub} \tag{6.3}$$

$$\sigma_b^2 = 1 / \text{taub}$$

and

$$\sigma_d^2 = 1 / \text{taud}$$

The four sigmas are the variance components, where σ_a^2 is the between replication variance component, because it measures the variation of the SUV values between the two replications, and in a similar way, σ_b^2 measures the variation between the three radiologists, and is called the between readers variance component. σ_d^2 is the variance component for lesions, and σ_w^2 is the so-called error variance component, that is, it measures the variation of what is not explained by the other (replications, readers, and lesions) factors in the model. Equation 6.1 is often referred to as a random model and is interpreted as follows. The replications, readers, and lesions are thought of as random samples from the relevant populations, thus the three readers are thought of as a random sample from some population of readers, as well as the patients in the study, who are selected at random from the population of newly diagnosed non-small-cell lung cancer patients.

Of course, there are many versions of this scenario, referred to as mixed models, because such models have both fixed and random effects, and one could consider the two replications as fixed, that is, one is interested in only the two replications, and they are not considered random samples from a population of replications thus, inferences about replications refer to only the two replications of the study. It is also important to note that the assumption of normality for the SUV values.

BUGS CODE 6.2 listed below includes the SUV values for three readers, two replications, and 20 patients. See the first list statement which includes the SUV values. The data is in six groups, where the first group corresponds to the 20 SUV values of the first replication and the first reader, while the sixth group included the 20 SUV values for the second replication and the third reader.

BUGS CODE 6.2

```
model;
# Marom SUV
{
# N is number or reps
# M is number of readers
# O is number of patients
for(i in 1:N){for(j in 1:M){ for( k in 1:O){
y[i,j,k] ~ dnorm(mu[i,j,k], tau.w)
mu[i,j,k] <- theta + a[i] + b[j] + d[k]}}}
# theta is overall mean
theta ~ dnorm(0.0,.00001)
for(i in 1:N){ a[i] ~ dnorm(0.0, taua)}
for(j in 1:M){ b[j] ~ dnorm(0.0,taub)}
for( k in 1:O){d[k] ~ dnorm(0.0,taud)}
```

```
taua ~ dgamma(0.00001,.00001)
taub ~ dgamma(0.00001,.00001)
taud ~ dgamma(0.00001,.00001)
tau.w ~ dgamma(0.00001,.00001)

sigma.w < -1/tau.w
sigmaa < -1/taua
sigmab < -1/taub
sigmad < -1/taud

# sigmaa is variance component for reps
# sigmab is variance component for readers
# sigmad is variance component for patients
# sigma.w is variance component for error
}
list(N = 2,M = 3,O = 20,

y = structure(.Data = c(

11.70,5.00,21.70,21.00,33.00,20.30,11.80,4.00,13.50,8.50,21.00,13.20,6.30,19.50,
13.40,10.90,14.70,3.60,4.30,15.50,

11.50,5.00,21.00,24.00,30.90,19.40,11.80,4.00,10.20,8.50,22.40,12.30,6.30,19.50,
13.40,8.50,13.50,3.50,4.80,17.00,

10.30,5.00,19.30,18.70,28.00,20.70,9.60,4.60,10.00,8.50,25.60,12.80,5.90,15.80,18.40,
9.30,10.60,3.60,4.60,15.60,

11.50,5.00,18.80,14.50,31.40,20.80,11.80,4.60,12.30,8.50,24.40,13.20,6.30,19.50,
11.00,11.00,14.60,3.60,4.70,16.00,

11.50,5.00,19.90,24.00,31.40,19.40,11.80,4.00,12.30,9.80,19.60,13.20,6.30,19.60,13.50,
12.70,13.80,3.30,4.80,15.60,

11.90,5.00,21.00,22.40,30.60,18.90,11.80,4.00,8.90,9.00,20.60,12.20,6.30,16.80,11.00,
12.70,12.10,3.30,4.40,18.30),.Dim = c(2,3,20)))

list(tau.w = 1, a = c(0,0),b = c(0,0,0),d = c(0,0,0,0,0,0,0,0,0,0,0,0,0,0,0,0,0,0,0,0),
theta = 0, taua = 1,taub = 1,taud = 1))
```

The box plots of Figure 6.4 shows the interobserver and intraobserver variation, and it is apparent that according to the median SUV, there is good agreement between the three, and that the "largest" intraobserver error is for rater 3, and that the smallest inter quartile range is for the first reader at the second rep.

A posterior analysis based on Equation 6.2 and BUGS CODE 6.2 with 845,000 observations generated from the joint posterior distribution, with a burn in of 5000, and a refresh of 100 appears in Table 6.4. Estimates of the various variance components are quite interesting, because of the very small

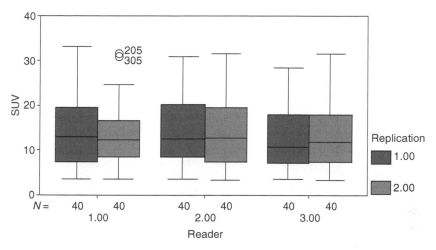

FIGURE 6.4
Box plots, SUV of three readers (two replications).

TABLE 6.4

Posterior Analysis for Three Readers, Two Replications, and 20 Patients

Parameter	Mean	SD	2½	Median	97½
σ_w^2	2.304	0.3338	1.741	2.273	3.044
σ_a^2	3.675	435.8	0.00001018	0.0017	3.059
σ_b^2	0.1522	2.653	0.00001115	0.002991	0.8384
σ_a^2	57.54	21.18	29.56	53.32	110.4
θ	13.26	1.802	9.744	13.26	16.8
$a[1]$	−0.002542	0.5852	−0.5373	0.000128	0.4975
$a[2]$	−0.006703	0.5854	−0.552	−0.000287	0.4806
$a[1]$	0.02711	0.2059	−0.3019	0.00288	0.4764
$b[2]$	0.04115	0.2199	−0.2585	0.00451	0.5243
$b[3]$	−0.07265	0.2355	−0.6402	−0.00886	0.1749

values for replications and readers and their right-skewed distributions. Because of the skewness, the posterior medians are preferred as measures of location in lieu of the posterior mean.

The estimates follow a typical pattern for inter and intraobserver studies, namely the largest component is between patients with a median of 53.32, followed by the error variance component with a median of 2.273, and the smallest component is the between replications with a median of 0.0017 with the next to smallest component for readers with a median of 0.002991. Therefore, expressed as percentages of a total variability of 55.59761, the error component is 4.088%, the replication component is 0.0030577%, the between readers component is 0.0053797%, while the patient component is 96.9034% of the total. One would expect the largest variation to be between patients and the smallest to be within observers, followed by between raters. The posterior means of the $a[i]$ are interesting, in that there is very little difference between them, and the same is true, but not to the same extreme, between the posterior means of the $b[j]$.

6.4 Intraclass Correlation Coefficient for Agreement

There are several ways to investigate rater agreement with continuous scores, and some of those approaches have been discussed in the previous sections, and to that end, regression and correlation methods, as well as analysis of variance techniques, were introduced. Along with analytical techniques, graphical procedures, including box plots and scatter plots, were introduced as visual methods of displaying inter and intraobsever disagreement. Graphical displays were also employed to check model assumptions, such as normality of the rater scores. The intraclass correlation coefficient is another way to study rater agreement, in that, higher correlations imply better agreement.

Recall the estimation of the intraclass correlation in Chapter 3 and Chapter 4 for discrete rater scores. For continuous observations, the approach is very similar. The model is based on normality of the rater scores and assumes the correlations are the same for all pairs of raters, thus, if there are four raters, the six correlations corresponding to the six pairs of raters are the same. This is somewhat of a restrictive assumption, and should be checked along with the normality of the observations. Normality will be checked with either PP or QQ plots, and the constant correlation assumption will be checked by estimating the Pearson correlations between all pairs of rater scores.

The analysis of variance models of the previous section are the basis for the intraclass correlation, which is a function of the variance components of the random factors of the mixed model. With the analysis of variance approach, the total variance of an observation is partitioned into several sources of variation, and the sources of variation in turn are determined by the layout of the design of the study, which dictate the assumptions that make up the model. The intraclass correlation coefficient goes one step further by using the variance components to define a common correlation between all pairs of rater scores, and in fact, the intraclass correlation is a ratio of sums of variance components, where the numerator and denominator are sums of the variance components of the model.

Several examples serve to illustrate the intraclass correlation, and the first two involve different imaging modalities: for the first example, MRI images were assigned tumor sizes by the raters, while the second example is another CT study involving cancer patients where CT images were taken of the liver (the disease has metastasized to the liver) and the blood flow and blood volume measured.

For all examples, the approach is to: (a) first check for normality of the observations and the constant correlation assumption; (b) define a model that takes into account the design of the study. The model is defined in terms of the sources of variation and assumptions about the distribution of the terms of the model. There are basically two types of studies considered, namely those with and without replication. By replication is meant the same image is viewed more than one time, and rater scores are assigned to each replication. A typical design of an agreement study is that all raters see each patient and assign scores to the images taken on each patient. Such studies are usually quite complicated and executed according to a protocol that details every facet of the study. What is presented here is very much simplified. For studies involving no replication there are three sources of variations, namely, between patients, between readers, and error, and for those with replication, in addition to the three named previously, there is a between replication source of variation. The description was used for the Marom et al.[3] study of the previous section, where replication was taken into account.

Kundra et al.[4] performed a study of interobserver agreement, where the lesion size of liver lesions were estimated with the aid of MRI and their interpretation by three radiologists. The study was quite involved using several MRI sequences and had two replications, but only the first replication is considered here and the MRI sequence is fast spin echo. See Broemeling[1] and Brant and Helms[5] for more information on the fundamentals of MRI and information about various sequences.

To begin the analysis, consider the descriptive statistics for the three radiologists given by Table 6.5. A marked skewness is detected for the tumor sizes of all three, but is quite obvious for reader 1 and 3.

There appears to be a small deviation from normality, however, normality will be assumed for the model. The next graphical display is Figure 6.5, a box plot of the three reader scores and shows the interrater variation is smallest between readers 2 and 3, and it also shows that reader 1 has the smallest interquartile range. What would one expect for the correlation between the three pairs of raters? I calculated the Pearson correlation between readers 1 and 2, 1 and 3, and 2 and 3 and found 0.740, 0.964, 0.754, respectively. Is it reasonable to assume a common correlation model?

In view of the graphical and descriptive evidence, the following normal model is proposed.

$$y[i,j] = 0 + a[i] + e[i,j] \qquad (6.4)$$

TABLE 6.5

Descriptive Statistics for Tumor Sizes of Liver Lesions (MRI Images)

Reader	N	Minimum	Maximum	Mean	SD	Median	Skew
1	19	10	116	37.21	26.01	33	1.72
2	20	8	75	32.75	18.54	26.5	0.544
3	19	9	104	33.57	24.47	27	1.528

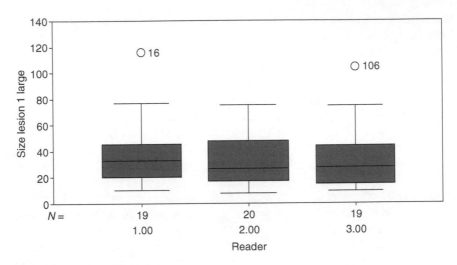

FIGURE 6.5
Box plots, tumor size (three radiologists).

where $y[i, j]$ is the tumor size for the i-th patient and representing the j-th reader, $i = 1, 2, \ldots, 22$, and $j = 1, 2, 3$. In addition, θ is a constant, the mean of the observations, the $a[i]$ are independent and normally distributed with mean 0 and variance σ_a^2, while the $e[i, j]$ are independent and normally distributed with mean 0 and variance σ_w^2. Also, all random variables are jointly independent.

Consider the covariance between $y[i, j]$ and $y[i, j']$, that is, between the tumor scores of the j-th and j'-th reader of the i-th lesion, then

$$\text{cov}(y[i, j], y[i, j']) = \text{cov}(\theta + a[i] + e[i, j], \theta + a[i] + e[i, j']) \qquad (6.5)$$

Note the covariance is conditional on the parameters and it can be shown that the covariance is σ_a^2, consequently, the intraclass correlation is

$$\rho = \sigma_a^2 / (\sigma_a^2 + \sigma_w^2) \qquad (6.6)$$

BUGS CODE 6.3 is to be executed in order to estimate the intraclass correlation of the Kundra study.

BUGS CODE 6.3

```
# program for Kundra inter reader agreement
# three readers
# 22 patients
# response is lesion size in mm
# intraclass correlation coefficient
```

```
model
        {
            for( i in 1 : images) {
                m[i] <- theta + a[i]
                for( j in 1 : radiologists) {
                    y[i,j]~dnorm(m[i], tauw)
        }
          }

          for( i in 1:images){ a[i]~dnorm(0,taua)}
          sigmaw <- 1/tauw
          sigmaa <- 1/taua
          tauw~dgamma(0.001, 0.001)
          taua~dgamma(0.001, 0.001)
          theta~dnorm(0.0, .001)
          icc <- (sigmaa)/( sigmaa + sigmaw)
          }

        list(images = 22, radiologists = 3,
          y = structure(.Data = c( 26.00, 22.00, 10.00,
        NA, 25.00,25.00,
        NA, NA,10.00,
NA,    17.00, 19.00,
20.00, 17.00, NA,
30.00, 28.00, NA,
39.00, 44.00, 40.00,
33.00, 53.00, 40.00,
46.00, 47.00, 27.00,
26.00, 25.00, 26.00,
50.00, 51.00, 49.00,
44.00, 48.00, 47.00,
77.00, 75.00, 74.00,
116.00, 44.00, 104.00,
11.00, 8.00, 12.00,
10.00, 12.00, 9.00,
12.00, 12.00, 15.00,
61.00, 56.00, 53.00,
21.00, 21.00, 27.00,
13.00, 13.00, 15.00,
```

36.00, NA,NA,

36.00, 37.00, 36.00), .Dim = c(22, 3)))

list(theta = 0 ,tauw = 1, taua = 1)

The statements are self explanatory, where the first list statement consists of 22 rows of three columns of tumor sizes, where the rows correspond to lesions and the columns to the three readers. The intraclass correlation is labeled as icc, while σ_a^2 is referred to sigma$a$. The variance of errors σ_w^2 is called sigma$w$. Note that the overall mean θ is given an uninformative normal distribution with mean 0 and precision 0.001. In a similar way, the lesion effects $a[i]$ are given means 0 and precision tau$a = 1/\sigma_a^2$ and the error terms means 0 and common precision tau$w = 1/\sigma_w^2$. In order to complete the specification for the prior distributions, the precision parameters are given uninformative gamma distributions, with first and second parameters 0.0001. The posterior analysis was done with 45,000 observations generated from the joint posterior distribution, with burn in of 5000, and refresh 100 (Table 6.6).

   Agreement between readers is fairly good and estimated as 0.7908 with the posterior mean and a 95% credible interval of (0.6285, 0.9044), while the mean of the observations is estimated as 32.53. Recall that the Pearson correlation coefficients were estimated (using SPSS version 11) as 0.740, 0.964, and 0.754 for readers 1 and 2, 1 and 3, and 2 and 3 respectively, thus, the estimated intraclass correlation is approximately equal to the mean of the three estimated Pearson correlation coefficients! Note that the posterior means of the variance components σ_a^2 and σ_w^2 are quite 'large', which is very reasonable in light of the actual tumor sizes appearing in the list statement of BUGS CODE 6.3.

   For the second example of the intraclass correlation, the Marom et al.[3] study is again considered because two replications were conducted by the three readers, who were measuring, with aid of PET, the SUV of lung cancer lesions. Refer to Table 6.4 for the posterior analysis of the study, which implies good agreement. The table provides the posterior mean, standard deviation, median, and 2½ and 97½ percentiles of the overall mean θ and the four variance components $\sigma_a^2, \sigma_b^2, \sigma_d^2,$ and σ_w^2 for the between replications, between readers, between patients, and error, respectively.

   Recall that the analysis of variance model is

$$y[i, j, k] = \theta + a[i] + b[j] + d[k] + e[i, j, k] \tag{6.2}$$

TABLE 6.6

Posterior Analysis for Kundra MRI Study

Parameter	Mean	SD	2½	Median	97½
ρ	0.7908	0.0711	0.6285	0.7994	0.9044
σ_a^2	446.5	167.3	219.2	414.5	862.6
σ_w^2	108.3	26.95	67.51	104.3	172.1
θ	32.53	4.676	23.19	32.56	41.62

TABLE 6.7

Posterior Analysis for Agreement of Marom Study

Parameter	Mean	SD	2½	Median	97½
ρ	0.9556	0.0197	0.9178	0.9582	0.9806
σ_a^2	1.986	162.4	0.00001015	0.0018	2.998
σ_b^2	0.1288	2.32	0.00001122	0.0029	0.8003
σ_d^2	57.31	20.97	29.37	53.15	109.4
σ_w^2	2.305	0.3339	1.741	2.274	3.047
θ	13.14	1.736	9.656	13.19	16.50

where θ is a constant, the $a[i] \sim n(0, \text{tau}a)$, $b[j] \sim n(0, \text{tau}b)$, $d[k] \sim n(0, \text{tau}d)$, and the $e[i,j,k]$ are distributed $n(0, \text{tau}w)$, and the variance components are the reciprocals of the precision parameters, namely: $\sigma_a^2 = 1/\text{tau}a$, $\sigma_b^2 = 1/\text{tau}b$, $\sigma_d^2 = 1/\text{tau}d$, and $\sigma_w^2 = 1/\text{tau}w$. All random variables are assumed to be independent, which implies that the intraclass correlation is

$$\rho = (\sigma_a^2 + \sigma_d^2)/(\sigma_a^2 + \sigma_b^2 + \sigma_d^2 + \sigma_w^2). \tag{6.7}$$

BUGS CODE 6.2 is amended with the statement

icc < - (sigmaa + sigmad)/(sigmaa + sigmab + sigmad + sigma.w)

for the intraclass correlation, and when executed gives Table 6.7.

The results are fairly typical of an agreement study, in that the largest component is between patients σ_d^2, and the smallest is between replications σ_a^2, as measured by the posterior medians. Note, the posterior distributions of the variance components are skewed to the right, thus the medians are more reliable for estimating the location of the distribution. Very good inter rater agreement is implied by the posterior median of 0.9582 and a 95% credible interval of (0.9178, 0.9806) for the intraclass correlation. 70,000 observations are generated from the joint posterior distribution, with a burn in of 5000, and a refresh of 100.

There have been many studies of the Bayesian approach to variance components and the intraclass correlation, for example, see Broemeling[1], Box and Tiao[6], and Palmer and Broemeling[7].

6.5 Agreement with Covariates

The theme of the previous section is continued with the Marom et al.[3] study, but now a covariate is taken into consideration and the intraclass correlation coefficient estimated with a Bayesian approach. Recall that this experiment involved three readers who are reading PET images, and are outlining the lesion as it appears on the screen. From the outlined lesion, the computer calculates the SUV which is an indication of the malignancy of the lesion. A potential covariate is the size of the lesion (measured in centimeters), which has a potential effect on the SUV value. Does it affect the intraclass correlation coefficient? A scatter plot of the SUV versus the lesion size, would show some effect, namely as the size of the lesion increases there is a tendency for the SUV values to increase. Will it have an effect on the estimated intraclass correlation? The model is Equation 6.2, but with a covariate x, where

$$y[i,j,k] = \theta + a[i] + b[j] + d[k] + \beta x[i,j,k] + e[i,j,k], \qquad (6.8)$$

and $i = 1, 2; j = 1, 2, 3;$ and $k = 1, 2, \ldots, 20$.

The constant β measures the effect of lesion size on SUV, where $x[i,j,k]$ is the size of the lesion corresponding to the i-th replication, the j-th reader, and the k-th patient. The goal is to estimate the intraclass correlation, using Equation 6.7.

With 90,000 generated from the joint posterior distribution, a burn in of 10,000, and a refresh of 100, the posterior analysis for the SUV study is portrayed in Table 6.8.

The effect of lesion size, as measured by β, is somewhat muted, as can be seen by Table 6.8 and the 95% credible interval (−0.1515, 2.135), but it does have some effect on the estimated variance components. Comparing Table 6.8 with Table 6.7 reveals a large change in θ, from a median of 13.19, when the lesion size was not considered, to a median of 9.155 when the lesion size is included in the model, however, there was practically no change in the posterior mean of ρ, which is the main parameter of interest. Also, there was a noticeable change in the posterior mean of σ_a^2, but not in the corresponding posterior median.

As the second example of an agreement study with a covariate, the Kundra[4] study serves as a good example. Recall that this study involved three raters, who are measuring the tumor size of liver lesions of mice with six MRI sequences, but what is reported here is for the fast spin echo sequence with two replications. The disease has metastasized to the liver, where there

TABLE 6.8

Posterior Analysis for Agreement of Marom Study

Parameter	Mean	SD	2½	Median	97½
ρ	0.9509	0.0242	0.9089	0.954	0.9789
σ_a^2	0.829	36.59	0.00001028	0.0016	1.902
σ_b^2	0.1646	3.007	0.00001094	0.00266	0.8054
σ_d^2	52.18	19.64	26.47	48.19	100.7
σ_w^2	2.303	0.3327	1.742	2.272	3.044
θ	9.078	2.88	3.216	9.115	14.77
β	0.9853	0.5727	−0.1515	0.9809	2.135

are usually multiple nodules that can vary from 0 to 11 per mouse, thus, since an increasing number of lesions usually indicates the severity of disease, it is taken as a covariate that perhaps will influence the dependent variable, which is the size of the largest liver lesion. It is important to remember that the size of the tumor and the number of nodules that are actually recorded depends on the MRI sequence, of which there are six, and the radiologist, who is interpreting the images!

The descriptive statistics are provided by Table 6.9A and 6.9B. The number of nodules also varies from reader to reader, within and between radiologists. Obviously, the distributions for the tumor sizes are skewed to the right, while those for the nodule size are skewed to the left!

With a scatter plot, one may show the effect of the number of lesions on the tumor size of the largest lesion, and the Lowess curve would show as the number of nodules (lesions) increase, the size of the tumor increase. However, it remains to be seen if it does indeed affect the posterior analysis of the intraclass correlation. For each mouse, the number of liver nodules is counted (the count varies with respect to the MRI setting as well as the radiologist reading the image) and the nodule with the largest size is the value of the dependent variable.

Suppose 290,000 observations are generated from the joint posterior distribution of the parameters, with a burn in of 100,000, and a refresh of 100, then the resulting posterior analysis is given by Table 6.10. A "large"

TABLE 6.9A

Descriptive Statistics for Lesion Size of Kundra Study

Replication	Reader	N	Minimum	Maximum	Mean	Median	SD
1	1	19	10	116	37.21	33	26.01
1	2	20	8	75	32.7	26.5	18.54
1	3	19	9	104	33.57	27	24.47
2	1	18	10	101	37.7	34	23.9
2	2	18	12	70	35.6	38	17.5
2	3	18	9	88	34.5	28.5	21.7

TABLE 6.9B

Descriptive Statistics for Number of Nodules of Kundra Study

Replication	Reader	N	Minimum	Maximum	Mean	Median	SD
1	1	21	1	12	8.09	10	3.30
1	2	20	1	10	7.55	10	3.56
1	3	20	0	10	7.50	10	3.80
2	1	21	1	11	7.85	10	3.56
2	2	22	0	10	7.45	10	3.87
2	3	21	0	10	7.71	10	3.64

TABLE 6.10

Posterior Analysis for Agreement of Kundra Study with Covariate

Parameter	Mean	SD	2½	Median	97½
ρ	0.7994	0.0670	0.6621	0.8056	0.9056
σ_a^2	104.2	62940	0.00001271	0.01054	50.94
σ_b^2	9.063	617	0.00001269	0.01051	14.56
σ_d^2	368.4	137	184.1	341.7	707.4
σ_w^2	83.65	12.91	62.08	82.37	112.5
θ	24.33	7.696	9.953	24.26	39.15
β	1.15	0.7563	−0.3472	1.153	2.622

simulation size is necessary, because the MCMC error is too "large" for the smaller sample sizes.

Large estimated variance components are evident with the posterior mean, but the posterior means show the usual order in agreement, where the smallest component are the between replications, σ_a^2, and the between radiologists, σ_b^2, and the largest is the between lesions component σ_d^2. The relatively "large" error component posterior median of 82 implies that that the model, perhaps, is not a good fit to the tumor size information. Note, that Equation 6.8 was employed, and that the posterior medians give a better estimate of location than the posterior mean. Note for σ_a^2, the mean is 104 compared to a median of 0.0105! The main concern is the number of lesions and their effect on the tumor size and the estimates of the other parameters. Also, what is the implication for agreement as measured by ρ?

6.6 Other Considerations with Continuous Scores

6.6.1 Introduction

When the rater scores are continuous, the common statistical methods of regression, correlation, and the analysis of variance are the most popular. With regression, normality of the observations is assumed, and the scores of one reader are regressed on the other. Good agreement is implied if the true regression line goes through the origin with a slope of 1, and this is implemented for all pairs of raters. This is most likely the best way to check for agreement. It is important to note that computing the correlation between all pairs of rater scores has its limitations, in that the correlations can be high, but the raters have poor agreement! It is usual to find some pairs of raters that have good agreement, while other pairs do not. The Bland–Altman procedure is the most popular technique for checking reader agreement, however, it is not recommended. The regression approach is preferred by the author and will be illustrated with a Bayesian approach in a later section. The Erasmus et al.[2] CT study with five readers measuring the tumor size of lung cancer patients illustrated the regression technique of assessing agreement experiments.

When referring to the analysis of variance, one means that the total variance of an observation is partitioned into several sources, and the percentage of each source is estimated by the appropriate variance component. Normality for the observations is usually assumed and the various factors of the model are assumed to have random effects, which are modeled as unobservable normal random variables, with mean zero and with a variance called the variance component for that particular factor. This approach involves using a very restrictive model, and the model assumptions should be checked. Later sections will explore use of mixed (those containing both fixed and random effects) models for estimating variance components. Also to be examined, is a Bayesian approach to estimating variance and covariance components when there are multiple dependent variables.

The intraclass correlation coefficient is the common correlation between all pairs of observers and is estimated as a ratio of the sum of variance components of a random model, thus, it is a natural extension of estimating the variance components of the model.

The Maron et al.[3] and Kundra[4] studies were referred to as good examples for estimating variance components and the intraclass correlation coefficient. In addition, these studies served as good illustrations when studying the effect of patient covariates on agreement. For example, the Marom et al. SUV study had the lesion size as a covariate, while the Kundra experiment included the number of lesions in the liver as a covariate. These ideas will be discussed further in the next section, when the type of the primary tumor is employed as a discrete covariate in another CT study that images blood flow and blood volume of the liver of cancer patients.

6.6.2 Bayesian Bland–Altman

I have not seen a Bayesian version of the Bland–Altman[8] approach to agreement, but it can be viewed as an extension of regression method, where the scores of one reader are regressed on those of another, and this is repeated for all pairs of readers. Briefly there are three steps to Bland–Altman: (1) for all pairs, regress the scores of one reader on those of the other, and assess agreement of the fitted simple linear regression, by checking to see if the "true" line goes through the origin with a slope of 1; (2) for each pair plot differences in the two paired scores, versus the sum or mean of the two paired scores. Of course, the more they agree, the smaller the differences. Look for systematic patterns in the association between the vertical variable (the paired differences) versus the horizontal variable (the sums of the paired observations); and (3) if there are no systematic patterns, summarize the plots by computing the mean and standard deviation of the differences.

The Erasmus et al.[2] study illustrated the regression approach to inter and intraobserver agreement, and will be the basis of the Bland–Altman method, where the focus will be on the intraobserver agreement. That is, the major objective is to estimate the difference between the rater's scores for the first and second replications. A scatter plot of reader 1 first replication scores versus the second replication score would show good intraobserver agreement.

The intrareader agreement of observer 1 seems fair, however there does appear to be a bias because the line does not appear to go through the origin.

Incidentally, the mean (SD) of the tumor size for replications 1 and 2 are 3.92 (1.612) and 4.132 (1.663) respectively, and the respective maximum values are 8 and 9. In addition, 1.5 and 1.4 are the respective minimum values, which implies fairly good agreement, but the fitted line is more informative.

Using BUGS CODE 6.2 (see Table 6.18 of Exercise 2 for the data), where the first list statement has the y and x vectors containing the rep 1 and rep 2 scores of reader 1, and using 20,000 observations generated from the joint posterior distribution (with a burn in of 5000 and a refresh of 100), the posterior analysis for the linear regression appears in Table 6.11.

Fair agreement is implied by the table, because the 95% credible interval for the intercept contains zero, and the interval for the slope contains 1 (just barely), and the sigma parameter is the variance about the regression line.

The second part of Bland–Altman is to plot the differences between the replications scores of reader 1 against the sum of the replication scores. The mean (SD) of the differences is -0.2125 (0.3817) with a minimum difference of -1 and a maximum difference of 0.5, thus, the replication 2 scores seem to be larger than the replication scores, but not by much. One may show a slight downward trend in the differences as the sums increase, however it is quite small. See Exercise 17. If there is a definite downward trend the range of the observations would be a function of the mean, which would violate an assumption of the simple linear regression, but this does not appear to be the case, at least, there is not enough evidence at this point. Of course, to fully implement the Bland–Altman approach, the remaining four plots for reader 2, 3, 4, and 5 must be executed, but this is left as an exercise, see Exercise 16.

TABLE 6.11

Intraobserver Agreement of Reader 1

Parameter	Mean	SD	2½	Median	97½
Intercept beta[1]	0.0212	0.1639	−0.3023	0.0210	0.3438
Slope beta[2]	0.9434	0.0368	0.8715	0.9433	1.016
Sigma	0.1483	0.0360	0.0935	0.143	0.2327

It should be remembered that the regression techniques for agreement assume that the rater scores are normally distributed, thus, this normality should be checked. In Section 6.2, the assumption of normality of the Erasmus study was verified by a QQ plot of the scores of reader 1. Another approach is to employ robust regression procedures, see for example Rousseeuw and Leroy[9] for a good reference on the subject.

6.6.3 More on Variance Components

For the analysis of variance components, several models were proposed and included the usual random effects models with and without covariates. One area that was not considered was to include interactions into the model.

For example, for the Marom et al.[3] study the analysis was presented in two stages: (1) the first was to include variance components for replications, readers, and lesions. Recall for this study the dependent variable was measured amount of SUV determined by a nuclear medicine procedure that used PET to count the amount of radioactivity emitted by the tumor, and there were two replications, three readers, and 20 lesions. It was assumed that the SUV could be modeled as

$$y[i,j,k] = \theta + a[i] + b[j] + d[k] + e[i,j,k] \qquad (6.2)$$

with the four sources of variation: between reps, between readers, between lesions, and error (the component that accounts for what is left over after accounting for those four sources), and the variation was expressed in terms of the corresponding four variance components. In order to account for the reader by lesion interaction, the model includes the interaction term $f[j,k]$, where

$$y[i,j,k] = \theta + a[i] + b[j] + d[k] + f[j,k] + e[i,j,k] \qquad (6.9)$$

The $f[j,k]$ are independent and $n(0, \text{tau}f)$ random variables, where $\text{tau}f$ is the precision component for the reader by lesion interaction, and the corresponding variance component is defined as $\text{sigma}f = 1/\text{tau}f$. The five random factors are given prior normal distributions with mean 0 and precision 0.00001, and the precision components are given independent prior gamma distributions with first and second parameter equal to 0.00001. Thus, the prior information is considered uninformative, which produces the posterior analysis of Table 6.12. Note, as measured by the posterior median, the largest source of variation is the between lesions component at 53.28, followed by the error component with a median of 2.215, then the interaction with a median 0.00619, then the reader median at 0.00281, and lastly the replication with a median of 0.001643. The usual order of the sources of variation as measured by the variance component is followed for this analysis (see Table 6.12).

TABLE 6.12

Posterior Analysis for Interaction of the Marom Study

Parameter	Mean	SD	2½	Median	97½
σ_a^2	7.299	2440	0.00001018	0.001643	2.69
σ_b^2	0.178	7.072	0.00001086	0.00281	0.858
σ_d^2	57.5	21.12	29.57	53.28	110.1
σ_j^2	0.0893	0.1818	0.00001266	0.006192	0.6457
σ_w^2	2.243	0.3465	1.639	2.215	3
θ	13.22	1.722	9.846	13.23	16.74

The posterior analysis was based on 2,000,000 observations generated from the joint posterior distribution, with a burn in of 500,000, and a refresh of 100. See Exercise 17 and Exercise 18.

6.6.4 Fixed Effects for Agreement

It should be pointed out that the terms random fixed, and mixed effects are relevant to the non Bayesian approach to statistics. In that situation, the term random implies that the model effects are considered random variables and one is making inference to a population of effects, while the term fixed means the effects are considered unknown constants, while the term mixed implies that that model contains both fixed and random factors. Of course, in the Bayesian approach, the terms in the model are all considered random, as is the case of Equation 6.2. For the sampling theory approach, one would consider θ as a fixed unknown constant, and the remaining terms, the $a[i]$, the $b[j]$, and $d[k]$, as well as the $e[i, j, k]$ as normal random variables with means 0 and unknown variances. But, in the Bayesian formulation of Equation 6.2, θ is considered a random variable with mean 0 and precision 0.00001, which produces a posterior mean of theta as an estimate of the overall mean of the observations. How would one define a so-called fixed model in the Bayesian context?

To answer this question, the Ng[10] CT study is analyzed with a fixed model. This study involved four radiologists using CT images of the liver of cancer patients. With the aid of CT, the blood flow (cubic mm/unit time) and blood volume (cubic mm) portal vein of the liver can be computed. The patient's disease has metastasized to the liver, and the primary tumor occurred at four sites: (1) the colon; (2) the kidneys; (3) the gastrointestinal tract; and (4) of carcinoid origin. The descriptive statistics for this study are presented in Table 6.13.

This study is stratified by type of primary tumor, and it appears that within type of primary tumor, the four readers are quite consistent.

Referring to Table 6.14, the average blood flow across readers varies from 73.90 for reader 3 to 81.89 for reader 1, and the standard deviation based on

TABLE 6.13

Blood Flow Through the Portal Vein of Cancer Patients by Reader and Primary Tumor

Primary tumor	Reader	Minimum	Maximum	Mean	SD	N	Cell
Colon	1	14.80	196.62	47.39	48.38	13	1
Colon	2	13.52	141.5	41.89	34.72	13	2
Colon	3	12.75	83.71	34.50	20.31	13	3
Colon	4	13.64	235	49.3	58.62	13	4
Kidney	1	14.15	323.35	127.98	73.05	14	5
Kidney	2	15.04	304.7	127.02	66.67	14	6
Kidney	3	15.28	305.35	123.11	66.10	14	7
Kidney	4	16.23	296.31	123.23	65.03	14	8
GI	1	12.36	20.36	16.35	5.66	2	9
GI	2	11.64	21.96	16.81	7.31	2	10
GI	3	10.94	16	13.47	3.57	2	11
GI	4	11.14	20.82	15.98	6.84	2	12
Carcinoid	1	16.25	16.25	16.25		1	13
Carcinoid	2	16.9	16.9	16.9		1	14
Carcinoid	3	18.01	18.01	18.01		1	15
Carcinoid	4	17.73	17.73	17.73		1	16

TABLE 6.14

Blood Flow by Reader

Reader	Minimum	Maximum	Mean	SD	N
1	12.35	323.35	81.89	73.25	30
2	11.64	304.7	79.11	68.0	30
3	10.94	305.35	73.90	65.9	30
4	11.14	296.31	80.54	71.13	30

30 observations averages about 68 cubic mm/minute. On the other hand, according to Table 6.15, the average variation across primary tumor types varies from a low of 17.22 for carcinoid to 125.34 for a kidney tumor.

The results for a conventional analysis of variance appears in Table 6.16. I ran a conventional analysis of variance and the results appear in Table 6.16, where 120 observations were used in the analysis. Thus, there is no significant difference ($P=0.945$) between readers, but between primary tumors, there is a significant ($P=0.001$ with SPSS version 11) difference. The analysis of variance is followed by a Scheffe multiple comparison procedure with $\alpha=0.05$, and there is no significant differences detected between the four radiologists, but on the other hand, for differences among the four primary tumor types, renal cell carcinoma (kidney) differed from the other three, namely, the colon, GI, and carcinoid.

TABLE 6.15

Liver Blood Flow by Primary Tumor

Primary tumor	Minimum	Maximum	Mean	SD	N
Colon	12.75	235.09	43.28	42.11	52
Kidney	14.15	323.35	125.34	65.95	56
GI	10.94	21.98	15.65	4.75	8
Carcinoid	16.25	18.01	17.22	.8018	4

TABLE 6.16

ANOVA for Blood Flow of Ng Study

Source	Type III SS	DF	MS	F	P-value
Corrected model	235075.53	6	39179.25	13.46	0.000
Intercept	98529.86	1	98529.86	38.86	0.000
Reader	1099.82	3	366.608	33.86	0.945
Primary tumor	233975.71	3	77991.90	26.80	0.000
Error	328761.83		113	2909.39	
Total	1310234.66	120			

A Bayesian analysis is initiated with a one-way fixed effects model that investigates the inter rater agreement.

$$y[i] = b[1]X1[i] + b[2]X2[i] + b[3]X3[i] + b[4]X4[i] + e[i] \qquad (6.10)$$

where $y[i]$ corresponds to the average blood flow of the i-th observation and $i = 1, 2, \ldots, 120$. In addition, $X1[i]$ is the i-th row of the 120×1 column vector corresponding to the first reader, $X2[i]$ is the 120×1 vector corresponding to the second reader and $b[2]$ is the corresponding effect of reader 2, etc. In addition, the $e[i]$ are independent $n(0, \text{tau})$, where tau is the common precision of an observation. In the non Bayesian sense, the coefficients of Equation 6.10 are regarded as unknown constants. The design matrix columns of Equation 6.10 are defined in BUGS CODE 6.4, where the X1 corresponds to the first reader, etc. The coefficients $b[i]$ are given non informative normal (0, 0.00001) prior distributions, and the common precision of the independent $e[i]$ are given a gamma(0.00001, 0.0001) prior distribution. See BUGS CODE 6.4 for additional information about Equation 6.10. Note that b12 is the difference in the effect of reader 1 minus the effect of reader 2, while b34 is the difference in the effect of reader 3 minus the effect of reader 4. The y vector is the column of 120 observations, the columns of the design matrix are denoted by X1, X2, X3, and X4, where X1 corresponds to the first rater, X2 to the second reader, X3 to the third, and X4 to the fourth reader. The 120 observations are paired, thus the first observation of y is paired with the first observations of the five columns of the design matrix. In addition, be aware that the first list statement defines the observations and design columns, while the second specifies the initial values for the MCMC simulation.

BUGS CODE 6.4

```
model;
{
for(i in 1:120){ y[i]~dnorm(mu[i],tau)
mu[i] < - b[1]*X1[i] + b[2]*X2[i] + b[3]*X3[i] + b[4]*X4[i]}
for (i in 1:4){ b[i]~dnorm(0,.0001)}
tau~dgamma(.00001,.00001)
sigma < - 1/tau
b12 < - b[1] - b[2]
b23 < - b[2] - b[3]
b34 < - b[3] - b[4]
}
list(
y = c(14.80,20.72,20.67,196.92,62.76,59.89,29.29,29.37,42.54,60.98,46.52,
16.59,15.10,93.62,144.17,323.35,198.57,141.81,14.15,130.02,93.27,111.05,61.49,94.6
5,73.16,156.94,155.60,12.35,20.36,16.25,13.52,19.29,19.17,141.50,68.06,54.14,27.67,
28.69,43.26,51.60,46.58,15.85,15.29,93.74,147.72,304.70,185.08,147.52,15.04,146.02,
102.76,103.53,62.181,12.31,80.61,136.11,141.05,11.64,21.98,16.90,12.91,21.68,19.22,
83.71,42.49,50.12,27.65,29.16,37.56,55.44,39.74,16.19,12.75,103.67,132.41,305.35,
173.94,131.29,15.28,140.96,75.10,88.79,61.74,99.25,115.52,133.87,
146.49,10.94,16.00,18.01,13.64,20.13,17.32,235.09,64.42,54.91,26.28,28.14,43.93,
59.93,44.86,17.22,15.42,104.18,154.87,296.31,162.98,123.52,16.23,145.85,95.21,94.17,
62.01,93.62,75.49,157.55,143.35,11.14,20.82,17.73),

X1 = c(1,1,1,1,1,1,1,1,1,1,1,1,1,1,1,1,1,1,1,1,1,1,1,1,1,1,1,1,1,1,0,0,0,0,0,0,0,0,0,0,0,0,0,0,
0,0,0,0,0,0,0,0,0,0,0,0,0,0,0,0,0,0,0,0,0,0,0,0,0,0,0,0,0,0,0,0,0,0,0,0,0,0,0,0,0,0,0,0,0,0,0,0,0,
0,0,0,0,0,0,0,0,0,0,0,0,0,0,0,0,0,0,0,0,0,0,0,0,0,0),

X2 = c(0,0,0,0,0,0,0,0,0,0,0,0,0,0,0,0,0,0,0,0,0,0,0,0,0,0,0,0,0,0,1,1,1,1,1,1,1,1,1,1,1,1,
1,1,1,1,1,1,1,1,1,1,1,1,1,1,1,0,0,0,0,0,0,0,0,0,0,0,0,0,0,0,0,0,0,0,0,0,0,0,0,0,0,0,0,0,0,0,0,0,
0,0,0,0,0,0,0,0,0,0,0,0,0,0,0,0,0,0,0,0,0,0,0,0,0,0,0,0,0),

X3 = c(0,0,0,0,0,0,0,0,0,0,0,0,0,0,0,0,0,0,0,0,0,0,0,0,0,0,0,0,0,0,0,0,0,0,0,0,0,0,0,0,0,0,0,0,
0,0,0,0,0,0,0,0,0,0,0,0,0,0,1,1,1,1,1,1,1,1,1,1,1,1,1,1,1,1,1,1,1,1,1,1,1,1,1,1,1,1,1,1,1,1,0,
0,0,0,0,0,0,0,0,0,0,0,0,0,0,0,0,0,0,0,0,0,0,0,0,0,0,0,0),

X4 = c(0,0,0,0,0,0,0,0,0,0,0,0,0,0,0,0,0,0,0,0,0,0,0,0,0,0,0,0,0,0,0,0,0,0,0,0,0,0,0,0,0,0,0,0,
0,0,0,0,0,0,0,0,0,0,0,0,0,0,0,0,0,0,0,0,0,0,0,0,0,0,0,0,0,0,0,0,0,0,0,0,0,0,0,0,0,0,0,0,0,0,0,0,0,
0,0,1,1,1,1,1,1,1,1,1,1,1,1,1,1,1,1,1,1,1,1,1,1,1,1,1,1,1,1))

list(b = c(0,0,0,0), tau = 1)
```

The program was executed with 45,000 observations generated from the joint posterior distribution, with a burn in of 5000 observations, and a refresh of 100.

From Table 6.17, note that $b[1]$ estimates the effect of reader 1 on the average blood flow, with a posterior mean of 80.49 and a 95% credible interval of (55,105), the other three readers have effects that do not differ much from reader 1 and from each other. This is made clear by referring to the posterior distributions of the contrasts $b12$, $b23$, and $b34$, all of which have 95% credible intervals centered at 0.

6.6.5 Multivariate Techniques

The Ng study of CT images for blood flow and volume provides a good example of multivariate response, and the responses will be examined for rater agreement with a fixed effects version of the Bayesian approach that employs multivariate techniques. There are many references, including Box and Tiao[6], Zellner[11], and Press[12], and Broemeling[13], and the basis for the analysis is the one-way multivariate linear model

$$y_{ij} = \theta + a[i] + e[i,j] \tag{6.11}$$

where y_{ij} is a p vector of observations, where i denotes the i-th group and j the j-th observation from the i-th group, θ represents the mean of the observations, $a[i]$ denotes the main effect of the i-th group, and the $e[i,j]$ are independent error terms. N is the total number of observations where

$$N = \sum_{i=1}^{i=q} n_i$$

where n_i is the number of observations in the i-th group.

The following presentation is due to Press[12] and is a transformation to the so-called regression form. Accordingly, let

TABLE 6.17

Posterior Analysis of Fixed Effects Model of Ng Study

Parameter	Mean	SD	2½	Median	97½
$b[1]$	80.49	12.75	55.4	80.47	105.6
$b[2]$	74.5	12.83	49.18	74.59	99.64
$b[3]$	72.6	12.79	47.51	72.52	97.68
$B[4]$	79.15	12.8	53.86	79.22	104.2
σ^2	4960	663.5	3828	4899	6433
b12	5.988	18.03	−29.56	5.977	41.52
b23	1.892	18.12	−33.5	1.758	37.52
b34	−6.543	17.99	−41.69	−6.447	28.76

$Y' = [y_{11}, y_{12}, \dots, y_{1n_1}; \dots ; y_{q1}, y_{q2}, \dots, y_{qn_q}]$ is a $p \times N$ matrix,
$U' = [u_{11}, u_{12}, \dots, u_{1n_1}; \dots ; u_{q1}, u_{q2}, \dots, u_{qn_q}]$ is a $p \times N$ matrix,
$\theta[i] = \theta + a[i]$ is a $p \times 1$ vector,
$B' = (\theta[1], \dots, \theta[q])$, is a $p \times q$ matrix, and
X is a $N \times q$ quasi-diagonal matrix,

where the i-th diagonal element is a $n_i \times 1$ vector of ones, $i = 1, 2, \dots, q$, and the model is expressed as

$$Y = XB = U \tag{6.12}$$

Suppose the u_{ij} are independent and distributed as a p-dimensional multivariate normal with mean vector zero and variance-covariance matrix Σ, where Σ is a positive definite matrix, thus the likelihood function is

$$l(Y / X, B, \Sigma) = [\det(\Sigma)]^{-N/2} \exp tr[V + (B - \hat{B})'S(B - \hat{B})]\Sigma^{-1}/2 \tag{6.13}$$

The least squares estimator of B is

$$\hat{B} = (X'X)^{-1}X'Y, \text{ also}$$

$S = (X'X)$, and
$V = (Y - X\hat{B})'(Y - X\hat{B})$ is the matrix of residuals. For additional details see Press[12].

For the Bayesian analysis, a prior distribution is assigned to B and Σ, and let B and Σ be independent, where the density of Σ is

$$f(\Sigma) \propto [\det(\Sigma)]^{-v/2} \exp - tr(\Sigma^{-1}H)/2$$

which is an inverse Wishart distribution with parameters v and H. As for B, let $B' = (\theta[1], \dots, \theta[q])$ and assume the $\theta[i]$ are independent and normally syllables with mean vector ξ and dispersion matrix Ψ, then the prior density is

$$g(B/\xi, \Psi) = [\det('\Psi')]^{-q/2} \exp - \sum_{i=1}^{i=q} (\theta[i] - \xi)' \Psi^{-1}(\theta[i] - \xi) / 2 \tag{6.14}$$

Combining the prior density of the parameters with the likelihood function, the joint posterior distribution is

$$p(B, \Sigma/\xi, \Psi, X, Y) \propto [\det(\Psi)]^{-q/2} \exp - tr(B - B^*)(B - B^*)' \Psi^{-1}/2 \tag{6.15}$$

$$[\det(\Sigma)]^{-(N+v)/2} \exp - tr[(V + H) + (B - \hat{B})'S(B - \hat{B})]\Sigma^{-1}/2$$

and when Σ is eliminated (via the properties of the inverted Wishart distribution), one can show that the marginal distribution of B is

$$h(B/\xi, \Psi, X, Y) \propto [\det(\Psi)]^{-q/2} \exp \sum_{i=1}^{i=q} (\theta[i] - \xi)' \Psi^{-1}(\theta[i] - \xi)/2 \qquad (6.16)$$

$$\{\det[(V + H) + (B - \hat{B})' \; S(B - \hat{B})]\}^{\kappa/2}$$

where $\kappa = N + v - p - 1$.

The marginal posterior distribution of B follows a matrix t-distribution, and as such, the columns of B follow a multivariate t-distribution, and Press[12] provides more details about posterior inferences regarding B.

Note the joint and marginal distributions depends on the hyper parameters v, H, ξ and Ψ, of which there are many (depending on the value of p), and values need to be assigned to them in order to implement a posterior analysis! If results from a prior related study are available, the hyper parameters could be estimated from the likelihood function.

If results from a previous experiment are not available, one can adopt a non-informative prior, where B and Σ are independent and the density of B is a constant and the density of Σ is

$$g(\Sigma) \propto [\det(\Sigma)]^{-(p+1)/2}$$

where Σ is a positive symmetric matrix of order p. If this is the case the marginal posterior density of B is

$$p(B / X, Y) \propto [\det(V + (B - \hat{B})' \; S(B - \hat{B})]^{-N/2} \qquad (6.17)$$

where V is the matrix of residuals and $\hat{B}$ is the matrix of least squares estimates of B.

Exercises

1. Refer to Table 6.2 and complete the table for pairs 3 and 4, 3 and 5, and 4 and 5. Use BUGS CODE 6.1 with a burn in of 5000, a refresh of 100, and generate 35,000 observations from the joint posterior distribution. Does your analysis imply strong agreement among the five readers? If not, explain.

2. With a Bayesian approach, regress reader i rep 1 scores against reader i rep two scores, for $i = 1, 2, 3, 4, 5$. Perform a posterior analysis similar

to Table 6.2 by computing the posterior mean, standard deviation, the 2½ and 97½ percentiles, and the median. Use the data in Table 6.18, where columns 1–5 correspond to the 40 tumor scores for the readers 1–5 for the first rep, and the last five columns correspond to the tumor scores for readers 1–5 for rep 2.

3. Using the tumor sizes in Table 6.18, plot the reader *i* rep1 scores versus the reader *i* rep 2 scores and comment on the intrareader agreement. Also, compute the correlation between the reader *i* rep 1 scores and reader *i* rep 2 scores, for $i = 1, 2, 3, 4, 5$.

4. Referring to Exercise 2 and Exercise 3, what is your overall conclusion about the intrareader agreement for the lung cancer tumor size information?

5. The judges' score for Cabernet are provided by Table 1.8 of Chapter 1, where there are 10 judges and 11 wines. Using a Bayesian approach, execute a posterior analysis that estimates the variance components for three sources of variation: (a) between wines; (c) between judges; and (c)

TABLE 6.18

Tumor Unidimensional Scores for Five Readers and Two Replications

Readers Replication 1					Readers Replication 1				
1	2	3	4	5	1	2	3	4	5
3.5	3.3	3.8	3.8	4	3.8	2.2	3.6	3.9	4
3.8	3.8	3.8	3.9	3.9	4	3.8	4	3.9	4
2.2	2	2.8	2.4	2.1	2	1.8	2.8	2.5	2
1.5	1.2	2.2	2.3	2	1.8	1.2	2	2	2
3.8	4	4.8	3.7	4.2	4	3.9	4.8	4.2	4.1
3.5	3.3	4.8	3.6	3.9	3.5	3	4.8	3.5	3.2
4.2	4	4.8	4.1	4.2	4.2	4	4.2	4.2	4.8
5.4	5.2	5.8	6.5	5	5	5.3	5.8	5.9	5.1
7.6	7.8	7	8.1	8.2	8	7.5	7	8.4	8.5
2.8	3	3.2	3.4	3.1	3	3	3.3	3.4	3
5	5.5	5.2	5.1	5.5	5.7	5.3	5.2	5.1	5
2.3	2.7	3.2	2.5	2.9	2.5	2.3	3	2.5	2.5
4.4	4.5	4.5	4.5	4.2	4.5	4.4	4.5	4.4	4.4
2.5	1.8	3.2	3	3.2	3.2	1	3	2	2.8
5.2	4	6	5.7	5.1	5.4	3.8	4.8	5.3	5.5
1.7	2	5	4.6	3.8	2	2	4.8	3.9	4.8
4.5	5.8	4.2	4.8	4.5	4.8	4.5	4.5	4.8	4.7
4.5	3.8	5.2	4.4	4.5	4.5	3.6	4.6	4.2	4.5
6	5.8	6.8	6.2	6	6	5.8	6.2	6.5	6
3.3	2.5	3.2	4.5	3.5	3.5	3	3.7	3.6	3.2
7	7	7.8	7.6	6.9	7	6.8	7.8	7.1	6.8
4	2.2	3.2	4.5	4.6	4.2	2	3.6	3.2	4.5
4.8	5	5	5.5	4.2	5	4.3	4.6	4.5	4.4

(continued)

TABLE 6.18 (Continued)

	Readers Replication 1					Readers Replication 1			
1	2	3	4	5	1	2	3	4	5
5	5	6.2	5.3	5	5.6	4.8	5.8	5.1	5
2.7	2.5	2.8	2.5	2.5	3	2.8	3	2.6	2.8
3.7	4	4.2	4.1	3.9	4	4	4	4	4
1.8	2	36	2.9	2.5	1.8	2	2.8	2.4	2.5
6.3	4.5	6	6.1	6.5	6	3.8	6	7.5	7
4	4.3	4.6	4.1	4	4.3	4.2	4.4	4.1	4
1.5	1.5	2.8	1.5	1.7	1.4	1.5	2.8	1.4	1.5
2.2	2.3	1.5	2.6	2.5	3.7	2	1.5	2.6	2.4
2.2	4.2	3.8	3.6	3.5	2.4	3	3.8	3.8	2
2.6	2.8	3	2.8	3	3	2.8	2.8	2.6	3
3.7	2	4.8	3.7	3	4	2	4	2.7	2.5
2.5	2.2	3	3.7	3.6	2.5	2.5	2.8	3.3	2.2
4.8	5	3.	5.9	4	5.7	3.8	3.8	5.9	3.8
8	5	9	9	9	9	5	8	8.5	9
4	3	4	4.4	3.3	3.5	3.2	4.2	4	3.8
3.5	3.5	4	3.4	3.1	3.2	3.3	3.6	3.1	3.2
4.8	4.2	4.8	4.3	5	4.6	4.2	4.4	4.4	4.6

error. I found that the largest component is error with a posterior mean of 9.89 and a posterior median of 9.74. The posterior distributions for the other two components are highly skewed to the right. Write the code and verify Table 6.19, using 95,000 observations generated from the joint posterior distribution, with a burn in of 5,000, and a refresh of 100.

TABLE 6.19

Variance Components of Wine Tasting Example

Parameter	Mean	SD	2½	Median	97½
Wines	2.473	2.138	0.0135	1.985	7.77
Judges	0.0694	0.2163	0.0000202	0.00354	0.5977
Error	9.894	1.504	7.396	9.742	13.26
Theta	11.75	0.5671	10.61	11.75	12.88

6. Refer to Table 1.7 for the scores of the Chardonnay wines and execute a Bayesian analysis that is similar to Exercise 5. What are the estimates of the variance components for the error, between wines, and between judges? How many observations did you generate from the joint posterior distribution?

7. Suppose in lieu of Equation 6.4 the model for the Kundra study is now specified as

$$y[i,j] = \theta + a[i] + b[j] + e[i,j]$$

where θ is a constant, the $a[i]$ and $e[i,j]$ are defined as in Equation 6.4, and the $b[j]$ are independent and normally distributed with mean 0 and variance σ_b^2, and is called the between reader variance component. (a) What is the correlation between two distinct tumor sizes, corresponding to two different readers of the same lesion? (b) Modify BUGS CODE 6.3 and execute a posterior analysis in order to estimate the intraclass correlation coefficient. Use 75,000 observations generated from the joint posterior distribution, with a burn in of 6000, and a refresh of 200.

8. Is the common correlation model (Equation 6.4) valid for the analysis of the Kundra data? See the list statement for BUGS CODE 6.3 for the actual data!

9. Refer to BUGS CODE 6.2, where the first list statement provides the SUV data for the three readers and the two replications. The first three columns refer to the SUV values of readers 1, 2, and 3 of the first replication, while the last three columns refer to the SUV values of readers 1, 2, and 3 of the second replication. (a) Compute the correlation between the six columns, and (b) in light of the correlations, is the estimated intraclass correlation of 0.95 reasonable? (c) Is the constant correlation model valid? Explain your answers.

10. Using BUGS CODE 6.2, verify Table 6.7, the posterior analysis for the Marom study and produce plots of the posterior densities of the variance components σ_a^2, σ_b^2, σ_d^2, and σ_w^2. In what way do the plot support the inferential implications of Table 6.7? What do the plots imply about the agreement between the radiologists?

11. I performed a posterior analysis of the tumor sizes of the Erasmus et al.[2] study and the results are reported in Table 6.2. This was analyzed in Section 6.1, but now the information includes two replications with five readers. The data set for the Erasmus study is given by Table 6.18. This information is in the form of a $2 \times 5 \times 40$ matrix, where there are two replications, five readers, and 40 lesions. Note that the right skewness of the distributions! Do the results imply good agreement between the five radiologists? Produce a plot of the posterior density of the intraclass correlation coefficient. Produce box plots of the tumor sizes that show the inter and intraobserver variation. For your analysis, generate 1,580,000 observations from the joint posterior distribution, with a refresh of 100 and a burn in of 500,000. Why did I use such a large sample size for the simulation? Examine the error of estimation of σ_a^2! I estimated the variance components via MLE and found: 2.165 (0.243) for the patient component, 0.077 (0.003) for the reader variation, 0.002 (0.00002429) for replications, and 0.297 (0.00001) for the error component. The MLE estimates are of the same order as the posterior medians, and I used SPSS (version 11).

12. Describe the effect of the lesion size on SUV, and refer to Table 6.7 and Table 6.8.

13. Validate Table 6.10. Is lesion size a needed covariate? Please explain in depth your answer. Execute your program with 2,900,000 observations generated from the joint posterior distribution, with a burn in of 100,000, and a refresh of 100, and produce a plot of the posterior density of the beta coefficient. Note the relevant model is Equation 6.8. Is the sample size for the simulation sufficiently large? Below is the BUGS CODE 6.5 for the Kundra study. The y matrix is $2 \times 3 \times 22$ for the tumor sizes, where there are two replications, three readers, and 22 mice. The x matrix is for the number of nodules and is of the same dimension as y, and the elements of the two matrices are paired.

BUGS CODE 6.5

```
model;
# Kundra liver lesions sizes are given by y
# Number of nodules is the covariate x
{
# N is number or reps=2
# M is number of readers=3
# O is number of patients=22
for(i in 1:N){for(j in 1:M){ for( k in 1:O){
y[i,j,k]~dnorm(mu[i,j,k], tau.w)
mu[i, j, k] <- theta + a[i] + b[j] + d[k] + beta*x[i,j,k]}}}
# theta is overall mean
# beta is the effect of nodule size on SUV
theta~dnorm(0.0,.0001)
beta~dnorm(0,.00001)
for(i in 1:N){ a[i] ~ dnorm(0.0, taua)}
for(j in 1:M){b[j] ~ dnorm(0.0,taub)}
for(k in 1:O){d[k] ~ dnorm(0.0,taud)}
taua ~ dgamma(0.00001,.00001)
taub ~ dgamma(0.00001,.00001)
taud ~ dgamma(0.00001,.00001)
tau.w ~ dgamma(0.00001,.00001)
sigma.w <- 1/tau.w
 sigmaa <-1/taua
 sigmab <-1/taub
 sigmad <-1/taud
```

icc < - (sigmaa + sigmad)/(sigmaa + sigmab + sigmad + sigma.w)

sigmaa is variance component for reps

sigmab is variance component for readers

sigmad is variance component for patients

sigma.w is variance component for error

}

list(N = 2,M = 3,O = 22,

y = structure(.Data = c(26.00,NA,NA,NA,20.00,30.00,39.00,33.00,46.00, 26.00,50.00,44.00,

77.00,116.00,11.00,10.00,12.00,61.00,21.00,13.00,36.00,36.00,22.00,25.00,NA, 17.00,17.00,28.00,

44.00,53.00,47.00,25.00,51.00,48.00,75.00,44.00,8.00,12.00,12.00,56.00,21.00, 13.00,NA,37.00,

10.00,25.00,10.00,19.00,NA,NA,40.00,40.00,27.00,26.00,49.00,47.00,74.00, 104.00,12.00,9.00,

15.00,53.00,27.00,15.00,NA,36.00,38.00,NA,NA,20.00,21.00,31.00,38.00, 33.00,44.00,27.00,48.00,

43.00,79.00,101.00,NA,11.00,10.00,65.00,21.00,14.00,NA,35.00,48.00,26.00, NA,17.00,14.00,28.00,

43.00,39.00,54.00,NA,45.00,41.00,70.00,43.00,NA,14.00,12.00,65.00,30.00, 15.00,NA,37.00,NA,

25.00,12.00,18.00,NA,30.00,39.00,35.00,27.00,27.00,48.00,47.00,73.00,88.00, NA,9.00,13.00,61.00,

22.00,15.00,NA,33.).,.Dim = c(2,3,22)),

x = structure(.Data − c(7.00,10.00,4.00,1.00,1.00,5.00,10.00,10.00,10.00,10.00 ,10.00,12.00,

10.00,10.00,3.00,10.00,10.00,10.00,10.00,10.00,6.00,7.00,8.00,10.00,1.00,1.00, 2.00,6.00,10.00,10.00,

10.00,7.00,10.00,10.00,10.00,10.00,1.00,10.00,10.00,10.00,10.00,10.00,8.00,5.00 ,10.00,10.00,2.00,

1.00,.00,3.00,10.00,10.00,10.00,7.00,10.00,10.00,10.00,10.00,1.00,3.00,10.00, 10.00,10.00,10.00,4.00,

6.00,7.00,10.00,2.00,1.00,1.00,5.00,10.00,10.00,10.00,11.00,10.00,10.00,10.00,1 0.00,1.00,10.00,10.00,

10.00,10.00,10.00,5.00,7.00,10.00,10.00,1.00,1.00,1.00,5.00,10.00,10.00,10.00,1 0.00,10.00,9.00,10.00,

10.00,2.00,10.00,10.00,10.00,10.00,10.00,0.00,5.00,10.00,10.00,2.00,1.00,0.00, 7.00,10.00,10.00,10.00,

10.00,10.00,10.00,10.00,10.00,2.00,3.00,10.00,10.00,10.00,10.00,6.00,7.00),. Dim = c(2,3,22)))

list(tau.w = 1, a = c(0,0), b = c(0,0,0), d = c(0,0,0,0,0,0,0,0,0,0,0,0,0,0,0,0,0,0,0,0,0,0,0,0,0), theta = 0, taua = 1, taub = 1, taud = 1, beta = 0))

14. (a) From the data contained in the first list statement of the above BUGS CODE 6.5, perform a PP plot of the three readers; (b) Do the plots indicate normality? (c) Perform a posterior analysis similar to Table 6.10, but using a log transform of the tumor sizes; and (d) How does the analysis differ for the one reported in Table 6.10? Note that the posterior analysis includes the effect of the covariate.

15. Refer to Section 6.6.2, and verify Table 6.12 for the intraobserver agreement of reader 1. In addition, complete the Bland–Altman analysis for the intraobsever agreement of readers 2, 3, 4, and 5, by revising BUGS CODE 6.1, where the dependent variable y is the difference in the replication 1 minus replication 2 scores of a reader and the independent variable x is the sum of the replications and 2 scores of the reader. The tumor sizes of the Erasmus study for the five readers and both replications are given Table 6.18.

16. Refer to section 6.6.2 and Table 6.18 and plot the difference in the scores of replications 1 and 2 of reader 1 versus the sum of the tumor sizes for replications 1 and 2 of reader 1. This is a Bland–Altman Plot. What does the plot reveal? Is there an association between the differences versus the sum?

17. Using the BUGS CODE 6.6, verify Table 6.12, the posterior analysis for the Marom study that included the lesion by reader interaction term in Equation 6.9.

BUGS CODE 6.6

model;

```
# Marom SUV with Interaction
{
# N is number or reps
# M is number of readers
# O is number of patients
for( i in 1:N){for(j in 1:M){ for(k in 1:O){
y[i,j,k] ~ dnorm(mu[i,j,k], tauw)
mu[i,j,k] <- theta + a[i] + b[j] + d[k] + f[j,k]}}}
# theta is overall mean
theta ~ dnorm(0.0,.00001)
for(i in 1:N){ a[i] ~ dnorm(0.0, taua)}
for(j in 1:M){ b[j] ~ dnorm(0.0,taub)}
for(k in 1:O){d[k] ~ dnorm(0.0,taud)}
```

```
for(j in 1:M){for( k in 1:O){f[j,k] ~ dnorm(0,tauf)}}

taua ~ dgamma(0.00001,.00001)
taub ~ dgamma(0.00001,.00001)
taud ~ dgamma(0.00001,.00001)
tauf ~ dgamma(0.00001,.00001)
tauw ~ dgamma(0.00001,.00001)

sigmaw <- 1/tauw
 sigmaa <-1/taua
 sigmab <-1/taub
 sigmad <-1/taud
 sigmaf <-1/tauf

# sigmaa is variance component for reps
# sigmab is variance component for readers
# sigmad is variance component for patients
# sigmaf is the variance component for the lesion by reader interaction
# sigmaw is variance component for error
}
list(N=2,M=3,O=20,

y=structure(.Data=c(
11.70,5.00,21.70,21.00,33.00,20.30,11.80,4.00,13.50,8.50,21.00,13.20,6.30,19.50,13.40,10.90,14.70,3.60,4.30,15.50,

11.50,5.00,21.00,24.00,30.90,19.40,11.80,4.00,10.20,8.50,22.40,12.30,6.30,19.50,13.40,8.50,13.50,3.50,4.80,17.00,

10.30,5.00,19.30,18.70,28.00,20.70,9.60,4.60,10.00,8.50,25.60,12.80,5.90,15.80,18.40,9.30,10.60,3.60,4.60,15.60,

11.50,5.00,18.80,14.50,31.40,20.80,11.80,4.60,12.30,8.50,24.40,13.20,6.30,19.50,11.00,11.00,14.60,3.60,4.70,16.00,

11.50,5.00,19.90,24.00,31.40,19.40,11.80,4.00,12.30,9.80,19.60,13.20,6.30,19.60,13.50,12.70,13.80,3.30,1.80,15.60,

11.90,5.00,21.00,22.40,30.60,18.90,11.80,4.00,8.90,9.00,20.60,12.20,6.30,16.80,11.00,12.70,12.10,3.30,4.40,18.30),.Dim=c(2,3,20)))

list(tauw = 1,a = c(0,0),b = c(0,0,0),d = c(0,0,0,0,0,0,0,0,0,0,0,0,0,0,0,0,0,0,0,0),

f=structure(.Data=c(0,0,0,0,0,0,0,0,0,0,0,0,0,0,0,0,0,0,0,0,0,0,0,0,0,0,0,0,0,0,0,0,0,0,0,0,0,0,0,0,0,0,0,0,0,0,0,0,0,0,0,0,0,0,0,0,0,0,0,0),.Dim=c(3,20)),

theta=0, taua = 1,taub = 1,taud = 1,tauf = 1)))
```

18. Refer to the previous problem, but now estimate the intraclass correlation coefficient

$$\rho = (\sigma_a^2 + \sigma_d^2) / (\sigma_a^2 + \sigma_b^2 + \sigma_d^2 + \sigma_f^2 + \sigma_w^2)$$

with a Bayesian approach and modify BUGS CODE 6.6 and generate 2,000,000 observations generated from the joint posterior distribution, with a burn in of 500,000, and a refresh of 100.

19. With regard to the Marom et al.[3] SUV study, compare the posterior analyses of Table 6.4 and Table 6.12, where the former is based on the random model (Equation 6.2) without interaction, and the latter on a model that includes the lesion by reader interaction terms. Does the addition of an interaction appreciably change the posterior estimates of the other variance components? Should the interaction terms be excluded from the analysis?

20. Using BUGS CODE 6.4 and the one-way fixed effects model (Equation 6.10), validate Table 6.17. Does the posterior analysis agree with Table 6.14, the descriptive statistics by reader?

21. Assume a fixed effects model for the Ng study with 120 observations and consider two factors: namely the site of primary tumor with four levels, and the reader factor with four levels. Thus, revise Equation 6.10 and BUGS CODE 6.4 and execute the posterior analysis with 100,000 observations generated from the joint posterior distribution with a burn in of 5000, and a refresh of 100. There are four sets of 120 paired observations below: the first set corresponds to the patient label, while the second corresponds to the reader identification, the third identifies the primary tumor, and the fourth consists of the 120 blood flows. Note that the first observation of the four sets are paired, as is the second observation, and the last, or 120th.

    Refer to Table 6.13, Table 6.14, and Table 6.15 for the descriptive statistics of the blood flow information and note the unbalanced design with a different number of observations for the four primary tumor sites, namely 52, 56, 8, and 4 for the colon, kidney, GI, and carcinoid types, respectively.

22. For the Ng information of Table 6.20 perform a Bayesian analysis, but assume a random effects model with three factors: (a) the primary tumor factor with four effects (site of primary tumor); (b) the reader factor with four effects; and (c) the patient factor with 120 effects, and assume:

$$y[i,j,k] = \theta + a[i] + b[j] + d[k] + e[i,j,k]$$

TABLE 6.20

Ng Study Data for Blood Flow and Volume by Reader and Primary Tumor

Patient

1.00, 2.00, 17.00, 18.00, 19.00, 20.00, 21.00, 22.00, 23.00, 24.00, 25.00, 26.00, 27.0, 1.00, 2.00, 17.00,
18.00, 19.00, 20.00, 21.00, 22.00, 23.00, 24.00, 25.00, 26.00, 27.00, 1.00, 2.00, 17.00, 18.00, 19.00,
20.00, 21.00, 22.00, 23.00, 24.00, 25.00, 26.00, 27.00, 1.00, 2.00, 17.00, 18.00, 19.00, 20.00, 21.00,
22.00, 23.00, 24.00, 25.00, 26.00, 27.00, 3.00, 4.00, 5.00, 6.00, 7.00, 8.00, 9.00, 10.00, 11.00, 12.00,
13.00, 14.00, 15.00, 16.00, 3.00, 4.00, 5.00, 6.00, 7.00, 8.00, 9.00, 10.00, 11.00, 12.00, 13.00, 14.00,
15.00, 16.00, 3.00, 4.00, 5.00, 6.00, 7.00, 8.00, 9.00, 10.00, 11.00, 12.00, 13.00, 14.00, 15.00, 16.00,
3.00, 4.00, 5.00, 6.00, 7.00, 8.00, 9.00, 10.00, 11.00, 12.00, 13.00, 14.00, 15.00, 16.00, 28.00, 29.00,
28.00, 29.00, 28.00, 29.00, 28.00, 29.00, 30.00, 30.00, 30.00, 30.00

Reader

1.00, 1.00, 1.00, 1.00, 1.00, 1.00, 1.00, 1.00, 1.00, 1.00, 1.00, 1.00, 1.00, 2.00, 2.00, 2.00, 2.00, 2.00,
2.00, 2.00, 2.00, 2.00, 2.00, 2.00, 2.00, 2.00, 2.00, 3.00, 3.00, 3.00, 3.00, 3.00, 3.00, 3.00, 3.00, 3.00, 3.00,
3.00, 3.00, 3.00, 4.00, 4.00, 4.00, 4.00, 4.00, 4.00, 4.00, 4.00, 4.00, 4.00, 4.00, 4.00, 4.00, 1.00, 1.00,
1.00, 1.00, 1.00, 1.00, 1.00, 1.00, 1.00, 1.00, 1.00, 1.00, 1.00, 2.00, 2.00, 2.00, 2.00, 2.00, 2.00,
2.00, 2.00, 2.00, 2.00, 2.00, 2.00, 2.00, 2.00, 3.00, 3.00, 3.00, 3.00, 3.00, 3.00, 3.00, 3.00, 3.00, 3.00,
3.00, 3.00, 3.00, 3.00, 4.00, 4.00, 4.00, 4.00, 4.00, 4.00, 4.00, 4.00, 4.00, 4.00, 4.00, 4.00, 4.00,
1.00, 1.00, 2.00, 2.00, 3.00, 3.00, 4.00, 4.00, 1.00, 2.00, 3.00, 4.00

Primary site tumor

1.00, 1.00, 1.00, 1.00, 1.00, 1.00, 1.00, 1.00, 1.00, 1.00, 1.00, 1.00, 1.00, 1.00, 1.00, 1.00, 1.00, 1.00,
1.00, 1.00, 1.00, 1.00, 1.00, 1.00, 1.00, 1.00, 1.00, 1.00, 1.00, 1.00, 1.00, 1.00, 1.00, 1.00, 1.00, 1.00,
1.00, 1.00, 1.00, 1.00, 1.00, 1.00, 1.00, 1.00, 1.00, 1.00, 1.00, 1.00, 1.00, 1.00, 1.00, 1.00, 2.00, 2.00,
2.00, 2.00, 2.00, 2.00, 2.00, 2.00, 2.00, 2.00, 2.00, 2.00, 2.00, 2.00, 2.00, 2.00, 2.00, 2.00, 2.00, 2.00,
2.00, 2.00, 2.00, 2.00, 2.00, 2.00, 2.00, 2.00, 2.00, 2.00, 2.00, 2.00, 2.00, 2.00, 2.00, 2.00, 2.00, 2.00,
2.00, 2.00, 2.00, 2.00, 2.00, 2.00, 2.00, 2.00, 2.00, 2.00, 2.00, 2.00, 2.00, 2.00, 2.00, 2.00, 2.00, 2.00,
3.00, 3.00, 3.00, 3.00, 3.00, 3.00, 3.00, 3.00, 4.00, 4.00, 4.00, 4.00

Blood flow

14.80, 20.72, 20.67, 196.92, 62.76, 59.89, 29.29, 29.37, 42.54, 60.98, 46.52, 16.59, 15.10, 13.52,
19.29, 19.17, 141.50, 68.06, 54.14, 27.67, 28.69, 43.26, 51.60, 46.58, 15.85, 15.29, 12.91, 21.68,
19.22, 83.71, 42.49, 50.12, 27.65, 29.16, 37.56, 55.44, 39.74, 16.19, 12.75, 13.64, 20.13, 17.32,
235.09, 64.42, 54.91, 26.28, 28.14, 43.93, 59.93, 44.86, 17.22, 15.42, 93.62, 144.17, 323.35, 198.57,
141.81, 14.15, 130.02, 93.27, 111.05, 61.49, 94.65, 73.16, 156.94, 155.60, 93.74, 147.72, 304.70,
185.08, 147.52, 15.04, 146.02, 102.76, 103.53, 62.18, 112.31, 80.61, 136.11, 141.05, 103.67, 132.41,
305.35, 173.94, 131.29, 15.28, 140.96, 75.10, 88.79, 61.74, 99.25, 115.52, 133.87, 146.49, 104.18,
154.87, 296.31, 162.98, 123.52, 16.23, 145.85, 95.21, 94.17, 62.01, 93.62, 75.49, 157.55, 143.35,
12.35, 20.36, 11.64, 21.98, 10.94, 16.00, 11.14, 20.82, 16.25, 16.90, 18.01, 17.73

Blood volume

2.70, 3.91, 3.60, 15.15, 6.02, 6.87, 4.38, 4.35, 6.75, 6.64, 6.49, 2.72, 2.52, 2.71, 4.16, 3.62, 10.01, 5.82,
6.61, 4.52, 4.32, 5.90, 6.83, 6.79, 2.73, 2.63, 2.97, 4.27, 3.62, 7.32, 5.11, 6.52, 4.14, 3.83, 5.54, 5.63,
5.55, 2.08, 1.78, 3.73, 4.49, 2.88, 16.88, 5.85, 6.96, 4.71, 4.43, 6.23, 6.74, 7.27, 2.85, 2.38, 14.17,
16.61, 21.58, 17.52, 14.58, 4.73, 10.94, 9.10, 12.46, 5.71, 10.62, 9.67, 16.12, 17.55, 13.31, 16.91,
21.20, 17.44, 15.14, 4.87, 10.83, 10.38, 12.77, 6.04, 12.31, 9.89, 14.47, 15.33, 13.98, 12.96, 21.46,
16.67, 15.0, 5.00, 10.63, 9.58, 12.24, 6.02, 12.94, 9.61, 13.65, 19.97, 14.53, 18.50, 21.29, 17.11,
14.37, 4.47, 11.00, 10.38, 11.85, 6.11, 11.90, 10.17, 15.48, 16.74, 2.23, 4.43, 2.21, 4.50, 2.32, 4.03,
2.21, 4.67, 3.38, 3.32, 3.44, 3.59

where $y[i,j,k]$ is the blood flow corresponding to the i-th primary tumor, the j-th reader, and the k-th patient. In addition, θ is the average blood flow, $a[i]$ is the effect of the i-th primary site, $i = 1, 2, 3, 4$, $b[j]$ is the effect of the j-th reader, $j = 1, 2, 3, 4$, and $d[k]$ is the effect of the k-th patient. The range of k (the patient id) depends on the primary site i, thus this is an unbalanced design. Additional assumptions include $\theta \sim n(0,0.00001)$, the $e[i]$ are independent and distributed $n(0, \sigma_a^2)$, the $b[j]$ are independent and distributed $n(0, \sigma_b^2)$, the $d[k]$ are independent and distributed $n(0, \sigma_d^2)$, and lastly the $e[i,j,k]$ are independent and distributed $n(0, \sigma_e^2)$. For the Bayesian analysis, assume the four variance components have independent gamma(0.00001, 0.00001) prior distributions. Display your code and execute it with 1,000,000 observations generated from the joint posterior distribution, with a burn in of 200,000 observations, and a refresh of 100. What is your conclusion about rater agreement? Should an interaction term (between readers by primary site) be included? How well does the model fit?

I estimated the variance components by maximum likelihood using SPSS (version 11) and the results are found in Table 6.21.

TABLE 6.21

Maximum Likelihood Estimates of Variance Components for Ng Study

Parameter	MLE	Estimated SD
Primary site	1613.74	1642
Patient	2891.99	809
Reader	5.576	10.01
Error	198.10	30.12

The usual order of variation is present for this study, because the smallest variation is between readers, and the next largest between patients, followed by the primary tumor site. Does this conform to Table 6.13, Table 6.14, and Table 6.15? It also follows that the MLE of the intraclass correlation is 0.9567, which implies very good rater agreement! How does the Bayesian analysis compare to the above conventional analysis?

23. Refer to Section 6.6, and explain how the hyper parameters v, ξ, H, and Ψ of the prior distribution (see Equation 6.14) can be estimated from the data of a related study. By related, I mean the design of the past experiment is the same as that of the future planned experiment, and that there are q groups, and that the M observations are $p \times 1$ vectors.

24. Perform a Bayesian one-way MANOVA with information from the Ng study included in Exercise 20. As dependent variables use the

blood flow and the blood volume data and assume there are four groups, namely the four readers. Assume a non informative prior for B and Σ. (a) Find the posterior density of B, see Equation 6.17. (2) Do the four raters have the same effect on the average values of the two dependent variables, the blood flow and volume? (c) What is your overall conclusion about rater agreement?

As a guide to the Bayesian approach, I computed the 'classical' MANOVA and found that all four (Pillai, Wilks, Hotelling, and Roy) multivariate tests imply very good agreement between the four readers. The null hypothesis being tested is that the readers are having the same effect on the average blood flow and volume, and the four P-values are in excess of 0.91, indicating there is not sufficient evidence to reject the null hypothesis. Calculations were executed with SPSS (version 11) and reported in Table 6.22.

TABLE 6.22

MAVOVA for Ng Study. The Effect of Readers on Blood Flow and Volume

Effect	Test	Value	F	Hypothesis df	Error df	P-value
Intercept	Pillai	0.782	206.125	2	115	0
	Wilks	0.218	206.125	2	115	0
	Hotelling	3.585	206.125	2	115	0
	Roy	3.585	206.125	2	232	0
Reader	Pillai	0.006	0.124	6	230	0.993
	Wilks	0.994	0.122	6	230	0.994
	Hotelling	0.006	0.121	6	228	0.994
	Roy	0.004	0.173	3	116	0.914

25. Similar to Exercise 24, execute a two-way Bayesian MANOVA, with two factors: (1) the reader factor with four levels, and (2) the primary tumor factor with four levels. See the Ng study data of Exercise 22. Are there any differences in the four primary tumors? Is there a difference in the four readers? Use WinBUGS with 2,000,000 observations generated from the joint posterior distribution, with a burn in of 500,000, and a refresh of 100. Refer to problem 6.5 of Press[12].

References

1. Broemeling, L. D. 2007. *Bayesian biostatistics and diagnostic medicine*. New York: Chapman & Hall/CRC.
2. Erasmus, J. J., G. W. Gladish, L. Broemeling, B. S. Sabloff, M. T. Truong, R. S. Herbst, and R. F. Munden. 2003. Interobserver and intraobserver variability in measurement of non-small-cell carcinoma lung lesions: Implications for assessment of tumor response. *J. Clin. Oncol.* 21(13), 2574.

3. Marom, E. M., R. F. Munden, M. T. Truong, G. W. Gladish, D. A. Podoloff, O. Mawlawi, L. D. Broemeling, J. F. Bruzzi, and H. M. Macapinlac. 2006. Interobserver and intraobserver variability of standardized uptake values measurements in non-small-cell lung cancer. *J. Thorac. Imaging* 21, 205.
4. Kundra, V. 2006. Personal communication. The University of Texas MD Anderson Cancer Center.
5. Brant, W. E., and C. A. Helmes. 1999. *Fundamentals of diagnostic imaging,* 2nd ed. New York: Lippincot, Williams, and Wilkins.
6. Box, G. E. P., and G. C. Tiao. 1973. *Bayesian inference in statistical inference.* Reading, MA: Addison-Wesley.
7. Palmer, J. L., and L. D. Broemeling. 1990. A comparison of Bayes and maximum likelihood estimation of the intraclass correlation coefficient. *Comm. Statist. Theory Meth* 19(3), 953.
8. Bland, J. M., and D. G. Altman. 1986. Statistical methods for assessing agreement between two methods of clinical measurement. *Lancet,* 307–310.
9. Rousseeue, P. J., and A. M. Leroy. 1987. *Robust regression and outlier detection.* New York: John Wiley & Sons.
10. Ng. 2003. Personal communication. The University of Texas MD Anderson Cancer Center.
11. Zellner, A. 1971. *An introduction to Bayesian inference in econometrics.* New York: John Wiley & Sons.
12. Press, J. S. 1989. *Bayesian statistics: Principles, models, and applications.* New York: John Wiley & Sons.
13. Broemeling, L. D. 1984. The *Bayesian analysis of linear models.* New York: Marcel-Dekker.

7

Sample Sizes for Agreement Studies

7.1 Introduction

The last chapter of the book is focused on a very important topic, namely the determination of the sample size for studies of agreement between several raters. Of course, sample size determination is only one part of the design of a study, but is probably the most important and is essential to achieve the objectives of the study. If insufficient information is employed, the conclusions to the study will be compromised, but on the other hand, if the sample size is too large, valuable resources will be wasted. For clinical studies, ethical issues play an important role in sample size choice. Also, in the case of clinical studies, when several readers are involved, the sample size is even more of a challenge because it not only involves determining the number of subjects, but also the number of raters necessary to achieve a given level of rater agreement.

Beginning with a review of the classical approaches to sample size estimation and power analysis for the standard problems, Bayesian methods are introduced, which is followed by sample size determinations for specific problems of interest to agreement. This includes those methods for Kappa and the G coefficient, methods for logistic-linear models, for variance components and the intra class correlation coefficient. Lastly, those Bayesian techniques of sample size estimation for continuous rater scores are described, including those for simple linear regression and correlation and for the fixed effects analysis of variance models.

When referring to the classical approach non Bayesian methods are meant that do not rely on Bayes theorem for incorporating prior information for the power analysis. The classical approach is based on the type I and II errors for the null and alternative hypotheses, that is, the sample size is chosen to achieve a given level of power for certain alternative instances of the alternative hypothesis for a given type I error alpha.

Null and alternative hypotheses are stated in terms of the parameter belonging to certain regions of the parameter space, and the null hypothesis is rejected in favor of the alternative if the sample belongs to a subset of sample space called the critical region. Significance levels or the alpha value is the probability of rejecting the null hypothesis, assuming it is true, while on the other hand, assuming the alternative hypothesis is true, the probability of rejecting the null hypothesis for certain alternative cases generates the power curve of the test. It is important to remember that a particular sample size formula is based on a set of assumptions that are more or less true in practice. It is also important to stress that the sample size determination is based on prior information from previous related experiments, and that this information is put into the calculations for the sample size, however, it is also important to remember that the uncertainty that is inherent in this prior information is not usually formally taken into account when performing the sample size and power calculations.

A hybrid Bayesian/classical method for sample size calculation will be used, where the process is initiated with the conventional sample size formula, conditional on the parameters of the model. Based on the formula, which is a function of the unknown parameters and the type I and II errors, the prior distribution of the parameters, available from previous related studies, induces a prior distribution for the sample size, from which the sample size can be estimated along with the corresponding power function.

Spiegelhalter, Abrams, and Myles[1] were among the first to suggest this approach. I found it very appealing and consequently have adopted it for the book. The conventional sample size thus serves as a background reference upon which Bayesian estimates of the sample size can be judged. It has the obvious advantage that it utilizes prior information from previous related experiments. The disadvantage of classical approach is in the use of prior information, where the uncertainty in the hypothesized values of the parameters under the null and alternative hypotheses is often not a part of the analysis.

As before, interesting examples introduced in the previous chapters will play an important role providing prior information for Bayesian sample size estimation. For example, for designing a future study involving kappa, the example of agreement between a gold standard (pathology) and a radiologist using ultrasound to diagnose prostate cancer for 235 patients is introduced in Table 2.21a of Chapter 2. This was a case of training a novice, but one suspects the trainee has improved her diagnostic ability and a future study is planned.

The kappa index based on the previous study was estimated by the posterior mean (SD) as 0.655 (0.0492), and using this information a future study is planned. What is the required sample size and what is the best use of this previous information? There are two alternatives to consider, one based on a formula of Donner and Eliasziw[2] and reported in Shoukri[3], and the other based on a formula developed by Cantor[4] and reported in von Eye and Alexander[5], but both formulas are based on prior information, which is easily included in the Bayesian analysis.

Recall that the logistic linear model compares two or more raters in the way they assign scores to subjects, where the coefficients in the model measure the effect of a rater on a score contrast. Such an example is portrayed in Table 5.2 of Chapter 5, where two psychiatrists are assigning depression scores (1, 2, or 3) to 129 patients, and the main interest was comparing the way the two raters are measuring the effect of a score of 1 minus a score of 3 (no depression versus clinical depression). Thus, two coefficients of the model are to be compared and as will be seen the results of the previous study can easily be included in the sample size estimation.

Table 6.1 of Chapter 6 provides another interesting example of Bayesian sample size estimation, where the Erasmus et al.[6] study involves five radiologists measuring the lesion size of lung via CT of lung cancer patients, and agreement was assessed by regressing the scores of one reader on those of another. If the true lines goes through the origin with a slope of one, the agreement is judged to be good.

Obviously, hypothesis testing is in order when the null hypothesis is that the slope is 1 versus the alternative, it is not 1. Another hypothesis involving the intercept is also of interest. One could also base the sample size on estimation principles, where one would want to achieve a given error of estimation for the slope and intercept. In any case, the prior information from the Erasmus study plays a valuable role in planning a future experiment. Well known formulas for regression can be employed in a Bayesian way to estimate the sample size and power!

Chapter 6 also provides some valuable information when planning studies to estimate the intraclass correlation coefficient. Consider the Kundra[7] study where three radiologists utilized MRI to measure the size of the largest liver lesion and the size of the lesion was modeled with a random effects model, and the overall goal was to estimate the intraclass correlation coefficient. In fact the posterior mean of the correlation was 0.79. How can this be used to estimate the sample size of a future experiment? The conventional sample size formula is reported in Shoukri[3] and is the basis of planning the future experiment. In fact there are two formulas reported by Shoukri, one based on the Fisher normalizing transformation, and another developed by Walter, Eliasziw, and Donner[8]. The latter is the most general and estimates the number of subjects for a given number of raters. One is assuming that a given number of raters are assigning scores to all of a given number of subjects. The former formula reported in Shoukri is restricted to two raters, and the number of subjects to be scored is estimated by a sample size formula, however, the Walter et al. approach is asymptotic.

Chapter 7 consists of a review of the classical approach to sample size estimation, Bayesian sample size estimation for the standard problems, a section on the planning studies to estimate kappa and other indices of agreement, a part devoted to designing future experiments modeled by the linear logistic model with a focus on comparing raters, a description of power studies for agreement using continuous scores, a discussion of calculating the number of subjects necessary to estimate the intraclass correlation coefficient, and lastly a presentation for planning future studies where the analysis is based on the fixed effects analysis of variance.

7.2 The Classical and Bayesian Approaches to Power Analysis

An important feature of sample size estimation is testing hypotheses. Often in agreement studies, the scientific hypothesis of that study can be expressed in statistical terms and a formal test implemented. Suppose $\Omega = \Omega_0 \cup \Omega_1$ is a partition of the parameter space, then the null hypothesis is designated as H: $\theta \in \Omega_0$ and the alternative by A: $\theta \in \Omega_1$, and a test of H versus A consists of rejecting H in favor of A if the observations $x = (x_1, x_2, \ldots, x_n)$ belong to a critical region C. In the usual approach, the critical region is based on the probabilities of type I errors, namely $\Pr(C/\theta)$, where $\theta \in \Omega_0$ and of type II errors $1-\Pr(C/\theta)$, where $\theta \in \Omega_1$. This approach to testing hypothesis was developed formally by Neyman and Pearson and can be found in many of the standard references, such as Lehmann[9], and Lee[10] presents a good elementary introduction to testing in a Bayesian context.

Once the null and alternative hypotheses have been specified and the type I and type II errors stated, it is often the case that the sample size can be expressed in closed form. This will be illustrated with many examples of agreement studies in later sections of the chapter. Contrasted with the above conventional approach, the Bayesian way to test hypotheses is defined below.

In the Bayesian approach, the decision to reject the null hypothesis is based on the probability of the alternative hypothesis

$$\varsigma_1 = \Pr(\theta \in \Omega_1 / x), \tag{7.1}$$

and the probability of the null hypothesis

$$\varsigma_0 = \Pr(\theta \in \Omega_0 / x).$$

Thus, the larger ς_1, the more the evidence that H is false.

If π_0 and π_1 denote the prior probabilities of the null and alternative hypotheses respectively, the Bayes factor is defined as

$$B = (\varsigma_0 / \varsigma_1) / (\pi_0 / \pi_1), \tag{7.2}$$

where the numerator is the posterior odds of the null hypothesis, and the denominator is the prior odds.

Suppose θ is scalar and that H: $\theta < \theta_0$ and A: $\theta > \theta_0$, where $\theta \in \Omega$, and Ω is a subset of the real numbers, then H and A are one-sided hypotheses. This situation can occur when there are so called nuisance parameters in the model. For example, if θ is the mean of normal population, then the standard deviation would be considered a nuisance parameter, since the primary focus is on the mean.

On the other hand, suppose H: $\theta = \theta_0$ and A: $\theta \neq \theta_0$, then the alternative is two-sided, and the null is referred to as a sharp null hypothesis. Note, in this situation θ can be multi dimensional and nuisance parameters can be present. For a sharp null hypothesis, special attention to the prior must be given to the null hypothesis. Let π_0 denote the probability of the null hypothesis, let $\pi_1 = 1 - \pi_0$, and suppose $\pi_1 \xi_1(\theta)$ is the prior density of θ when $\theta \neq \theta_0$. The marginal density of x is

$$f(x) = \pi_0 f(x / \theta_0) + \pi_1 f_1(x), \qquad (7.3)$$

where

$$f_1(x) = \int \xi_1(\theta) f(x / \theta) d\theta.$$

The posterior probabilities of the null and alternative hypotheses are

$$\varsigma_0 = [\pi_0 f(x / \theta_0)] / [\pi_0 f(x / \theta_0) + \pi_1 f_1(x)] \qquad (7.4)$$

and $\varsigma_1 = 1 - \varsigma_0$. See Lee[10] for additional details and examples of the Bayesian approach to testing hypotheses, however, that approach will not be taken, but instead will be based on the prior distribution of the conventional sample size formula for a particular application.

7.3 The Standard Populations: Classical and Bayesian Approaches

The classical and Bayesian approaches to sample size estimation is illustrated with two standard test problems: (1) testing for the difference in two proportions; and (2) testing for differences in the means of two normal populations with a common precision.

Comparing two binomial populations is a common problem in statistics and involves the null hypothesis H: $\theta_1 = \theta_2$ versus the alternative A: $\theta_1 \neq \theta_2$, where θ_1 and θ_2 are parameters from two Bernoulli populations. The two Bernoulli parameters might be the sensitivities of two diagnostic modalities.

Assuming the prior probability of the null hypothesis is π and assigning independent uniform priors for the two Bernoulli parameters, it can be shown that the Bayesian test rejects H in favor of A if the posterior probability P of the alternative hypothesis satisfies

$$P > \gamma, \qquad (7.5)$$

where

$$P = D_2 / D,$$

and $D = D_1 + D_2$.

It can be shown that

$$D = \left\{\pi \binom{n_1}{x_1}\binom{n_2}{x_2} \Gamma(x_1 + x_2 + 1)\Gamma(n_1 + n_2 - x_1 - x_2 - 1)\right\} \div \Gamma(n_1 + n_2 + 2), \quad (7.6)$$

where Γ is the gamma function.

$D_2 = (1 - \pi)(n_1 + 1)^{-1}(n_2 + 1)^{-1}$, and π is the prior probability of the null hypothesis. X_1 and X_2 are the number of responses from the two binomial populations with parameters (θ_1, n_1) and (θ_2, n_2), respectively. In order to choose sample sizes n_1 and n_2, one must calculate the power function

$$g(\theta_1, \theta_2) = \Pr_{x_1, x_2/\theta_1, \theta_2}[P > \gamma / x_1, x_2, n_1, n_2], \quad (\theta_1, \theta_2) \in (0,1) \times (0,1), \quad (7.7)$$

where P is given by Equation 7.5 and the outer probability is with respect to the conditional distribution of X_1 and X_2, given θ_1 and θ_2.

The classical approach is to reject H in favor of A if

$$abs(X_1 - X_2) \geq c(\alpha), \quad (7.8)$$

where $c(\alpha)$ depends on the probability of a type I error α, such that

$$P[abs(X_1 - X_2) \geq c(\alpha) / \theta_1 = \theta_2] \leq \alpha. \quad (7.9)$$

The power of the test at (θ_1, θ_2) is the probability of rejecting the null hypothesis, namely

$$g(\theta_1, \theta_2) = P[abs(X_1 - X_2) \geq c(\alpha) / (\theta_1, \theta_2)] \quad (7.10)$$

Note that for a given value of the parameters, Equation 7.9 is a function of the sample sizes and can be chosen in such a way that $g(\theta_1, \theta_2)$ has a given power at (θ_1, θ_2). Fortunately, a sample size formula can be derived, and according to Fleiss, Levin, and Paik[11] is as follows.

$$n(\theta_1, \theta_2, \alpha, \beta) = [Z_{\alpha/2}(2\bar{\theta}\bar{\phi})^{1/2} + Z_\beta\sqrt{\theta_1\phi_1 + \theta_2\phi_2}]^2 / [\theta_2 - \theta_1]^2, \quad (7.11)$$

where $\phi_i = 1 - \theta_i$, $i = 1, 2$, $\bar{\theta} = (\theta_1 + \theta_2)/2$, and $\bar{\phi} = (\phi_1 + \phi_2)/2$. Note that the sample size is the sample size for each sample and depends on the parameters, but only as a difference $\theta_1 - \theta_2$, the type I error α, and the type II error β. Thus for $\alpha = 0.05$ and $\beta = 0.2$, $Z_{\alpha/2} = 1.96$ and $Z_\beta = 0.843$, and one should refer to Fleiss et al. for the assumptions and the details of the derivation of Equation 7.11.

If prior information is available from previous studies, the prior distribution of the parameters induces a prior distribution for the sample size, via Equation 7.11, which will be the basis of choosing the sample size of a future study. As will be seen, this formula will be the basis for choosing sample sizes for agreement studies based on the G coefficient.

It is important to remember that the sample size formula is based on conventional frequentist ideas, and that one of the most important assumptions is that the sample size applies to a particular test of hypothesis for the binomial distribution, thus when used in a Bayesian sense, the corresponding Bayesian critical region (the one formed by rejecting the null hypothesis when the Equation 7.5 should be equivalent to the conventional critical region given by Equation 7.9). In this case, equivalent means the critical region should be of the form $abs(X_1 - X_2) \geq$ constant.

If the null hypothesis is H: $\theta_1 \leq \theta_2$ and the alternative is A: $\theta_1 > \theta_2$, then the sample size formula is revised so that $Z_{\alpha/2}$ is replaced by Z_α.

It also should be emphasized that the sample size determined by Equation 7.11 assumes that θ_1 and θ_2 are known, but of course they are not, thus, when the formula is used in practice estimates for them are substituted into the formula, and usually a range of values is used for the difference $\theta_1 - \theta_2$. If prior relevant studies are available, then estimates of the parameters based on that information are substituted into the formula, often without taking into account the uncertainty of the estimates. The Bayesian approach has a solution for this: in the Bayesian use of the formula, the posterior distribution of the parameters, based on previous studies, induces a posterior distribution for $n(\theta_1, \theta_2, \alpha, \beta)$. Based on the posterior distribution of n, the sample size is selected and a posterior distribution of the power function computed.

The sample size formula can be revised by taking into account a continuity correction and is

$$m(\theta_1, \theta_2, \alpha, \beta) = (n / 4)[1 + \sqrt{1 + 4 / (n \ abs(\theta_1 - \theta_2))}]^2, \qquad (7.12)$$

where n is given by Equation 7.11.

Suppose the null hypothesis is H: $\theta_1 \leq \theta_2$ versus the alternative A: $\theta_1 > \theta_2$ and that previous studies with 100 patients gives 0.10 as an estimate of the response rate θ_1 for the standard therapy, and it is anticipated that a new therapy will improve the response rate to 0.25. A panel of experts expects hypothetically that a future study of 100 patients would result in 25 who would respond to the new therapy. In the conventional sense with a significance level of 0.05 and a power of 0.80, 78 patients will be required with Equation 7.11, but with the continuity correction (Equation 7.12), the sample size is estimated as 91.

The posterior analysis for sample size appears in Table 7.1. The posterior medians of $m = 89.9$ and $n = 77$ agree quite well with 91 and 78, respectively based on the conventional approach. The uncertainty in the past and future

TABLE 7.1

Posterior Analysis of Sample Size

Parameter	Mean	SD	2½	Median	97½
Power	0.7624	0.0406	0.7236	0.7654	0.7964
m	127500	2.6×10^7	36.67	89.9	799
n	127500	2.6×10^7	29.17	77	760
θ_1	0.1001	0.0299	0.0493	0.0973	0.1659
θ_2	0.2499	0.04301	0.1708	0.2481	0.3395
Ln(n)	4.512	0.906	3.375	4.345	6.634
Exp(ln(n))	91	2.474	29.22	77.09	760.5

values of the response rate is taken into account, where the 95% credible intervals for θ_1 and θ_2 are (0.0493, 0.1659) and (0.1708, 0.3395), respectively, which produces an uncertainty of (29.17, 758) and (36.67, 799) in the estimated sample sizes for n and m respectively, as measured by 95% credible intervals. The posterior distributions for n and m are extremely skewed, thus one should refer to the medians for estimators of location! I have used the exponential of the log of n which is reported in the last row. This distribution is much more stable showing a mean sample size of 91 and a median of 77.09. The analysis is executed via BUGS CODE 7.1, where 70,000 observations are generated from the joint posterior distribution with a burn in of 5000 and a refresh of 100. Note the power refers to the average power of the conventional test (with a significance level of 0.05 and a power of 0.8), where the average is taken with respect to the posterior distribution of θ_1 and θ_2, and provides a posterior mean of 0.7624 and a median of 0.7654. What is your estimate of the sample size?

BUGS CODE 7.1 follows closely the form of the sample size formulas given in Equation 7.11 for n and Equation 7.12 for m, and the code for the power is based on Equation 4.17 of Fleiss, Levin, and Paik[11].

BUGS CODE 7.1

model;

```
# one-sided test
# alpha = .05
# beta = .20
# refer to Fleiss, Levin, and Paik[11]

{
# null hypothesis is th1 = .10

# alternative is at th2 = .25

th1 ~ dbeta(10,90)
```

th2 $\sim$ dbeta(25,75)

```
# th1 is prior information based on 100 patients
# th2 is prior information based on a hypothetical future study of 100
patients
```

ph1<-1-th1
ph2<-1-th2

mth<-(th1+th2)/2

mph<-(ph1+ph2)/2

num<-1.645*sqrt(2*mth*mph)+.843*sqrt(th1*ph1+th2*ph2)

num2<-num*num

den<-(th1-th2)*(th1-th2)

```
# num2 is the numerator
```

```
# den is the denominator
```

n<-num2/den

```
# n is sample size for given alternative p
```

m1<-(1+sqrt(1+4/(n*abs(th1 − th2))))

m<-(n/4)*m1*m1

```
# m is adjusted sample size based on the continuity correction
```

zbn<-(th2-th1)*sqrt(m)-1.96*sqrt(2*mth*mph)

zbd<-sqrt(th1*ph1+th2*ph2)

zb<-zbn/zbd

power<-phi(zb)

```
# power is the average power taken over the posterior distribution
}
```

list(th1 = .5,th2 = .5)

As a second example of classical and Bayesian power analysis, the difference in the means of two normal populations is considered.

How are the means of two normal populations, $n(\theta_1,\tau^{-1})$ and $n(\theta_2,\tau^{-1})$ compared? Suppose that the null hypothesis is H: $\theta_1 = \theta_2$ versus the alternative A: $\theta_1 \neq \theta_2$ and the common precision is a nuisance parameter. The conventional sample size for each of the two populations is

$$n(\theta_1,\theta_2,\sigma^2,\alpha,\beta) = 2\sigma^2(Z_{\alpha/2}+Z_\beta)^2 / (\theta_1 - \theta_2)^2 , \qquad (7.12)$$

where $\sigma^2 = 1/\tau$ is the common variance and Z_β is that value of the standard normal distribution such that the probability to the right of it is β.

If a previous related study is available, then the posterior distribution of the parameters θ_1, θ_2, and σ^2 will induce a posterior distribution for n, via Equation 7.12 and is the basis for choosing an appropriate sample size. Suppose from a previous study the θ_1 value is estimated as 0 with a variance of 0.1, and that for a future experiment with mean θ_2 it is anticipated the mean will be 0.5 with a variance of 0.1. From a previous study the sample variance was 1 and this estimate is given a variance of 0.0342, and it is assumed the variance of the second population will be the same as that of the previous study. See BUGS CODE 7.2 and how the formula for the sample size is expressed, where θ_1 is the represented as th1, while th2 represents θ_2, th1 is given a normal distribution with mean 0 and precision 10, and th2 a normal with mean 0.5 and precision 10 (variance 0.1). In addition, the common variance is designated as sigma and its inverse by tau and is given a gamma distribution with first and second scale parameters of 31 and 30, which implies a mean of 1 and a variance of 0.0344. The posterior analysis is given in Table 7.2.

Note the instability in the posterior distribution of n, thus, working with the log of n produces a much more stable distribution resulting in a posterior mean of 76.78 compared to a posterior median of 46. See the last two rows of the table. Also, note that the posterior median based on the exponential of the log sample size is the same as the posterior median computed directly from the posterior distribution of n. The posterior median of n is 46.04 compared to the classical estimate of 49.2. I used a type I error of 0.05 and a power of 0.80 at $\theta_2 - \theta_1 = 0.5$, which implies a sample size of 49.2 for each of the two populations. However, with the Bayesian approach the average posterior power is 0.9964, which is much larger than the nominal value of 0.80 used in Equation 7.12. 40,000 observations were generated from the joint posterior distribution with a burn in of 5000 and a refresh of 100.

TABLE 7.2

Posterior Analysis for Sample Size for Difference in Two Normal Populations

Parameter	Mean	SD	2½	Median	97½
n	1.085×10^6	1.217×10^8	6.14	46.04	18220
Power	0.9964	0.0034	0.9871	0.9975	0.9999
Sigma	1	0.1852	0.6999	0.9792	1.423
$\theta_2 - \theta_1$	0.4984	0.3162	−0.1211	0.4984	1.117
Tau	1.033	0.1855	0.7029	1.021	1.429
$\mathrm{Ln}(n)$	4.341	2.079	1.815	3.83	9.81
$\mathrm{Exp}(\mathrm{ln}(n))$	76.78	7.996	6.141	46	18214

BUGS CODE 7.2

```
model;

{

th1 ~ dnorm(0, 10)
```

the distribution of th1 is based on a previous study
```
th2 ~ dnorm(.5,10)
```
the distribution of th2 is hypothetical
```
th21<-th2-th1
```

```
tau ~ dgamma(31,30)
```
the distribution of tau induces a distribution for sigma

```
sigma<-1/tau
```

sigma is the common variance

```
n<-(2*sigma*2.48*2.48)/(th21*th21)
```
n is the sample size based on alpha = .05 and power .8

```
thsq<-th21*th21
```

```
power <-phi(sqrt((n*th21*th21/2*sigma)+1.96))
```

power is the power of the conventional t-test for the difference in two means with a common variance
the posterior distribution of power is the average power averaged over the distribution of th1,th2, and sigma.

```
}
list(th2=.5,th1=0, tau=1)
```

7.4 Kappa, the *G* Coefficient, and Other Indices

Sample size determinations for difference in two proportions and for difference in two normal means was introduced in the previous section, where the posterior distribution of the classical sample size formulas was the basis for choosing the appropriate value. It was shown how the information from previous related studies can be used to induce a posterior distribution for the sample size. This section will be the first encounter with estimating the appropriate sample size for agreement studies involving two raters and using the *G* coefficient to measure that agreement.

In Chapter 2 the *G* coefficient was introduced as

$$G = (\theta_{00} + \theta_{11}) + (1 - \theta_{00} - \theta_{11}) \tag{7.13}$$

for the 2×2 table (Table 1.1), where θ_{ij} is the probability that rater 1 gives a score of i and rater 2 a score of j, where $i, j = 0$ or 1, and n_{ij} is the corresponding number of subjects. The experimental results have the structure of a multinomial distribution. Obviously, the probability of agreement is the sum of the diagonal probabilities $\phi = \theta_{00} + \theta_{11}$, and G consequently is defined as

$$G = 2\phi - 1.$$

The main interest of this section is to test the null H: $G \leq G_0$ versus A: $G > G_1$, where $G_1 > G_0$, and to estimate the appropriate sample size. Since G depends only on the probability of agreement ϕ, the null and alternative hypotheses can be phrased as

$$\text{H: } \phi \leq \phi_0 \text{ versus A: } \phi > \phi_1.$$

Of course the main index of agreement is kappa which is defined as

$$\kappa = [(\theta_{00} + \theta_{11}) - (\theta_{0.}\theta_{.0} + \theta_{1.}\theta_{.1})] / [1 - (\theta_{0.}\theta_{.0} + \theta_{1.}\theta_{.1})] , \tag{7.14}$$

where the numerator is the probability of agreement minus the probability of agreement, assuming independence between readers. Chapter 2 and Chapter 3 were primarily devoted to kappa and related indices of agreement. There are several ways to estimate the sample size for an agreement study involving kappa and two of these are presented by Donner and Eliasziw[2] and Cantor[4], and are presented by Shoukri[3].

Consider the null hypothesis $\kappa \leq \kappa_0$ versus $\kappa > \kappa_0$, where κ_0 is based on previous related experiments. First to be considered is the Donner and Eliasziw method based on the formula

$$n(\alpha, \beta, \pi, \kappa_0, \kappa_1) = C(\alpha, \beta)[n_1 + n_2 + n_3]^{-1}, \tag{7.15}$$

where,

$$c(\alpha, \beta) = (Z_\alpha + Z_\beta)^2 ,$$

$$n_1 = [\pi(1 - \pi)(\kappa_1 - \kappa_0)]^2 / [\pi^2 + \pi(1 - \pi)\kappa_0] ,$$

$$n_2 = 2[\pi(1 - \pi)(\kappa_1 - \kappa_0)]^2 / [\pi(1 - \pi)(1 - \kappa_0)]$$

and

$$n_3 = [\pi(1 - \pi)(\kappa_1 - \kappa_0)]^2 / [(1 - \pi)^2 + \pi(1 - \pi)\kappa_0] .$$

The terms of n are usually unknown, however, there is some knowledge of κ_0 from previous related studies and there is some evidence that kappa has increased to κ_1 and a future study is planned to determine if kappa has indeed increased, therefore, a hypothetical experiment is envisaged where κ_1 is the future kappa value and π is future probability a subject will be assigned a 1. Thus, the distribution of κ_0 used in Equation 7.15 is the actual posterior distribution of kappa based on a past experiment, while the distribution of κ_1 to be used in 1 is based on a hypothetical 2×2 table, based on evidence or opinion that kappa has increased.

Suppose the two raters are the same radiologist looking at the same images where the early training of the radiologist was centered on an agreement value of κ_0, now one year later, there is evidence the intrarater variation of the radiologist has improved to $\kappa_1 > \kappa_0$. A study is planned to assess the improvement, if any, in the radiologist's interrater readings of the same images. How many images should be used to assess the radiologist?

The radiologist is being trained in CT and in particular in diagnosing non-small-cell lung cancer and has been assigning two scores to each image, either a 0 for no evidence of disease or a 1 indicating at least some evidence of the disease, and the scores for 78 patients were reported in the early phase of training.

Based on this information, the posterior mean (SD) of kappa is 0.1647 (0.1031), the probability of agreement is 0.577 (0.0556), the probability of chance agreement is 0.4939 (0.0168), and the posterior mean of π is 0.442 (0.0420). Assuming a uniform prior, I generated 40,000 observations from the joint posterior distribution, with a burn in of 5000 and a refresh of 100. As indicated by the value of kappa and the probability of agreement, overall agreement is deemed poor to fair.

Recent experience of the radiologist suggests an improvement in agreement as measured by kappa and the probability of agreement and a study is planned to assess the interrater variation of this radiologist, thus, a committee of radiology faculty have considered a hypothetical study with 100 patients and predicted the following outcomes reported in Table 7.3.

Based on this information, kappa is estimated with the posterior mean (SD) as 0.4895 (0.0865), which of course implies good improvement, thus, using this information and the information from the previous related study of Table 7.3, the sample size is estimated by substituting κ_0 and κ_1 from the previous and hypothetical experiments into Equation 7.15. In this way the uncertainty from those two sources can be included in the posterior distribution of the sample size n. BUGS CODE 7.3 demonstrates how this is

TABLE 7.3

Classification Table of a Radiologist Hypothetical CT Training for Non-Small-Cell Lung Cancer, a Future Study

First Look	Second Look		
	$Y=0$	$Y-1$	
$X=0$	59	19	78
$X=1$	3	19	22
	62	38	100

accomplished. The program consists of three parts: (a) the posterior distribution of κ_0 based on the results of the previous study in Table 7.3; (b) the posterior distribution of κ_1, based on Table 7.3; and (c) the posterior distribution of n, based on Equation 7.15.

BUGS CODE 7.3

model;

```
# Code for the sample size for kappa hypothesis test
# one-sided test
{
g00 ~ dgamma(27,2)
g01 ~ dgamma(23,2)
g10 ~ dgamma(10,2)
g11 ~ dgamma(18,2)

h<-g00+g01+g10+g11

th00<-g00/h
th01<-g01/h
th10<-10/h
th11<-g11/h

th0.<-th00+th01
th.0<-th00+th10
th1.<-th10+th11

th.1<-th01+th11

k0<-(agree0-cagree0)/(1-cagree0)

# k0 is the kappa based on Table 7.3

agree0<-th00+th11

cagree0<-th0.*th.0+th1.*th.1

a00 ~ dgamma(59,2)
a01 ~ dgamma(19,2)
a10 ~ dgamma(3,2)
a11 ~ dgamma(19,2)

b<-a00+a01+a10+a11

ph00<-a00/b
ph01<-a01/b
ph10<-a10/b
ph11<-a11/b

ph0.<-ph00+ph01
ph.0<-ph00+ph10
```

```
ph1.<-ph10+ph11
ph.1<-ph01+ph11
k1<-(agree1-cagree1)/(1-cagree1)
# k1 is the kappa based on hypothetical study Table 7.4
agree1<-ph00+ph11
cagree1<-ph0.*ph.0+ph1.*ph.1
pi<-(ph1.+ph.1)/2
n<-c/(n1+n2+n3)
# n is the sample size formula ( 7.15)
n1<-n11/n12
n2<-n21/n22
n3<-n31/n32
n11<-pi*(1-pi)*(k1-k0)*pi*(1-pi)*(k1-k0)
n12<-pi*pi +pi*(1-pi)*k0
n21<-2*pi*(1-pi)*(k1-k0)*pi*(1-pi)*(k1-k0)
n22<-pi*(1-pi)*(1-k0)
n31<-pi*(1-pi)*(k1-k0)*pi*(1-pi)*(k1-k0)
n32<-(1-pi)*(1-pi)+pi*(1-pi)*k0
}
list( c=6.19)
# c= (Za+Zb)(Za+Zb)
list( g00=1,g01=1,g10=1,g11=1,a00=1,a01=1,a10=1,a11=1)
```

Table 7.4 presents the posterior analysis for the sample size estimation based on Equation 7.15, and it shows the change in kappa from the information in the previous related study compared to the information of the hypothetical outcomes. The faculty really believes the radiologist has improved!

The sample size posterior median is 62.15 which is my estimate of the required sample size, and it is interesting to know that this value is quite close to the 'classical' estimate of 63.2, which is found by substituting 0.1645, 0.4895, and 0.300 for κ_0, κ_1, and π, respectively into Equation 7.15. Note that $\alpha = 0.05$ and $\beta = 0.20$ was used in the formula for the classical estimate of 43.1, thus using a power of 0.80. It should be emphasized that the posterior distribution for n is quite unstable showing extreme right skewness, however the median appears to be providing a stable estimate of the sample size. The instability in the distribution is due to the "small" values of the numerator

TABLE 7.4

Posterior Distribution of the Sample Size for Kappa

Parameter	Mean	SD	2½	Median	97½
κ_0	0.1645	0.1033	−0.0403	0.165	0.3647
κ_1	0.4895	0.0865	0.316	0.4911	0.6532
N	$1.347*10^7$	$7.669*10^9$	17.72	62.15	1544
$n1$	0.0496	0.0482	0.00113	0.0369	0.1732
$n2$	0.05855	0.0390	0.00244	0.05246	0.1497
$n3$	0.0109	0.0093	0.000293	0.0086	0.0354
π	0.300	0.0390	0.2268	0.2989	0.3794
$\mathrm{Ln}(n)$	4.362	1.203	2.875	4.13	7.342
$\mathrm{Exp}(\ln(n))$	78.41	3.33	17.72	62.17	1543

in Equation 7.15, thus the posterior distribution based on the exponential of the log of n is given in the last row of the table, where right skewness is discerned in the distribution of the sample size.

A related approach to the sample size is presented by Shoukri[3] and is based on controlling the width w of a 95% confidence interval for κ, which is given by

$$m(\alpha,\kappa,\pi) = (4Z_{\alpha/2}^2 / w^2)\{(1-\kappa)[(1-\kappa)(1-2\kappa) + \kappa(2-\kappa) / 2\pi(1-\pi)]\}, \quad (7.16)$$

where κ is the hypothetical future value of kappa that is believed will occur and π is the probability of a positive response (denoted by 1). In a Bayesian sense, the sample size m has a distribution induced by the prior distribution of a future hypothetical study.

What is the posterior distribution of m? It can be shown that the posterior mean (SD) of m is 1390 (185.6) with a median of 1392 and 95% credible interval (1019, 1748), thus approximately 1400 patients are needed to achieve a 95% confidence interval with a width of 0.10. On the other hand, for a confidence interval of width 0.2, the median of the posterior distribution is 348 subjects, and for one with a width of 0.3, the median sample size estimate is 154. Table 7.5 demonstrates the effect of the width of the 95% confidence interval on the posterior distribution of the sample size.

As the width decreases the posterior mean increases in a dramatic fashion, from a high of 1390 patients, for a width of 0.1, to a low of 87, corresponding to a width of 0.4. Therefore, if one chooses a width of $w = 0.2$, 357 patients will be recruited into a study to assess the diagnostic ability of the novice radiologist. Figure 7.1 depicts the posterior density of the sample size, when the width of the confidence interval, $w = 0.4$ and appears to be symmetric about the mean of 86.8, as do the other densities of the sample size.

TABLE 7.5

Effect of Width of Confidence Interval on Sample Size

Width	Mean	SD	2½	Median	97½
0.1	1390	186.1	1018	1393	1749
0.2	347	46.5	254.5	348.2	437.2
0.3	154	20.68	113.1	154.8	194.3
0.4	86.87	11.63	63.61	87.05	109.3

TABLE 7.6

Posterior Analysis of Depression Study with Two Psychiatrists and a Covariate of Paranoid Scores

Parameter	Mean	SD	2½	Median	97½
$b[1]$	−4.77	0.4164	−5.61	−4.76	−3.98
$b[2]$	0.3068	0.2001	−0.0749	0.3036	0.7076
$b[3]$	−2.781	0.3809	−3.56	−2.769	−2.07
$b[4]$	−0.7697	0.2356	−1.245	−0.7662	−0.3185
$b[5]$	−1.645	0.32	−2.317	−1.641	−1.06
$b[6]$	0.1162	0.0238	0.07047	0.116	0.1636

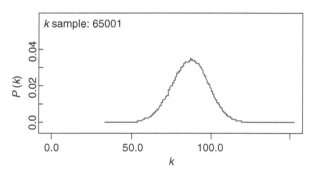

FIGURE 7.1
Posterior density of sample size for w = 0.4.

Cantor[4] developed another method of sample size estimation for binary scores and two readers, and for a more general method of sample size estimation that involves multiple readers and scores, see Indurkhya, Zayas, and Buka[12]. See Exercise 3 for additional information on the Cantor method.

7.5 The Logistic Linear Model

Chapter 5 presented the Bayesian approach to agreement using the logistic linear model in order to measure the effects of raters on the scores that are being assigned to the experimental units. In addition, the model was employed to assess the effects of other factors (in addition to the raters) on the agreement between readers, including patient and rater covariates. For example, patient covariates, such as their age and medical history, were taken into account in order to determine their effect on rater scores.

In order to design an agreement study, the number of experimental units to be included needs to be carefully done. When the main tool of the analysis is the logistic linear model the choice of sample size should include the study objectives. The protocol for an agreement study should include a section that describes in detail the study objectives and should rephrase the objectives in statistical terms. Often the goals of the study are stated as tests of hypotheses that allow one to reasonably select the sample size. In an agreement study using a logistic model, the objectives are often stated as test of hypotheses about the regression coefficients. For example, the study objectives are stated in such a way in order to want to compare the two raters in the way they assign scores to the patients. For example, in the first example of Chapter 5 when two psychiatrists are assigning depression scores (see Table 5.2) to patients, it was of interest to see if the two are assigning scores in the same way.

This was investigated by a posterior analysis reported in Table 5.3 when rater 1 was compared to rater 2 in the way they assigned scores 1 (a 1 denotes no depression) and 2 (a 2 designates mild depression). The comparison depends on the difference between two regression coefficients of the logistic model.

Suppose a future study is to be designed that will compare the depression scores of two psychiatrists in an investigation that has the same design as the one reported in Table 5.2. How many patients n should be included? The design of such an experiment is a random assignment of n patients into the 3×3 table and the number of patients "falling" into the cells follow a multinomial distribution. Recall that the logistic linear model is defined as

$$\log \text{it}(\phi_i) = \sum_{j=1}^{j=p} b[j]X_j[i], \tag{5.1}$$

where,

$$\log \text{it}(\phi_i) = \log[\phi_i / (1 - \phi_i)],$$

and $i = 1, 2, \ldots, 9$, the nine cells of the 3×3 Table 5.2 of depression scores. Note that ϕ_1 is the cell probability corresponding to the i-th cell of the above table, where i is read from top to bottom and from left to right. Thus, the model consists of p factors which might have some effect on the logit of the cell probabilities. It is obvious how to extend the model to more than two raters and scores. Also, the $b[j]$ are considered as scalar constants to be estimated, and the X_j are known independent variables, the effect of which are on the logits as measured by the corresponding $b[j]$.

By convention, the column X_1 is a 9×1 vector of ones, while the other $p-1$ vectors of the design matrix are specified by the user, where $X_j[i]$ is the i-th row of the vector X_j. There are p vectors that represent the effects of the two raters as well as the effects of various covariates on the cell probabilities and their corresponding logits.

The choice of a sample size will be illustrated with the example of depression scores, where the factors are the two raters and a covariate of patient paranoid scores. In this case the columns of the design matrix are:

$$X_1 = (1,1,1,1,1,1,1,1,1)',$$

$$X_2 = (1,1,1,0,0,0,-1,-1,-1)',$$

$$X_3 = (0,0,0,1,1,1,-1,-1,-1)',$$

$$X_4 = (1,0,-1,1,0,-1,1,0,-1)',$$

$$X_5 = (0,1,-1,0,1,-1,0,1,-1)', \text{ and}$$

$$X_6 = (17,27,3,16,45,14,1,3,3).$$

Note that the first column is a vector of ones which gives the effect of the overall mean of the logits, the second vector corresponds to a contrast of score 1 minus score 3 for rater 1, the third corresponds also to rater 1, but for the contrast of score 2 minus score 3. In a similar fashion, the fourth column contrasts score 1 minus score 3 for rater 2, while the fifth contrast score 2 minus 3 for rater 2. The last column is the vector of covariate which gives the average value of a paranoid score for the nine cells of the table. Attention is focused on $b[6]$, which gives the effect of the paranoid scores on the logits, in that, do the paranoid scores, affect the agreement between the two raters, i.e. is $b[6]=0$?

A Bayesian analysis is performed using non informative prior distributions for the $b[.]$, coefficients of Equation 5.1, with the results shown in Table 7.6.

The results of this study will be considered as those of a hypothetical future study and will be used to estimate the sample size. A look at Table 7.6 and Figure 7.2 shows a symmetric density about the mean of 0. 1162 with a standard deviation of 0.0238 and a 95% credible interval of (0.0704,0.1636), which implies that the posterior distribution of $b[6]$ is approximately normal.

Recall that the sample size of 129 patients for the depression study are partitioned into the cells of the 3×3 Table 5.2 as 11, 2, 19, 1, 3, 3, 0, 8, 82, and these cell values affect the estimates of the coefficients of the model in a complicated way. For example, see Woolson[13], who presents the dependence

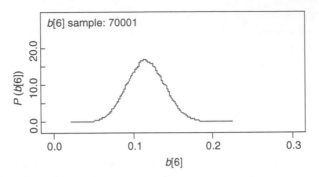

FIGURE 7.2
Posterior density of the effect of the covariate.

of the estimated regression coefficients and their standard errors on the cell frequencies for a simple linear logistic model. It is quite difficult to derive the dependence of the estimated coefficients on the cell frequencies, thus, the sample size will be determined indirectly.

The sample size will be chosen by trial and error with hypothetical future cell frequencies and using the width of the 95% credible intervals of the posterior distribution of the coefficients. For example, with the cell frequencies of Table 5.1, the 95% credible interval for $b[6]$ is (0.0704,0.1636), corresponding to a width of 0.0932, or a 95% error of estimation of 0.0466. Thus it appears that the paranoid scores are associated with the logits. Now suppose a future study is planned with twice the number of patients, namely 259 patients, and the patients are partitioned into the cells as: 22, 4, 38, 2, 6, 6, 1, 16, 164. What will be the effect on the width of the 95% credible interval for $b[6]$? The 95% credible interval is (0.0816,0.1465) with a width of 0.0649, a reduction of some 30%. Thus a doubling of the sample size (with a doubling in each cell) reduces the error of estimation from 0.0466 to 0.0324. Will it be worth doubling the sample size in order to reduce the width of credible interval by 30%? Of course, the sample size will not actually have to be doubled, but instead, repeated with 129 patients, and the results combined with the past study. It is important to remember that in a future experiment with 259 patients, the cell frequencies will not be the double those in the original study, if the patients are selected at random and one waits to see what happens!

The Bayesian analysis is performed with 140,000 observations, a burn in of 5000, and a refresh of 100. See BUGS CODE 7.4:

BUGS CODE 7.4

Model;

{

von eye and mun example page 50

for (i in 1:N){y[i] ~ dbin(p[i],259)}

```
for(i in 1:N) {logit(p[i])<- b[1]*X1[i]+b[2]*X2[i]+b[3]*X3[i]+b[4]*X4[i]+
b[5]*X5[i]+b[6]*X6[i] }
```

```
# b[2] is the effect of the contrast (1-3) for rater 1
# b[3] is the effect of the contrast (2-3) for rater 1
# b[4] is the effect of the contrast (1-3) for rater 2
# b[5] is the effect of the contrast (2-3) for rater 2
# b[6] is the effect of the paranoid scores
```

```
}
```

```
# prior distrubution for the coefficients
 for(i in 1:6){ b[i] ~ dnorm(0.0,.000001)}
```

```
}
```

```
list(N=9, y=c(11,2,19,1,3,3,0,8,82),
X1=c(1,1,1,1,1,1,1,1,1), X2=c(1,1,1,0,0,0,-1,-1,-1),
X3=c(0,0,0,1,1,1,-1,-1,-1), X4= c(1,0,-1,1,0,-1,1,0,-1),
X5 =c(0,1,-1,0,1,-1,0,1,-1), X6=c(17,27,3,16,45,14,1,3,3))
```

```
# X1 is the vector for the constant
# X2 is the contrast between the score 1 and 3 for rater 1
# X3 is the contrast between the score 2 and 3 for rater 1
# X4 is the contrast between the score 1 and 3 for rater 2
# X5 is the contrast between the score 2 and 3 for rater 2
# X6 is the vector of paranoid scores
```

```
list(b=c(0,0,0,0,0,0))
```

It is somewhat arbitrary to base the sample size estimation on the width of $b[6]$. One could base the sample size on the width of the credible interval for the posterior distribution of $b[2] - b[4]$, which compares the 1 minus 3 contrast of rater 1 with that of rater 2. See Exercise 7.

7.6 Regression and Correlation

For continuous or quantitative scores, Chapter 6 presents several approaches to reader agreement including the analysis of variance, with fixed and random effects, and regression and correlation methods. It is the latter that will be emphasized here.

Consider the regression approach to agreement, where the scores of one reader are regressed on those of another, and agreement is assessed by estimating the slope and intercept of the fitted simple linear regression. If the "true" line goes through the origin with a slope of one, the agreement is deemed very good, however, this is difficult to determine. The approach in Chapter 6 was to determine the 95% confidence intervals for the slope and intercept and see if the interval for the intercept contains zero and if interval for the slope contained one. Recall the Erasmus study of Section 6.2, which involves five readers who are estimating the lesion size of lung cancer patients, and where Figure 6.3 shows the scores of reader 1 regressed on those of reader 2. Does the "true" line go through the origin with a slope on 1?

A normal theory simple linear regression model (Equation 6.1, $y[i] = \beta_1 + \beta_2 x[i] + e[i]$) is employed, where $y[i]$ is the tumor size of i-th patient measured by one reader, $x[i]$ is the tumor size of the i-th patient as measured by another rater (for the same replication), and the $e[i]$ are independent and distributed as $n(0, tau)$, where tau is the precision of $e[i]$, $i = 1, 2,..., 20$. The intercept and slope are the unknown parameters β_1 and β_2, respectively. With the Bayesian approach the parameters are given uninformative prior distributions, i.e. β_1 and β_2 are given independent $n(0, 0.00001)$ distributions, and tau is specified as a gamma(0.00001,0.00001).

For the regression of reader 1 versus reader 2, the 95% credible interval for the intercept is (−0.175, 1.292), and for the slope is (0.725, 1.09), thus one would conclude that readers 1 and 2 are in fair to good agreement. Note that the credible interval for the slope almost does not include 1.

Based on Equation 6.1, one can show that the $1 - \alpha$ confidence interval for the intercept is

$$\hat{\beta}_1 \pm t_{1-\alpha/2}(n-2)\hat{\sigma}_{\beta_1}, \qquad\qquad (7.17)$$

where $\hat{\beta}_1$ is the least squares estimate of the intercept, and the estimated variance of the estimated intercept is

$$\hat{\sigma}^2_{\beta_1} = \sigma^2_{y/x}[1/n + \bar{X}^2 / \sum_{i=1}^{i=n}(x_i - \bar{x})^2],$$

where $\sigma^2_{y/x}$ is the estimated variance about the regression line. See Woolson[13] for additional details about simple linear regression. In a similar way, the 95% confidence interval for the slope is

$$\hat{\beta}_2 \pm t_{1-\alpha/2}(n-2)\hat{\sigma}_{\beta_2}, \qquad\qquad (7.18)$$

where, $\hat{\beta}_2$ is the least squares estimate of the slope, $\hat{\sigma}_{\beta_2} = \sigma_{y/x}\sqrt{1/n S_x^2}$ is the estimated standard deviation of the estimated slope, and S_x^2 is the sample variance of the x variables.

Since the agreement between readers depends on the 95% credible intervals about the origin, the choice of the sample size will be based on the confidence intervals. For example, one may use the confidence interval for the intercept in order to select the sample size, by letting

$$w = t_{1-\alpha/2}(n-2)\hat{\sigma}_{\beta_1},$$

where w is the predetermined length of the confidence interval of the intercept, chosen by the experimenter, thus 'solving' for the sample size gives

$$n = [t_{1-\alpha/2}(n-2)]^2 \sigma^2_{y/x} / w^2 S^2_x. \qquad (7.19)$$

For "large" sample sizes, the t percentile $t_{1-\alpha/2}(n-2)$ may be replaced by $Z_{1-\alpha/2}$, say by 1.96, for $\alpha = 0.05$, which simplifies the way the sample size is selected. How does the Bayesian use this formula to choose a sample size in order to estimate the intercept? If the results of a previous study are available, one would use the posterior distribution of

$$\sigma^2_{y/x} = \sum_{i=1}^{i=n} (y[i] - \beta_1 - \beta_2 x[i])^2 / (n-2)$$

in order to induce a posterior distribution for the sample size! Note the y values are the scores of reader 1 and the x scores are those of reader 2. Based on the scores of reader 1 and reader 2 listed in BUGS CODE 7.1 and with a width of $w = 0.10$, $\alpha = 0.05$, 45,000 observations generated from the posterior distribution, a burn in of 10,000 and a refresh of 250, the posterior density of the sample size is shown in Figure 7.3.

Also, the posterior mean (SD) of n is 126 (30.78). The posterior median is 122.2 implying a right skewness, thus the median $122 = n$ is my estimate of the sample size based on a width of 0.10 for the 95% confidence interval of the intercept. In a similar way the sample size based on the confidence interval for the slope is:

$$m = t^2_{1-\alpha/2}(n-2)\sigma^2_{y/x}[1 + \bar{X}^2 / S^2_x] / w^2, \qquad (7.20)$$

where w is the desired width of the confidence interval. Replacing the t percentile by 1.96, and letting $w = 0.10$, the posterior distribution of the estimated

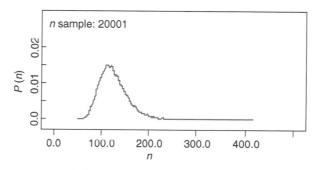

FIGURE 7.3
Posterior density of the sample size.

TABLE 7.7

Effect of Width of 95% Confidence Interval of Sample Size for Intercept and Slope

Width	Parameter	Mean	SD	2½	Median	97½
0.1	*m*	2025	489	1281	1953	3185
	n	126	30.6	80.1	122	199
0.3	*m*	225	54.7	141.6	217.3	354.3
	n	14.09	3.42	8.85	13.59	22.16
0.4	*m*	126.6	30.75	79.95	122	199
	n	7.9	1.923	4.98	7.63	12.46
0.8	*m*	31.66	7.69	19.94	30.52	49.82
	n	1.98	0.48	1.247	1.909	3.166

sample size based on the slope has a posterior mean (SD) of 2025 (491) with a 95% credible interval of (1277, 3183), a rather large sample size for an agreement study. The effect of the width w on the posterior distribution of the sample size is given by Table 7.7. The sample size m is based on the slope while n is based on the intercept.

The effect is rather pronounced with a median sample size of 1953 and 122, based on 95% confidence intervals for the slope and intercept respectively, when the width is 0.1, and a median of 30.52 and 1.9 for them when the width is 0.8. These calculations are executed with 40,000 observations generated from the joint posterior distribution, with a burn in of 5000, and a refresh of 100. See BUGS CODE 7.5:

BUGS CODE 7.5

model;

erasmus example tumor size

```
{
# Regression of y on x
for(i in 1:N) { y[i] ~ dnorm(mu[i],precy)
mu[i]<- beta[1] + beta[2]* × [i]
}
# prior distribution of betas
for ( i in 1:P){ beta[i] ~ dnorm(0.0,.000001)}
precy ~ dgamma(.00001,.00001)
# sigma is the variance about the regression line
sigma<-1/precy
# n is the sample size based on a width w for intercept CI
```

n<-m1/(w*w*sqx)

m<-m1*m2/m3

m is the sample size based in the width of slope CI

m1<-Z*Z*sigma

m2<-(1+xbsq/sqx)

m3<-w*w

}

values of y and x
the y values are for reader 1
the x values are for reader 2
N is the number of observations
sqx is the sum of squared deviations for the x values
xbsr is the square of the mean of the x values
Z is the z-value of the standard normal corresponding to alpha

list(N=40, P=2, w=.3, Z=1.96, sqx=2.30, xbsq=13.69, y=c(3.50,3.80,2.20,
1.50,3.80,3.50,4.20,5.40,7.60,2.80,5.00,2.30,4.40,2.50,5.20,1.70,4.50,4.50,6.00,3.30,
7.00,4.00,4.80,5.00,2.70,3.70,1.80,6.30,4.00,1.50,2.20,2.20,2.60,3.70,2.50,4.80,8.00,
4.00,3.50,4.80),

x=c(3.30,3.80,2.00,1.20,4.00,3.30,4.00,5.20,7.80,3.00,5.50,2.70,4.50,1.80,4.00,2.00,
5.80,3.80,5.80,2.50,7.00,2.20,5.00,5.00,2.50,4.00,2.00,4.50,4.30,1.50,2.30,4.20,2.80,
2.00,2.20,5.00,5.00,3.00,3.50,4.20))

initial values of betas

list(beta=c(0,0), precy=1)

Two choices for the sample size, one based on the slope, and the other on the intercept presents a problem. For a given width and the same confidence level, the sample size based on the slope is much larger than that for the intercept, thus, for example, when $w=0.3$, the posterior mean of m is 225 compared to 14 for n, the estimate based on the intercept. Choosing 224 refers to the sample size for the number of patients to be included in a future study, where each of the five readers look at all 224. This is probably too large for an agreement study, thus consider a larger width, say 0.4 for both intervals then $m=126$ and $n=179$, which is still probably too large. Remember the original study has 40 patients, and one would have to increase the width to probably 0.6 in order to achieve a reasonable sample size. Because of time and money constraints, sample sizes between 40 and 60 are feasible in an academic health science center.

Another alternative is to base the sample size using another criterion, such as estimating the mean tumor size by rater 1 at a given tumor size X_0 for rater 2. A 95% confidence interval for the mean of Y (reader 1 scores) at the reader 2 score of $X=X_0$ is

TABLE 7.8

Posterior Distribution of the Sample Size

Width	Parameter	Mean	SD	2½	Median	97½
0.3	p	32.37	7.848	20.41	31.21	50.88
0.2	p	72.48	17.66	45.92	70.23	114.5
0.1	p	291.3	70.63	183.7	280.9	457.9

$$\hat{\beta}_1 + \hat{\beta}_2 + t_{1-\alpha/2}(n-2)\sigma_{y/x}\sqrt{1/n + (X_0 - \bar{X})^2 / nS_x^2}$$

and from which one may show the sample size is

$$p = t_{1-\alpha/2}^2(n-2)[1 + (X_0 - \bar{X})^2 / S_x^2]/w^2. \qquad (7.21)$$

A special case is when $X_0 = \bar{X}$, and the sample size simplifies to

$$p = t_{1-\alpha/2}^2(n-2)/w^2, \qquad (7.22)$$

which estimates the sample size based on estimating the reader 1 score at the sample mean $\bar{X}$ of the reader 2 scores. For reader 1 scores regressed on reader 2 scores, the posterior distribution of the sample size based on the confidence interval for the mean of reader 1 scores at the mean tumor size $\bar{X} = 3.705$ of reader 2 scores, as in Table 7.8.

Thus, the posterior median for the sample size is 70.23 when $w = 0.2$ at the mean tumor score of 3.70 cm for reader 2. This appears reasonable for agreement studies in diagnostic radiology. Recall that the sample size estimate is based on a past study of five radiologists with 40 patients. The effect of the width is dramatic with the smallest posterior median of 31 for a width of 0.3 and the largest estimate of 280 when the width is 0.1. Choosing a width of 0.2 corresponds to a small error of estimation, because the tumor sizes range form 1.2 to 7.8 cm! The sample size estimate of 280 corresponding to an error of estimation of 0.1 is much too large.

Of course, other approaches to sample size estimation could have been adopted, for example, those based on hypothesis testing principals, but this is left for Exercise 5.

7.7 The Intraclass Correlation

Recall the approach to agreement using the random linear model

$$y[i,j] = \theta + a[i] + e[i,j], \qquad (7.23)$$

where $y[i,j]$ is the score assigned by the j-th reader to the i-th patient, $i=1$, $2,\ldots,k$ and $j=1, 2,\ldots, n$. Also, θ is a constant, the mean of the observations, the $a[i]$ are independent and normally distributed with mean 0 and variance σ_a^2, while the $e[i,j]$ are independent and normally distributed with mean 0 and variance σ_w^2, and all random variables are jointly independent.

Consider the covariance between $y[i,j]$ and $y[i,\,j']$, that is, between the scores of the j-th and $j'-$th reader of the i-th patient, then

$$\mathrm{cov}(y[i,j], y[i,j']) = \mathrm{cov}(\theta + a[i] + e[i,j], \theta + a[i] + e[i,j']).$$

Note, the covariance is conditional on the parameters and it can be shown that the covariance is σ_a^2, consequently, the intra class correlation is

$$\rho = \sigma_a^2 / (\sigma_a^2 + \sigma_w^2). \qquad (7.24)$$

One way to measure agreement is the intraclass correlation coefficient, which is the common correlation between all pairs of raters. The main focus of this section will be on choosing the sample size based on (a) testing hypotheses about the intraclass correlation, (b) based on cost considerations, and (c) based on the width of a confidence interval for the intraclass correlation.

Consider a test of H: $\rho \leq \rho_0$ versus A: $\rho > \rho_0$, then Shoukri[3] gives

$$k = (Z_{1-\alpha} + Z_{1-\beta})^2 / (\eta(\rho_0) - \eta(\rho_1))^2, \qquad (7.25)$$

as the sample size, where

$$\eta(\rho) = (1/2)\log((1+\rho)/(1-\rho)).$$

The sample size is based on normality and assumes that the number of readers is 2. For a more general technique, Walter et al.[8] presents the sample size as

$$k = 1 + 2n(Z_{1-\alpha} + Z_{1-\beta})^2 / (n-1)[\ln C_0]^2, \qquad (7.26)$$

where n is the number of raters and

$$C_0 = (1 + n\phi_0)/(1 + n\phi_1)$$

where

$$\phi_i = \rho_i / (1 - \rho_i)$$

with $i=0$ or 1 and $\rho_1 > \rho_0$, and ρ_1 is a value of ρ under the alternative hypothesis.

The Kundra[14] example was introduced in Chapter 6 and will illustrate the Bayesian approach to selecting the sample size based on Equation 7.26.

TABLE 7.9

Posterior Distribution of the Sample Size k

Parameter	Mean	SD	2½	Median	97½
k	7.33×10^5	5.78×10^7	21.04	172.9	76200
Ln(k)	5.655	2.141	2.988	5.147	11.24
ρ	0.7914	0.0707	0.6294	0.7999	0.904
θ	35.32	4.722	22.98	32.6	41.57
σ_a^2	448.5	169	220	416	868
σ_w^2	108.2	26.91	67.85	104.2	172

Kundra et al. performed a study of interobserver agreement, where the lesion size of liver lesions were estimated with the aid of MRI and their interpretation by three radiologists. The study was quite involved using several MRI sequences and had two replications, but only the first replication is considered here and the MRI sequence is fast spin echo. The main objective of the study is to estimate the inter and intra reader agreement, of three radiologists, who are measuring the size of the largest liver lesion. Recall from Section 6.4, that the Kundra study was examined in detail using a regression approach, where the $n = 22$ tumor sizes of one reader were regressed on those of another.

Suppose the Kundra results are considered a pilot study and that a future study is planned, where the future intraclass correlation will have a value of 0.85, in contrast to the prior value with a posterior mean (SD) of 0.7914 (0.0707) and a 95% credible interval of (0.629,0.904). Such values were determined with 55,000 observations generated from the posterior distribution of ρ, and assuming non informative prior distributions for the one-way random model (Equation 7.23). BUGS CODE 7.6 presents the Bayesian solution to the sample size problem, based on Equation 7.26, and it is executed with a burn in of 5000, a refresh of 100, and 50,000 observations generated from the posterior distribution of ρ. Table 7.9 presents the prior distribution of the sample size k, assuming a type I error of $\alpha = 0.05$, a power of 0.80, and an alternative value of $\rho_1 = 0.85$.

BUGS CODE 7.6

```
# Data is from Kundra inter reader agreement
# three readers
# 22 patients
# response is lesion size in mm
# intraclass correlation coefficient
# sample size (Equation 7.26)
model;
        {                       for( i in 1 : images) {
                                m[i] <- theta + a[i]
```

```
               for( j in 1 : radiologists) {
                      y[i,j] ~ dnorm(m[i], tauw)
               }}
```

```
               for( i in 1:images){ a[i] ~ dnorm(0,taua)}
# sigmaw is the within readers component
               sigmaw <-1 / tauw
# sigmaa is the between readers variance component
               sigmaa <-1 / taua
               tauw ~ dgamma(0.001, 0.001)
               taua ~ dgamma(0.001, 0.001)
               theta ~ dnorm(0.0, .001)
```

rho is the intraclass correlation coefficient

```
rho<-(sigmaa)/( sigmaa +sigmaw)
k<-1+k1/k2
# k is from formula (7.26)
```

```
        lk1<-log(k1)
        lk2<-log(k2)
```

lk is the log of the sample size k

```
        lk<-lk1-lk2
        k1<-2*n*n1
        k2<-(n-1)*log(C)*(n-1)*log(C)
        C<-c1/c2
        c1<-1+n*phi0
        c2<-1+n*phi1
        phi0<-rho/(1-rho)
```

```
# n1 is (Z 1-alpha+Z 1-beta) squared
# phi1 is the target correlation
# phi0 is the transform of rho, the past correlation based on a previous
experiment }
```

list(images = 22, radiologists = 3, n1 = 7.85, n = 3, phi1 = 5.66,

        y = structure(.Data = c(26, 22,10,

NA, 25,25,	NA,NA ,10,	NA,17, 19,
20,17,NA,	30,28,NA,	39, 44, 40,
33,53, 40,	46,47, 27,	26,25 ,26 ,
50, 51,49,	44, 48,47,	77,75, 74,
116, 44,104,	11, 8, 12,	10, 12 , 9,
12, 12, 15,	61, 56 ,53,	21,21,27,
13,13,15,	36,NA,NA,	36,37,36), .Dim = c(22, 3)))

list(theta = 0, tauw = 1, taua = 1)

Results of the analysis show the posterior distribution of the parameters
of the model, including theta which measures the average tumor size as

35.32 mm and the intraclass correlation coefficient of 0.7914. There are two sources of variation, between readers, measured by σ_a^2, and within readers, estimated by σ_w^2. The distribution of the intraclass correlation, given directly by k, is quite unstable, because of extremely small values in the denominator of k, but the distribution of $\log(k)$ is stable and determines the median sample size as $\exp(\log(k)) = \exp(5.147) = 171.9$, with a 95% credible interval of (19.84,76114).

Thus, in order to plan for a future experiment with an intraclass correlation of 0.85, 172 patients should be sufficient, based on the posterior median, or 286, based on the posterior mean of k. Upon substituting $\rho_0 = 0.8$ and $\rho_1 = 0.85$ into the classical formula, 236 patients should be enough for the future study.

If one uses $\rho_1 = 0.90$ as the alternative with the Bayesian approach via BUGS CODE 7.6, the posterior mean (SD) of k is 58.2 (4.72), a median of 40.76, and a 95% credible interval of (10.16,4117), which is a significant reduction in the sample size estimates. Recall that the Kundra study has 22 patients, with estimated ρ of 0.79 and it is believed the future study will have $\rho = 0.90$. Thus to pay for this improved value, more patients than the 22 of the Kundra study will be required.

Often it is difficult to specify a null and alternative values in a test of hypothesis of ρ, and another criterion is more appropriate for choosing the sample size, therefore, a second version of estimating the sample size of a future study will be based on the width of a $1 - \alpha$ confidence interval for ρ, namely,

$$w = 2Z_{\alpha/2}\sqrt{\text{var}(\hat{\rho})}, \tag{7.27}$$

where

$$\text{var}(\hat{\rho}) = 2(1-\rho)^2[1+(n-1)\rho]^2 / kn(n-1), \tag{7.28}$$

and k is the number of patients and n the number of raters. For a fixed n, the number of patients is found by isolating k in Equation 7.27, which gives

$$k = 8Z_{1-\alpha/2}(1-\rho)^2[1+(n-1)\rho]^2/[w^2n(n-1)], \tag{7.29}$$

where ρ, as it appears in Equation 7.29, is the desired value of the intraclass correlation in a future study. With the Bayesian approach, the posterior distribution of k is determined by the posterior distribution of a past relevant experiment, and the future study will be a replication of the previous one, therefore, the posterior distribution of a hypothetical future study will induce a prior distribution of k.

Marom et al.[14] investigated the interobserver error of three radiologists who are measuring the SUV activity of lung cancer lesions in 20 patients.

SUV values measure the metabolic activity of the lesions and indicate the degree of malignancy. The SUV is the amount of radiation emitted by the radionuclide which has been deposited in the tumor and is detected by a gamma camera, which is a PET imaging device. The intraclass correlation is estimated as 0.9528 and the planned future study is expected to have the same ρ. How many subjects should be included in the future study, when the intraclass correlation is expected to be 0.95? Suppose a width of $w = 0.1$ is desired for the 95% confidence interval for ρ, then using Equation 7.29, the posterior mean (SD) of k is 11.41 (18.33), a median of 8.042, and a 95% credible interval of (1.626,36.76). If one calculates the sample size k in the conventional manner by substituting $\rho = 0.95$ in Equation 7.29, one gets 11.21, which is quite close to the posterior mean of 11.41! With a plot of the posterior density, one may show the extreme right skewed density of k.

Cost considerations are important when planning a future agreement study. In the diagnostic division of a busy health science center, medical procedures are expensive and time is at a premium, and designing such a study is no simple matter. For diagnostic studies one must recruit patients, raters, and schedule the imaging times for the patients and the times at which the images will be reviewed by the readers.

Many other people are involved, including the medical technicians, who are actually performing the imaging, the information technologists, which includes the system and database personnel, and the administrative assistants, who have the responsibility of the scheduling and other clerical duties. First developed by Eliasziw and Donner[15], then followed by Flynn[16], and presented by Shoukri[3], a sample size estimation method that estimates n, the number of raters, and k, the number of patients, in such a way that the variance, var ($\hat{\rho}$), Equation 7.28, of the estimated correlation is minimized, subject to a cost constraint, where the overall cost is

$$C = c_0 + kc_1 + nkc_2 . \tag{7.30}$$

The components of the cost function C are c_0, the overall fixed cost (the expenses that involve all activities of the project that do not involve recruiting the patients and producing the image), c_1, the cost of recruiting a patient, and c_2 the cost of producing the image. Note the sample size k is

$$k = (C - c_0) / (c_1 + nc_2) , \tag{7.31}$$

when taking the cost into consideration and assuming the number of raters is fixed at n.

In diagnostic studies the number of raters n is usually "small", say five or less, and depends on the objectives of the study. Assuming n is known the value of k is determined via Equation 7.13. Assuming n is known, the value of k can be chosen using the posterior distribution as above, either with Equation 7.26, based on testing hypotheses, or by Equation 7.29, based on the width of the confidence interval, then once k chosen , the value can be compared to Equation 7.30, in order to see if the choice is realistic with regard to the cost of the study.

Consider the Maron et al.[14] study, where the sample size is chosen on the basis of the width $w = 0.10$ of 95% confidence interval for the slope of the one-way random model. It was shown that the posterior mean of k is 11.41. Is this is a reasonable choice of the sample size? Suppose the cost of recruiting a patient is $c_1 = \$100$, and the cost of imaging a patient is $c_2 = \$300$, then from Equation 7.31,

$$C - c_0 = 11.41 \ (\$100 + 3 \ (\$300)) = \$11{,}410.$$

This tells the investigator if the sample size of 11.41 patients is reasonable. Can the investigator afford the \$11,410? Of course, it depends on the fixed cost c_0.

7.8 Bayesian Approaches to Sample Size

Bayesian issues of sample size estimation was the focus of a special issue of the Statistician, (46, 2), in 1997. The choice of sample size is introduced by Lindley[17], who takes a formal decision–theoretic approach to the problem, and compares it to other methods presented in the special issue. In the decision–theoretic approach, a utility or loss function must be specified that characterizes the loss, in monetary terms, of making a bad decision, and the optimal decision (choice of sample size) is chosen as the one that minimizes the expectation of the loss function with respect to the posterior distribution of the parameters.

There are less formal Bayesian procedures which place a prior distribution on the parameter space, both under the null and alternative hypotheses, and the null hypothesis is rejected if the posterior probability of the alternative hypothesis is sufficiently large. This approach is demonstrated in A5.3 of Appendix A and is a topic described by Lee[10], where the technique is illustrated with the normal and binomial populations. The sample size can be chosen by trial and error using hypothetical future observations, and for each hypothetical sample size the posterior probability of the alternative hypothesis calculated. The chosen sample size is one where the probability of the alternative is sufficiently high.

A good review of the sample size choice is given in the special issue by Adcock[18], who reviews both Bayesian and non Bayesian approaches. With respect to the Bayesian, the decision–theoretic approach is taken, where a utility function is specified and the ideas demonstrated with sample size selection for the difference in the means of two normal populations. An interesting variation of sample size selection is presented by Joseph[19] who bases the selection on the length of a posterior credible interval.

The method of sample size selection taken for this chapter is a mixture of Bayesian and non Bayesian concepts. For example, when based on testing hypotheses, the significance level and power are taken into account by the classical sample size formula, then the prior distribution of sample size determined. Another hybrid method is based on the length of a confidence interval for the relevant parameter, and the sample size determined by the prior distribution of the sample size. As far as I am aware, Spiegelhalter et al.[1] were among the first to use this method, and is adopted for this chapter because the classical sample size is familiar to the reader and easily revised to include prior information, thus the Bayesian flavor. The sample size formula is a function of the unknown parameters, and estimates of those parameters are substituted into the formula, and those estimates are either from knowledge of past relevant studies or are hypothetical values that might occur in the future study. For the Bayesian, the prior distribution of the parameters induces a prior distribution for the sample size, from which a value is selected.

Exercises

1. Verify Table 7.2 employing BUGS CODE 7.2 and use 80,000 observations generated from the joint posterior distribution, with a burn in of 5000, and a refresh of 100.

2. Refer to Fleiss Levin, and Paik[11] and Equation 7.11 for the sample size for a binomial test and verify Table 7.1. Execute BUGS CODE 7.1 with 50,000 observations generated from the joint posterior distribution, with a burn in of 5000, and a refresh of 100.

3. There are several approaches to selecting the sample size, and the first to be presented was by Donner and Eliaszwi[2] given by Equation 7.15, and the other approach to be considered is by Cantor[4]. His is similar to Donner and Eliaszwi, and is based on the principles of testing hypotheses, where the null hypothesis is

$$H: \kappa \leq \kappa_0$$

and the alternative is

$$A: \kappa > \kappa_0,$$

and the type I and II errors determine the sample size

$$n(\alpha, \beta, \pi, \kappa_0, \kappa_1) = [(Z_\alpha \sqrt{Q_0} + Z_\beta \sqrt{Q_1}) / (\kappa_1 - \kappa_0)]^2.$$

Also,

$$Q = (1 - \theta_2)^{-4} (A1 + A2),$$

$$A1 = \{ \sum_{i-0}^{i=1} \theta_{ii}[(1 - \theta_2) - (\theta_{.i} + \theta_{i.})(1 - \theta_1)]^2,$$

and

$$A2 = (1 - \theta_2)^2 \sum_{i \neq j} \theta_{ij}(\theta_{.i} + \theta_{j.})^2 - (\theta_1\theta_2 - 2\theta_2 + \theta_1)^2.$$

Note that the θ_{ij}; $i, j = 0, 1, 2$ are the probabilities in the 2×2 Table 1.1,

$$\kappa = (\theta_1 - \theta_2)/(1 - \theta_2),$$

$$\theta_1 = \theta_{00} + \theta_{11},$$

and

$$\theta_2 = \theta_0\theta_0 + \theta_1\theta_1.$$

Kappa is expressed in terms of θ_1 and θ_2, where the former is the probability of agreement and the latter the probability of agreement, assuming independence between the raters. The Bayesian interpretation of the Cantor sample size (Equation 7.17) is as a parameter depending on the value of kappa, κ_0 under the null hypothesis, a value of kappa, κ_1, under the alternative, and the probabilities of Table 1.1.

As an example, the CT training example of the radiologist will be used, where the prior distribution of the sample size is induced from two sources: (a) the prior distribution of κ_0 and the cell probabilities of Q_0 estimated from Table 7.3, and (b) the prior distribution of κ_1 and the cell probabilities of Q_1, estimated from Table 7.4.

Two sources of information are available for the prior distribution of the sample size, namely, Table 7.3, the study in the early phase of training for the radiologist, and Table 7.4, the hypothetical outcomes predicted by a faculty committee of radiologists.

In the first case, the radiology student was in the initial phase of CT training for non-small-cell lung cancer, and the estimated kappa was 0.1645; in the latter study, the outcomes reflect the improvement in the diagnostic skills of the student as determined by the supervising faculty.

(a) From the above information, revise and ammend BUGS CODE 7.3 and estimate the sample size with the Cantor sample size formula using 70,000 observations generated from the joint posterior distribution, with a burn in of 5000 and a refresh of 100.

(b) Present a plot of the posterior density of the Cantor sample size.

4. Based on Exercise 3, compare the posterior distributions of the sample size given by Donner and Eliasziw (Equation 7.13) and that given by Cantor Exercise 3 and explain why they are different. In your comparison, use the same sample size for both methods and present the BUGS code you executed.

5. Consider a hypothesis testing approach to choosing the sample size for an agreement study based on simple linear regression, where the scores of one reader are regressed on those of another reader, and in particular, consider a test of the intercept , where the null hypothesis is H: $\beta_1 = 0$ versus the alternative $\beta_1 \neq 0$ with a type I error of 0.05. Develop a Bayesian test that has an approximate power of 0.80 when the alternative value of $\beta_1 = 0.3$. What is the prior distribution for parameters of the model?

6. (a) Duplicate the results of Table 7.8 using BUGS CODE 7.4. (b) Using the cell frequencies 22, 4, 38, 2, 6, 6, 1, 16, 164 for the y vector in the list statement of BUGS CODE 7.4, find a 95% credible interval for b[6], the effect of the paranoid scores on the logits of the cell frequencies, (c) What is width of the 95% credible interval and how does it compare to the width of b[6] in Table 7.8?

7. Refer to Table 7.8, which gives the posterior distribution of the coefficients of the logistic model based on the 129 observations with cell frequencies y = c(11,2,19,1,3,3,0,8,82) and determine the 95% credible interval for b[2] − b[4]. Using the cell frequencies 22, 4, 38, 2, 6, 6, 1, 16, 164, and BUGS CODE 7.4, find the posterior distribution of b[2] − b[4] and the corresponding 95% credible interval. Compare the widths of the 95% credible interval for b[2] − b[4] for the two scenarios. Execute the program with a burn in of 5000, a refresh of 100, and generate 55,000 observations from the posterior distribution. For the latter scenario with 259 patients, what does the 95% credible interval for b[2] − b[4] tell us about the way the contrast of score 1 minus 3 for rater 1 compares to the same contrast for rater 2?

8. Consider the simple linear logistic model

$$\text{logit}(\phi_i) = b[1] + b[2]\,X_2[i],$$

where $i = 1, 2, \ldots, k$. and ϕ_i is the probability of the i-th cell.

From the conventional viewpoint, one may show the dependence of the weighted least squares estimators of the coefficients and their standard errors on the cell frequencies and the total sample size n.

Suppose n subjects are selected at random and the number who respond in the i-th cell is designated as n_i and the number who do not respond is $n - n_i$, where $\sum_{i=1}^{i=k} n_i = n$.

Based on Grizzle, Starmer, and Koch[14], Woolson[13] presents the following. The weighted least squares estimator of the slope $b[2]$ is

$$\hat{b}[2] = \sum_{i=1}^{i=k} \delta_i(X_2[i] - \bar{X})(Y[i] - \bar{Y}) / \sum_{i=1}^{i=k} \delta_i(X_2[i] - \bar{X})^2,$$

where

$$Y[i] = \log(n_i/(n - n_i)), \ i = 1, 2, \dots, k.$$

Also, the estimator of the standard error of $\hat{b}[2]$ is

$$\sigma^2_{b[2]} = \sum_{i=1}^{i=k} \delta_i(X_2[i] - \bar{X})^2,$$

and the i-th weight is

$$\delta_i = [n_i^{-1} + (n - n_i)^{-1}]^{-1}.$$

Thus, an approximate $1 - \alpha$ confidence interval for $b[2]$ is

$$\hat{b}[2] \pm Z_{1-\alpha/2}\sigma^2_{b[2]}$$

and one may base the sample size selection on the width $w = Z_{1-\alpha/2} \times \sigma^2_{b[2]}$ of the confidence interval.

(a) Refer to BUGS CODE 7.4 and using the column vector X6 (the vector of paranoid scores) as the dependent variable X2 (as the vector corresponding to $b[2]$ in the model) and the vector of cell frequencies $Y = (22, 4, 38, 2, 6, 6, 1, 16, 164)$ as the dependent variable, what is the width of the confidence interval for the slope?

(b) Describe a Bayesian approach to choosing the sample size based on the formula for the width of the confidence interval based of the slope $b[2]$? By trial and error with various choices of hypothetical cell frequencies, the width of the confidence for the slope can be estimated.

9. Refer to Table 7.9, BUGS CODE 7.6, and Equation 7.26 for the sample size k for testing hypotheses about the intraclass correlation ρ. Table 7.9 presents the posterior analysis for the sample size of a future study when $\alpha = 0.05$, power $= 0.80$, and the alternative value of ρ is 0.80. The sample size formula for k appearing in BUGS CODE 7.6 depends on the posterior distribution of ρ of the Kundra study.

Revise the code using the alternative $\rho = 0.95$, and find the posterior distribution of k. Determine the posterior mean, standard deviation, median, and 95% credible interval for k, based on the code for lk (the log of k), and generate 65,000 observations from the posterior distribution, with a refresh of 100, and a burn in of 5000.

10. Refer to the sample size Equation 7.31, which is based on cost consideration. Shoukri[3] presents a way to choose n, the number of raters, and k, the number of patients, so that the variance (Equation 7.28) of the estimated interclass correlation is minimized, subject to the cost constraint (Equation 7.30). Develop a Bayesian approach of determining the posterior distribution of the sample size which is founded on the Shoukri presentation.

References

1. Spiegelhalter, D. J., K. R. Abrams, and J. P. Myles. 2004. *Bayesian approaches to clinical trials and health care evaluation.* Chichester, New York: John Wiley and Sons.
2. Donner, A., and M. Eliasziw. 1987. Sample size requirements for reliability studies. *Stat. Med.* 6, 441–48.
3. Shoukri, M. M. 2004. *Measures of interobserver agreement.* Boca Raton, London, New York: Chapman and Hall/CRC.
4. Cantor, A. B. 1996. Sample size calculations for Cohen's Kappa. *Psych. Meth.* 1, 150.
5. Von Eye, A., and Mun, E. Y. 2005. *Analyzing rater agreement, manifest variable methods.* Mahwah, NJ: Lawerence Erlbaum Associates.
6. Erasmus, J. J., G. W. Gladish, L. Broemeling, B. S. Sabloff, M. T. Truong, R. S. Herbst, and R. F. Munden. 2003. Interobserver and intraobserver variability in measurement of non-small-cell carcinoma lung lesions: Implications for assessment of tumor response. *J. Clin. Oncol.* 21 (13), 2574.
7. Kundra, V. 2006. Personal Communication, The University of Texas MD Anderson Cancer Center.
8. Walter, D. S., M. Eliasziw, and A. Donner. 1998. Sample size and optimal design for reliability studies. *Stat. Med.* 17, 101–10.
9. Lehmann, E. I. 1959. *Testing statistical hypotheses.* New York: John Wiley and Sons.
10. Lee, P., M. 1997. *Bayesian statistics, an introduction.* 2nd ed. London: Arnold, a member of the Hodder Headline Group.
11. Fleiss, J. L., B. Levin, and M. C. Paik. 2003. *Statistical methods for rates and proportions.* 3rd ed. New York: John Wiley and Sons.
12. Indurkhya, A., L. H. Zayas, and S. L. Buka. 2004. Sample size estimates for interrater agreement. *Meth. Psychol. Res.* online: www.hhpub.com/journals/methodology/journals.html.
13. Woolson, R. F. 1987. *Statistical methods for the analysis of biomedical data.* New York: John Wiley and Sons.

14. Marom, E. M., R. F. Munden, M. T. Truong, G. W. Gladish, D. A. Podoloff, L. D. Broemeling, J. F. Bruzzi, and H. A. Macapinlac. 2006. Interobserver and intraobserver variability of standardized uptake value measurements in non-small-cell lung cancer. *J. Thorac. Imaging.* 21 (3), 205.
15. Eliasziw, M., and A. Donner. 1987. A cost function approach to the design of reliability studies. *Stat. Med.* 6, 647–55.
16. Flynn, N. T., E. Whitley, and T. Peters. 2002. Recruitment strategy in a cluster randomized study-cost implications. *Stat. Med.* 21, 397–405.
17. Lindley, D. V. 1997. The choice of sample size. *The Statistician* 46 (2), 129–38.
18. Adcock. C. J. 1997. Sample size determination: A review. *The Statistician*, 46, 261–83.
19. Joseph, L., and P. Belisle. 1997. Bayesian sample size determination for normal means and differences between normal means. *The Statistician*, 46 (2), 209–26.

Appendix A: Bayesian Statistics

A1 Introduction

Bayesian methods will be employed to design and analyze agreement studies. Appendix A will introduce the theory that is necessary in order to describe Bayesian inference. Bayes theorem, the foundation of the subject, is first introduced and followed by an explanation of the various components of Bayes theorem: prior information from the sample given by the likelihood function, the posterior distribution which is the basis of all inferential techniques, and lastly the Bayesian predictive distribution. A description of the main three elements of inference, namely, estimation, tests of hypotheses, and forecasting future observations follows.

The remaining sections of the appendix refer to the important standard distributions for Bayesian inference, namely: the Bernoulli, Beta, the multinomial, Dirichlet, normal, gamma, and normal–gamma, multivariate normal, Wishart, normal–Wishart, and the multivariate t-distributions. As will be seen, the relevance of these standard distributions to inferential techniques is essential for understanding the analysis of agreement studies.

Of course, inferential procedures can only be applied if there is adequate computing available. If the posterior distribution is known, often analytical methods are quite sufficient to implement Bayesian inferences, and will be demonstrated for the binomial, multinomial, and Poisson populations, and several cases of normal populations. For example, when using a Beta prior distribution for the parameter of a binomial population, the resulting Beta posterior density has well known characteristics, including its moments. In a similar fashion, when sampling from a normal population with unknown mean and precision and with a vague improper prior, the resulting posterior t-distribution for the mean has known moments and percentiles which can be used for inferences.

Posterior inferences by direct sampling methods is easily done if the relevant random number generators are available. On the other hand, if the posterior distribution is quite complicated and not recognized as a standard distribution, other techniques are needed. To solve this problem, Monte Carlo Markov Chain (MCMC) techniques have been developing for the last 25 years and have been a major success in providing Bayesian inferences for quite complicated problems. This has been a great achievement in the field and will be described in later sections.

Minitab, S-Plus, and WinBUGS are packages that provide random number generators for direct sampling from the posterior distribution for many standard distributions, such as, binomial, gamma, Beta, and t-distributions. On occasion these will be used, however my preference is WinBUGS, because it has been well-accepted by other Bayesians. This is also true for indirect sampling, where WinBUGS, is a good package and is the software of choice for the book, and is introduced in Appendix B. Many institutions provide special purpose software for specific Bayesian routines. For example, at MD Anderson Cancer Center, where Bayesian applications are routine, several special purpose programs are available for designing (including sample size justification) and analyzing clinical trials, and will be described. The theoretical foundation for MCMC is introduced in the following sections.

Inferences for agreement experiments consist of testing hypotheses about unknown population parameters, estimation of those parameters, and forecasting future observations. When a sharp null hypothesis is involved, special care is taken in specifying the prior distribution for the parameters. A formula for the posterior probability of the null hypothesis is derived, via Bayes theorem, and illustrated for Bernoulli, Poisson, and normal populations. If the main focus is estimation of parameters, the posterior distribution is determined, and the mean, median, standard deviation, and credible intervals found, either analytically or by computation with WinBUGS. For example, when sampling from a normal population with unknown parameters and using a conjugate prior density, the posterior distribution of the mean is a t and will be derived algebraically. On the other hand in agreement studies, the experimental results are portrayed in a 2×2 table, such as Table 1.1, and follow a multinomial distribution, where the consequent posterior distribution is Dirichlet, and posterior inferences are provided both analytically and with WinBUGS. Of course, all analyses should be preceded by checking to determine if the model is appropriate, and this is where the predictive distribution comes into play. By comparing the observed results of the experiment with those predicted, the model assumptions are tested. The most frequent use of the Bayesian predictive distribution is for forecasting future observation in time series studies. Time series are part of many agreement studies and are analyzed with repeated measures type models.

A2 Bayes Theorem

Suppose X is a continuous observable random vector and $\theta \in \Omega \subset R^m$ is an unknown parameter vector, and suppose the conditional density of X given θ is denoted by $f(x/\theta)$. If $x = (x_1, x_2, \ldots, x_n)$ represents a random sample of size n from a population with density $f(x/\theta)$, and $\xi(\theta)$ is the prior density of θ, then Bayes theorem is given by

$$\xi(\theta/x) = c \prod_{i=1}^{i=n} f(x_i/\theta)\,\xi(\theta), \ x_i \in R \text{ and } \theta \in \Omega \qquad (A.1)$$

where the proportionality constant is c, and the term $\prod_{i=1}^{i=n} f(x_i/\theta)$ is called the likelihood function. The density $\xi(\theta)$ is the prior density of θ and represents the knowledge one possesses about the parameter before one observes X. Such prior information is most likely available to the experimenter from other previous related experiments. Note that θ is considered a random variable and that Bayes theorem transforms one's prior knowledge of θ, represented by its prior density, to the posterior density, and that the transformation is the combining of the prior information about θ with the sample information represented by the likelihood function.

"An essay toward solving a problem in the doctrine of chances" by the Reverend Thomas Bayes[1] is the beginning of our subject. He considered a binomial experiment with n trials and assumed the probability θ of success was uniformly distributed (by constructing a billiard table) and presented a way to calculate $\Pr(a \leq \theta \leq b/x = p)$, where x is the number of successes in n independent trials. This was a first in the sense that Bayes was making inferences via $\xi(\theta/x)$, the conditional density of θ given x. Also, by assuming the parameter as uniformly distributed, he was assuming vague prior information for θ. The type of prior, where very little is known about the parameter is called non informative or vague information, and both terms are used.

It can well be argued that Laplace[2] is the greatest Bayesian who made many significant contributions to inverse probability (he did not know of Bayes), beginning in 1774 with "Mémoire sur la probabilité des causes par les évènements', with his own version of Bayes theorem, and over a period of some 40 years culminating in "Theorie analytique des probabilités". See Stigler[3] and Chapters 9–20 of Hald[4] for the history of Laplace's contributions to inverse probability.

It was in modern times that Bayesian statistics began its resurgence with Lhoste[5], Jeffreys[6], Savage[7], and Lindley[8]. According to Broemeling and Broemeling[9], Lhoste was the first to justify non informative priors by invariance principals, a tradition carried on by Jeffreys. Savage's book was a major contribution in that Bayesian inference and decision theory was put on a sound theoretical footing as a consequence of certain axioms of probability and utility, while Lindley's two volumes showed the relevance of Bayesian inference to everyday statistical problems and was quite influential and set the tone and style for later books such as Box and Tiao[10], Zellner[11], and Broemeling[12]. Box and Tiao, and Broemeling were essentially works that presented Bayesian methods for the usual statistical problems of the analysis of variance and regression, while Zellner focused Bayesian methods primarily on certain regression problems in econometrics. During this period, inferential problems were solved analytically or by numerical integration. Models with many parameters (such as hierarchical models with many levels) were difficult to use because at that time numerical integration methods had limited capability in higher dimensions. For a good history of inverse probability see Chapter 3 of Stigler[3] and Hald[4], who present a comprehensive history and are invaluable as a reference. Dale[14] gives a complete and very interesting account of Bayes' life.

The last 20 years is characterized by the rediscovery and development of resampling techniques, where samples are generated from the posterior distribution via MCMC methods, such as Gibbs sampling. Large samples generated from the posterior make it possible to make statistical inferences and to employ multi-level hierarchical models to solve complex, but practical problems. See Leonard and Hsu[15], Gelman et al.[16], Congdon[17,18,19], Carlin and Louis[20], Gilks, Richardson, and Spiegelhalter[21], who demonstrate the utility of MCMC techniques in Bayesian statistics.

A3 Prior Information

Where do we begin with prior information, a crucial component of Bayes theorem rule? Bayes assumed the prior distribution of the parameter is uniform, namely

$$\xi(\theta) = 1, \quad 0 \le \theta \le 1$$

where θ is the common probability of success in n independent trials and

$$f(x/\theta) = \binom{n}{x} \theta^x (1-\theta)^{n-x} \tag{A.2}$$

where x is the number of successes $= 0, 1, 2, \ldots, n$. The distribution of X, the number of successes is binomial and denoted by $X \sim \text{Binomial}(\theta, n)$. The uniform prior was used for many years, however Lhoste[5] proposed a different prior, namely

$$\xi(\theta) = \theta^{-1}(1-\theta)^{-1}, \quad 0 \le \theta \le 1 \tag{A.3}$$

to represent information which is noninformative and is an improper density function. Lhoste based the prior on certain invariance principals, quite similar to Jeffreys[6]. Lhoste also derived a noninformative prior for the standard deviation σ of a normal population with density

$$f(x/\theta, \sigma) = (1/\sqrt{2\pi}\sigma)\exp{-(1/2\sigma)(x-\mu)^2}, \quad \mu \in R \quad \text{and} \quad \sigma > 0 \tag{A.4}$$

He used invariance as follows: he reasoned that the prior density of σ and the prior density of $1/\sigma$ should be the same, which leads to

$$\xi(\sigma) = 1/\sigma \qquad (A.5)$$

Jeffreys' approach is similar in that in developing noninformative priors for binomial and normal populations, but he also developed noninformative priors for multi-parameter models, including the mean and standard deviation for the normal density as

$$\xi(\mu, \sigma) = 1/\sigma, \quad \mu \in R \quad \text{and} \quad \sigma > 0. \qquad (A.6)$$

Non-informative priors where ubiquitous from the 1920s to the 1980s and were included in all the textbooks of that period, for example, see Box and Tiao, Zellner, and Broemeling. Looking back, it is somewhat ironic that non-informative priors were almost always used, even though informative prior information was almost always available. This limited the utility of the Bayesian approach, and people saw very little advantage over the conventional way of doing business. The major strength of the Bayesian way is that it a convenient, practical, and logical method of utilizing informative prior information. Surely the investigator knows informative prior information from previous related studies.

How does one express informative information with a prior density? For example, suppose one has informative prior information for the binomial population (Equation A.2). Consider

$$\xi(\theta) = [\Gamma(\alpha + \beta)/\Gamma(\alpha)\Gamma(\beta)]\theta^{\alpha-1}(1-\theta)^{\beta-1}, \quad 0 \le \theta \le 1 \qquad (A.7)$$

as the prior density for θ. The Beta density with parameters α and β has mean $[\alpha/(\alpha + \beta 0]$ and variance $[\alpha\beta/(\alpha + \beta)^2 (\alpha + \beta + 1)]$ and can express informative prior information in many ways. For example, suppose from a previous study with 20 trials, there were six successes and 14 failures, then the probability mass function for the observed number of successes $x = 6$

$$f(6/\theta) = \binom{20}{6}\theta^6(1-\theta)^{14}, \quad 0 \le \theta \le 1 \qquad (A.8)$$

As a function of θ, Equation A.8 is a Beta (7, 15) density and expresses informative prior information, which is combined with Equation A.3, via Bayes theorem, in order to make inferences (estimation, tests of hypotheses, and predictions) about the parameter θ. The Beta distribution is an example of a conjugate density, because the prior and posterior distributions for θ belong to the same parametric family. Thus the likelihood function based on previous sample information can serve as a source of informative prior information. The binomial and Beta distributions occur quite frequently in agreement studies, as for example in Phase II clinical trials, where a team of radiologists estimate the success or failure of a cancer therapy.

Of course, the normal density (Equation A.4) also plays an important role as a population model in agreement studies. For example, as was seen in the previous chapter in estimating the lesion size by several radiologists. The tumor size is considered a continuous measurement. How is informative prior information expressed for the parameters μ and σ? Suppose a previous study has m observations $X = (x_1 x_2, ..., x_m)$, then the density of X given μ and σ is

$$f(x / \mu, \sigma) \propto [\sqrt{m}/\sqrt{2\pi\sigma^2}] \exp{-(m/2\sigma^2)(\bar{x}-\mu)^2 [(2\pi)^{-(n-1)/2}\sigma^{-(n-1)}]}$$

$$\exp{-(1/2\sigma^2)\sum_{i-1}^{i=m}(x_i-\bar{x})^2}. \tag{A.9}$$

This is a conjugate density for the two-parameter normal family and is called the normal-gamma density. Note it is the product of two functions, where the first, as a function of μ and σ, is the conditional density of μ given σ, with mean $\bar{x}$ and variance σ^2/m, while the second is a function of σ only and is an inverse gamma density. Or equivalently, if the normal is parameterized with μ and the precision $\tau = 1/\sigma^2$, the conjugate distribution is as follows: (a) the conditional distribution of μ given τ is normal with mean $\bar{x}$ and precision $m\tau$, and (b) the marginal distribution of τ is gamma with parameters $(m+1)/2$ and $\Sigma_{i=1}^{i=m}(x_i - \bar{x})^2/2 = (m-1)S^2/2$, where S^2 is the sample variance. Thus, if one knows the results of a previous experiment, the likelihood function for μ and τ provides informative prior information for the normal population. Such prior information will be relevant when considering estimation of the interclass correlations for continuous agreement scoring information.

A4 Posterior Information

The preceding section explains how prior information is expressed in an informative or in a non-informative way. Several examples are given and will be revisited as illustrations for the determination of the posterior distribution of the parameters. In the Bayes example, where $X \sim \text{Binomial}(\theta, n)$, a uniform distribution for θ is used. What is the posterior distribution? By Bayes theorem,

$$\xi(\theta/x) \propto \binom{n}{x} \theta^x (1-\theta)^{n-x} \tag{A.10}$$

where x is the observed number of successes in n trials. Of course, this is recognized as a Beta $(x+1, n-x+1)$ distribution, and the posterior mean is $(x+1)/(n+2)$. On the other hand, if the Lhoste prior density (Equation A.3) is used, the posterior distribution of θ is Beta $(x, n-x)$ with mean x/n, the usual estimator of θ. The conjugate prior (Equation A.7) results in a Beta $(x+\alpha, n-x+\beta)$ with mean $(x+\alpha)/(n+\alpha+\beta)$. Suppose the prior is informative with a previous 10 successes in 30 trials, then $\alpha=11$ and $\beta=21$, and the posterior distribution is Beta $(x+11, n-x+21)$ If the current experiment has 40 trials and 15 successes, the posterior distribution is Beta $(26, 46)$ with mean $26/72=0.361$, compared to a prior mean of 0.343.

Consider a random sample $X=(x_1, x_2,..., x_n)$ of size n from a normal $(\mu, 1/\tau)$ population, where $\tau=1/\sigma^2$ is the inverse of the variance, and suppose the prior information is vague and the Jeffreys–Lhoste prior $\xi(\mu,\tau)\propto 1/\tau$ is appropriate, then the posterior density of the parameters is

$$\xi(\mu,\tau/x)\propto\tau^{n/2-1}\exp-(\tau/2)[n(\mu-\bar{x})^2+\sum_{i=1}^{i=n}(x_i-\bar{x})^2] \qquad (A.11)$$

Using the properties of the gamma density, τ is eliminated by integrating the joint density with respect to τ to give

$$\xi(\mu/x)=\{\Gamma(n/2)n^{1/2}/(n-1)^{1/2}S\pi^{1/2}\Gamma((n-10/2)\}$$

$$/[1+n(\mu-\bar{x})^2/(n-1)S^2]^{(n-1+1)/2} \qquad (A.12)$$

which is recognized as a t-distribution with $n-1$ degrees of freedom, location $\bar{x}$ and precision n/S^2. Transforming to $(\mu-\bar{x})\sqrt{n}/S$, the resulting variable has a Student's t-distribution with $n-1$ degrees of freedom. Note the mean of μ is the sample mean, while the variance is $[(n-1)/n(n-3)]S^2, n>3$. Eliminating μ from Equation 2.12 results in the marginal distribution of τ as

$$\xi(\tau/x)\propto\tau^{[(n-1)/2]-1}\exp-\tau(n-1)S^2/2, \quad \tau>0 \qquad (A.13)$$

which is a gamma density with parameters $(n-1)/2$ and $(n-1)S^2/2$. This implies the posterior mean is $1/S^2$ and the posterior variance is $2/(n-1)S^4$.

The Poisson distribution often occurs as a population for a discrete random variable with mass function

$$f(x/\theta)=e^{-\theta}\theta^x/x! \qquad (A.14)$$

where the gamma density

$$\xi(\theta)=[\beta^\alpha/\Gamma(\alpha)]\theta^{\alpha-1}e^{-\theta\beta} \qquad (A.15)$$

is a conjugate distribution that expresses informative prior information. For example, in a previous experiment with m observations, the prior density would be Equation A.12 with the appropriate values of alpha and beta. Based on a random sample of size n, the posterior density is

$$\xi(\theta/x) \propto \theta^{\sum_{i=1}^{n} x_i + \alpha - 1} e^{-\theta(n+\beta)} \qquad (A.16)$$

which is identified as a gamma density with parameters $\alpha' = \sum_{i=1}^{i=n} x_i + \alpha$ and $\alpha' = n + \beta$. Remember the posterior mean is α'/β', median $(\alpha' - 1)/\beta'$, and variance $\alpha'/(\beta')^2$.

A5 Inference

A5.1 Introduction

In a statistical context, by inference one usually means estimation of parameters, tests of hypotheses, and prediction of future observations. With the Bayesian approach, all inferences are based on the posterior distribution of the parameters, which in turn is based on the sample, via the likelihood function and the prior distribution. We have seen the role of the prior density and likelihood function in determining the posterior distribution, and presently will focus on the determination of point and interval estimation of the model parameters, and later will emphasize how the posterior distribution determines a test of hypothesis. Lastly, the role of the predictive distribution in testing hypotheses and in goodness of fit will be explained.

When the model has only one parameter, one would estimate that parameter by listing its characteristics, such as the posterior mean, media, and standard deviation and plotting the posterior density. On the other hand, if there are several parameters, one would determine the marginal posterior distribution of the relevant parameters and as above, calculate its characteristics (e.g. mean, median, mode, standard deviation, etc.) and plot the densities. Interval estimates of the parameters are also usually reported and are called credible intervals.

A5.2 Estimation

Suppose we want to estimate θ of the binomial example of the previous section. The posterior distribution is Beta (21, 46) with the following characteristics: mean $= 0.361$, median $= 0.362$, standard deviation $= 0.055$, lower 2½ percent point $= 0.254$, and $0.473 =$ upper 2½ percent point. The mean and median are the same, while the lower and upper 2½ percent points determine a 95% credible interval of (0.254,0.473) for θ.

Inferences for the normal (μ, τ) population are somewhat more demanding, because both parameters are unknown. Assuming the vague prior density $\xi(\mu,\tau) \propto 1/\tau$, the marginal posterior distribution of the population mean μ is a t-distribution with $n - 1$ degrees of freedom, mean $\bar{x}$, and precision n/S^2,

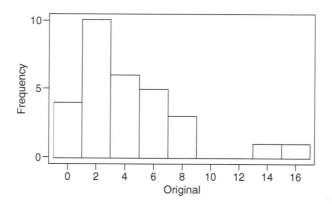

FIGURE A.1
Histogram of original observations.

thus the mean and the median are the same and provide a natural estimator of μ, and because of the symmetry of the t-density, a $(1-\alpha)$ credible interval for μ is $\bar{x} \pm t_{\alpha/2,n-1}S/\sqrt{n}$, where $t_{\alpha/2,\,n-1}$ is the upper 100 $\alpha/2$ percent point of the t-distribution with $n-1$ degrees of freedom. To generate values from the $t(n-1,\bar{x},n/S^2)$ distribution, generate values from Student's t-distribution with $n-1$ degrees of freedom, multiply each by $S/\sqrt{n}$, and then add $\bar{x}$ to each. Suppose $n=30$, $x=$ (7.8902,4.8343,11.0677,8.7969,4.0391,4.0024,6.6494,8.47 88,0.7939,5.0689,6.9175,6.1092,8.2463,10.3179,1.8429,3.0789,2.8470,5.1471,6.3730, 5.2907,1.5024,3.8193,9.9831,6.2756,5.3620,5.3297,9.3105,6.5555,0.8189,0.4713), then $\bar{x} = 5.57$ and $S = 2.92$, see Figure A.1.

Using the same dataset, the following WinBUGS code was used to analyze the problem.

BUGS CODE A.1

Model;

```
{ for( i in 1:30) {x[i]~dnorm(mu,tau) }
mu~dnorm (0.0,.0001)
tau ~dgamma(.0001,.0001)
sigma <- 1/tau }

list(x = c(7.8902,4.8343,11.0677,8.7969,4.0391,4.0024,6.6494,8.4788,0.7939,5.068
9,6.9175,6.1092,8.2463,10.3179,1.8429,3.0789,2.8470,5.1471,6.3730,5.2907,1.5024,
3.8193,9.9831,6.2756,5.3620,5.3297,9.3105,6.5555,0.8189,0.4713))

list( mu=0, tau=1)
```

Note, that a somewhat different prior was employed here, compared to previously, in that μ and τ are independent and assigned proper, but non informative distributions. The corresponding analysis gives Table A.1. Upper and lower refer to the lower and upper 2½ percent points of the posterior

TABLE A.1

Posterior Distribution of μ and $\sigma = 1/\sqrt{\tau}$

Parameter	Mean	SD	MC error	Median	Lower	Upper
μ	5.572	0.5547	0.003566	5.571	4.4790	6.656
σ	9.15	2.570	0.01589	8.733	5.359	15.37

distribution. Note a 95% credible interval for μ is (4.47, 6.65) and the estimation error is 0.003566. See Appendix A for the details on executing the WinBUGS statements above.

The program generated 30,000 samples from the joint posterior distribution of μ and σ using a Gibbs sampling algorithm, and used 29,000 for the posterior moments and graphs, with a refresh of 100.

A5.3 Testing Hypotheses

A5.3.1 Introduction

An important feature of inference is testing hypotheses. Often in agreement studies, the scientific hypothesis of that study can be expressed in statistical terms and a formal test implemented. Suppose $\Omega = \Omega_0 \cup \Omega_1$ is a partition of the parameter space, then the null hypothesis is designated as H: $\theta \in \Omega_0$ and the alternative by A: $\theta \in \Omega_1$, and a test of H versus A consists of rejecting H in favor of A if the observations $x = (x_1, x_2, ..., x_n)$ belong to a critical region C. In the usual approach, the critical region is based on the probabilities of type I errors, namely $\Pr(C/\theta)$ where $\theta \in \Omega_0$ and of type II errors, $1\text{-}\Pr(C/\theta)$, where $\theta \in \Omega_1$. This approach to testing hypothesis was developed formally by Neyman and Pearson and can be found in many of the standard references, such as Lehmann[22]. Lee[23] presents a good elementary introduction to testing and estimation in a Bayesian context.

In the Bayesian approach, the decision to reject the null hypothesis is based on the probability of the alternative hypothesis

$$\varsigma_1 = \Pr(\theta \in \Omega_1 / x) \tag{A.17}$$

and the probability of the null hypothesis

$$\varsigma_0 = \Pr(\theta \in \Omega_0 / x)$$

Thus, the larger ς_1, the more the indication that H is false. If π_0 and π_1 denote the prior probabilities of the null and alternative hypotheses respectively, the Bayes factor is defined as

$$B = (\varsigma_0 / \varsigma_1)/(\pi_0 / \pi_1) \tag{A.18}$$

where the numerator is the posterior odds of the null hypothesis, and the denominator is the prior odds.

Suppose θ is scalar and that H: $\theta < \theta_0$ and A: $\theta > \theta_0$, where $\theta \in \Omega$, and Ω is a subset of the real numbers, then H and A are one-sided hypotheses. This situation can occur when there are so called nuisance parameters in the model. For example, if θ is the mean of normal population, then the standard deviation would be considered a nuisance parameter, since the primary focus is on the mean.

On the other hand, suppose H: $\theta = \theta_0$ and A: $\theta \neq \theta_0$, then the alternative is two-sided, and the null is referred to as a sharp null hypothesis. Note, in this situation θ can be multi dimensional and nuisance parameters can be present. For a sharp null hypothesis, special attention to the prior must be given to the null hypothesis. Let π_0 denote the probability of the null hypothesis, let $\pi_1 = 1 - \pi_0$, and suppose $\pi_1 \xi_1(\theta)$ is the prior density of θ when $\theta \neq \theta_0$. The marginal density of x is

$$f(x) = \pi_0 f(x/\theta_0) + \pi_1 f_1(x) \tag{A.19}$$

where

$$f_1(x) = \int \xi_1(\theta) f(x/\theta) d\theta$$

The posterior probabilities of the null and alternative hypotheses are

$$\varsigma_0 = [\pi_0 f(x/\theta_0)] / [\pi_0 f(x/\theta_0) + \pi_1 f_1(x)] \tag{A.20}$$

and $\varsigma_1 = 1 - \varsigma_0$.

A5.3.2 A Binomial Example

A binomial example is considered in the context of a Phase II clinical trial, where the null and alternative hypotheses are one-sided. Consider a random sample from a Bernoulli population with parameters n and θ, where n is the number of patients and θ is the probability of a response. Let X be the number of responses among n patients, and suppose the null hypothesis is H: $\theta \leq \theta_0$ versus the alternative A: $\theta > \theta_0$. From previous related studies and the experience of the investigators, the prior information for θ is determined to be Beta (a, b), thus the posterior distribution of θ is Beta $(x+a, n-x+b)$, where x is the observed number of responses among n patients. The null hypothesis is rejected in favor of the alternative when

$$\Pr[\theta > \theta_0/x, n] > \gamma \tag{A.21}$$

where γ is usually some 'large' value as 0.90, 0.95, or 0.99. The above equation determines the critical region of the test, thus the power function is

$$g(\theta) = \Pr_{X/\theta}\{\Pr[\theta > \theta_0/x, n] > \gamma\} \qquad (A.22)$$

where the outer probability is with respect to the conditional distribution of X given θ.

A5.3.3 Comparing Two Binomials

Comparing two binomial populations is a common problem in statistics and involves the null hypothesis H: $\theta_1 = \theta_2$ versus the alternative A: $\theta_1 \neq \theta_2$, where θ_1 and θ_2 are parameters from two Bernoulli populations. The two Bernoulli parameters might be the sensitivities of two diagnostic modalities.

Assuming the prior probability of the null hypothesis is π and assigning independent uniform priors for the two Bernoulli parameters, it can be shown that the Bayesian test rejects H in favor of A if the posterior probability P of the alternative hypothesis satisfies

$$P > \gamma \qquad (A.23)$$

where $P = D_2/D$, and $D = D_1 + D_2$. It can be shown that

$$D_1 = \{\pi \binom{n_1}{x_1}\binom{n_2}{x_2}\Gamma(x_1 + x_2 + 1)\Gamma(n_1 + n_2 - x_1 - x_2 + 1)\} \div \Gamma(n_1 + n_2 + 2) \qquad (A.24)$$

where Γ is the gamma function.

$D_2 = (1-\pi)(n_1 + 1)^{-1}(n_2 + 1)^{-1}$, and π is the prior probability of the null hypothesis. X_1 and X_2 are the number of responses from the two binomial populations with parameters (θ_1, n_1) and (θ_2, n_2), respectively. In order to choose sample sizes n_1 and n_2, one must calculate the power function:

$$g(\theta_1, \theta_2) = \Pr_{x_1, x_2/\theta_1/\theta_2}[P > \gamma/x_1, x_2, n_1, n_2], \ (\theta_1, \theta_2) \in (0,1) \times (0,1) \qquad (A.25)$$

where P is given by Equation A.21 and the outer probability is with respect to the conditional distribution of X_1 and X_2, given θ_1 and θ_2.

A5.3.4 A Sharp Null Hypothesis for the Normal Mean

Let $N(\theta, \tau^{-1})$ denote a normal population with mean θ and precision τ, where both are unknown and suppose we want to test the null hypothesis H: $\theta = \theta_0$ versus A: $\theta \neq \theta_0$, based on a random sample of size n with sample mean $\bar{x}$ and variance S^2. Assume the prior probability of the null hypothesis is α and a non informative prior distribution

$$\xi(\theta, \tau) \propto 1/\tau$$

for θ and τ, then the Bayesian test is to reject the null in favor of the alternative if the posterior probability P of the alternative hypothesis satisfies

$$P > \gamma$$

where

$$P = D_2/D \qquad\qquad (A.26)$$

and

$$D = D_1 + D_2$$

It can be shown that

$$D_1 = \{\alpha \Gamma(n/2) 2^{n/2}\}/\{(2\pi)^{n/2}[n(\theta_0 - \bar{x})^2 + (n-1)S^2]^{n/2}\}$$

and

$$D_2 = \{(1-\alpha)\Gamma((n-1)/2) 2^{(n-1)/2}\}/\{(2\pi)^{(n-1)/2}[(n-1)S^2]^{(n-1)/2}\},$$

where α is the prior probability of the null hypothesis.

The above three examples involve standard one and two sample problems and will be applied in the context of diagnostic test accuracy and reader agreement studies and will be visited again when illustrating sample size estimation.

A5.3.5 Comparing Two Normal Populations

How are the means of two normal populations, $n(\theta_1, \tau^{-1})$ and $n(\theta_2, \tau^{-1})$ compared? Suppose the null hypothesis is H: $\theta_1 = \theta_2$ versus the alternative A: $\theta_1 \neq \theta_2$ and the common precision is a nuisance parameter. Suppose the prior probability of the null hypothesis is α and that the prior density of the parameters $\theta = \theta_1 = \theta_2$ and τ under the null hypothesis is $\xi_0(\theta,\tau) = 1/\tau$, $\theta \in R$ and $\tau > 0$.

Also, suppose under the alternative hypothesis, that the prior density of the parameters is

$$\xi_0(\theta_1, \theta_2, \tau) = 1/\tau, \ \theta_1, \theta_2 \in R \text{ and } \tau > 0$$

The null hypothesis is rejected when

$P > \gamma$, where

$$P = D_2/D \qquad\qquad (A.27)$$

$$D = D_1 + D_2$$

$$D_1 = \{\alpha\Gamma[(n_1 + n_2 - 1)/2]\} \div (2\pi)^{(n_1+n_2-1)/2}$$

$$\times [(n_1 - 1)S_1^2 + (n_2 - 1)S_2^2 + n_1 n_2(n_1 + n_2)^{-1}(\bar{x}_1 - \bar{x}_2)^2]^{(n_1+n_2-1)/2}$$

and

$$D_2 = \{(1-\alpha)\Gamma[(n_1 + n_2 - 2)/2]\} \div (2\pi)^{(n_1+n_2-2)/2}[(n_1 - 1)S_1^2 + (n_2 - 1)S_2^2]^{(n_1+n_2-2)/2}.$$

There are elements in the above expression that are familiar as the two-sample t-test. For example in the denominator of D_1, as the sample means become closer together, the denominator becomes smaller and D_1, the posterior probability of the null hypothesis, becomes larger. In addition, as the sample variances become larger, D_1 becomes smaller. Of course, the standard two sample t-test does not depend on a prior probability α of the null hypothesis, as does the Bayesian version.

As an example, consider two samples of size 5 from normal populations with mean 0 for the first and 3 for the second, and both with a variance of 1.

$$x^1 = (6.19739, 0.94196, 1.11206, 3.45737, 1.46134)$$

and

$$x^2 = (0.58176, 3.96194, 3.53676, 3.80614, 0.95053)$$

where $\bar{x}_1 = 2.634$, $\bar{x}_2 = 2.567$, $S_1^2 = 4.981$, and $S_2^2 = 2.742$.

A6 Predictive Inference

A6.1 Introduction

Our primary interest in the predictive distribution is to check for model assumptions. Is the adopted model for an analysis the most appropriate?

What is the predictive distribution of a future set of observations Z? It is the conditional distribution of Z given $X = x$, where x represents the past observations, which when expressed as a density is

$$g(z/x) = \int_\Omega f(z/\theta)\xi(\theta/x)d\theta, \; z \in R^m \tag{A.28}$$

where the integral is with respect to θ, and $f(x/\theta)$ is the density of $X = (x_1, x_2, \ldots, x_n)$, given θ. This assumes that given θ, that Z and X are independent. Thus the predictive density is posterior average of $f(z/\theta)$.

The posterior predictive density will be derived for the binomial and normal populations.

A6.2 The Binomial

Suppose the binomial case is again considered, where the posterior density of the binomial parameter θ is

$$\xi(\theta/x) = [\Gamma(\alpha+\beta)\Gamma(n+1)/\Gamma(\alpha)\Gamma(\beta)\Gamma(x+1)\Gamma(n-x+1)]\theta^{\alpha+x-1}(1-\theta)^{\beta+n-x-1}$$

a beta with parameters $\alpha+x$ and $n-x+\beta$, and x is the sum of the set of n observations. The population mass function of a future observation Z is $f(z/\theta) = \theta^z(1-\theta)^{1-z}$, thus the predictive mass function of Z, called the beta-binomial, is

$$g(z/x) = \Gamma(\alpha+\beta)\Gamma(n+1)\Gamma(\alpha+\sum_{i=1}^{i=n} x_i + z)\Gamma(1+n+\beta-x-z)$$

$$\div \Gamma(\alpha)\Gamma(\beta)\Gamma(n-x+1)\Gamma(x+1)\Gamma(n+1+\alpha+\beta) \tag{A.29}$$

where $z=0, 1$. Note this function does not depend on the unknown parameter, and that the n past observations are known, and that if $\alpha=\beta=1$, one is assuming a uniform prior density for θ.

A6.3 Forecasting from a Normal Population

Moving on to the normal density with both parameters unknown, what is the predictive density of Z, with a non informative prior density

$$\xi(\mu,\tau) = 1/\tau, \ \mu \in R \text{ and } \sigma > 0?$$

The posterior density is

$$\xi(\mu,\tau/x) = [(\tau^{n/2-1}/(2\pi)^{n/2}]\exp-(\tau/2)[n(\mu-\bar{x})^2 + (n-1)S_x^2]$$

where $\bar{x}$ and S_x^2 are the sample mean and variance, based on a random sample of size n, $x = (x_1, x_2, \ldots, x_n)$. Suppose z is a future observation then the predictive density of Z is

$$g(z/x) = \int\int [\tau^{(n)/2-1}/(2\pi)^{(n)/2}]\exp-(\tau/2)[n(\mu-\bar{x})^2 + (n-1)S_x^2] \tag{A.30}$$

where the integration is with respect to $\mu \in R$ and $\tau > 0$. This simplifies to a t density with $d = n - 1$ degrees of freedom, location $\bar{x}$, and precision

$$p = n/(n+1)\, S_x^2 \qquad (A.31)$$

Recall that a t density with d degrees of freedom, location $\bar{x}$, and precision p has density

$$g(t) = \{\Gamma[(d+1)/2]p^{1/2}/\Gamma(d/2)(d\pi)^{1/2}\}\{1 + (t - \bar{x})^2 p/d\}^{-(d+1)/2}$$

where $t \in R$, the mean is $\bar{x}$ and the variance is $d/(d-2)p$.

The predictive distribution can be used as an inferential tool to test hypotheses about future observations, to estimate the mean of future observations, and to find confidence bands for future observations.

A7 Checking Model Assumptions

A7.1 Introduction

It is imperative to check model adequacy in order to choose an appropriate model and to conduct a valid study. The approach taken here is based on many sources, including Gelman et al.[16], Carlin and Louis[20], and Congdon[17]. Our main focus will be on the likelihood function of the posterior distribution, and not the prior distribution, and to this end graphical representations such as histograms, box plots, various probability plots of the original observations will be compared to those of the observations generated from the predictive distribution. In addition to graphical methods, Bayesian versions of overall goodness of fit type operations are taken to check model validity. Methods presented at this juncture are just a small subset of those presented in more advanced works, including Gelman et al., Carlin and Louis, and Congdon. This topic is revisited in Chapter 5 on modeling. When one must choose one among a set of nested log linear models for an appropriate way to express agreement between two or more raters, taking into account other experimental factors such as subject covariates, etc.

Of course, the prior distribution is an important component of the analysis, and if one is not sure of the "true" prior, one should perform a sensitivity analysis to determine the robustness of posterior inferences to various alternative choices of prior information. See Gelman et al. or Carlin and Louis for details of performing a sensitivity study for prior information. Our approach is to either use informative or vague prior distributions, where the former is done when prior relevant experimental evidence determines the prior, or the latter is taken if there is none or very little germane experimental studies. In scientific studies, the most likely scenario is that there are relevant experimental studies providing informative prior information.

A7.2 Sampling from an Exponential, but Assuming a Normal Population

Consider a random sample of size 30 from an exponential distribution with mean 3.

$x =$ (1.9075,0.7683,5.8364,3.0821,0.0276,15.0444,2.3591,14.9290,6.3841,7.6572,5.9606,1.5316,3.1619,1.5236,2.5458,1.6693,4.2076,6.7704,7.0414,1.0895,3.7661,0.0673,1.3952,2.8778,5.8272,1.5335,7.2606,3.1171,4.2783,0.2930).

The sample mean and standard deviation are 4.13 and 3.739, respectively.

Assume the sample is from a normal population with unknown mean and variance, with an improper prior density $\xi(\mu,\tau) = 1/\tau$, $\mu \in R$ and $\tau > 0$, then the posterior predictive density is a univariate t with $n-1 = 29$ degrees of freedom, mean $\bar{x} = 3.744$, standard deviation 3.872 and precision $p = 0.0645$. This is verified from the original observations x and Equation A.31 for the precision. From the predictive distribution, 30 observations are generated:

$z =$ (2.76213,3.46370,2.88747,3.13581,4.50398,5.09963,4.39670,3.24032,3.58791,5.60893,3.76411,3.15034,4.15961,2.83306,3.64620,3.48478,2.24699,2.44810,3.39590,3.56703,4.04226,4.00720,4.33006,3.44320,5.03451,2.07679,2.30578,5.99297,3.88463,2.52737)

which gives a mean of $\bar{z} = 3.634$ and standard deviation $S = 0.975$. The histograms for the original and predicted observations are portrayed in Figure A.1 and Figure A.2, respectively.

The histograms obviously are different, where for the original observations, a right skewness is depicted, however, this is lacking for the histogram

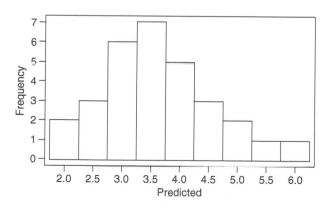

FIGURE A.2
Histogram of Predicted Observations.

of the predicted observations, which is for a *t*-distribution. Although the example seems trivial, it would not be for the first time that exponential observations were analyzed as if they were generated from a normal population! The sample statistics do not detect the discrepancy, because they are very similar; the mean and standard deviation for the exponential are 3.744 and 3.872, respectively, but are 3.688 and 3.795 for the predicted observations. Thus it is important to use graphical techniques to assess model adequacy.

It would be interesting to generate more replicate samples from the predictive distribution in order to see if these conclusions hold firm.

A7.3 Speed of Light Study of Newcomb

Gelman et al.[16] analyze the Newcomb measurements of the speed of light data and the sample mean is 26.21 with a standard deviation of 10.75. The original 66 observations are:

$x = (26,33,24,34,-44,27,16,40,-2,29,22,24,21,25,30,23,29,31,19,24$
$,20,36,32,36,28,25,21,28,29,37,25,28,26,30,32,36,26,30,22,36,23,27$
$,27,28,27,31,27,26,33,26,32,32,24,39,28,24,25,32,25,29,27,28,29,16,23,28)$

Assuming a non informative prior density, what is the posterior predictive density of a future observation? It is a t with 65 degrees of freedom, location 23.34, standard deviation 9.27, and precision 0.008524, and generating 66 observations, the future observations are:

$Z = (13.4959,20.6623,18.9337,15.8691,24.1766,-2.1886,9.6549,10.8169,5.3368,28.4$
$309,27.0961,33.0761,9.3243,13.1424,29.4281,36.0933,24.4052,22.1943,17.5710,34.16$
$64,26.4039,33.4919,20.5668,27.4045,19.1334,18.9078,20.1107,28.0435,32.1448,20.35$
$04,37.3514,27.4511,29.5548,29.0943,15.3733,32.0988,9.9439,24.4920,24.4597,16.947$
$7,20.5937,29.7578,26.3714,14.0564,24.3357,22.1057,18.5147,27.6458,33.4705,17.6636$
$,46.4281,37.5776,20.9465,24.5106,20.9145,35.4306,13.6199,25.6976,13.7429,40.5979,$
$19.4294,12.9588,24.6378,19.1301,47.9036,18.1050).$

Note that the predicted observations are to four decimal places compared to the original observations that were given to two digit accuracy. I used Minitab to generate the predicted observations from the *t*-distribution with 65 degrees of freedom, then divided each observation by the square root of the precision, and finally added the sample mean to each. The results are observations generated from a t with 65 degrees of freedom, location 23.35 and precision 0.008524.

Notice that the median indicates symmetry, however there are two negative observations, thus how should the original model be modified? The two negative observations were also confirmed by the histogram. Another approach to detect a discrepancy between the original and predicted is to compare the posterior distributions of

$$D[(\mu,\tau)/x] = \sum_{j=1}^{j=66} [x(j) - E(X(j)/\mu,\tau)]^2 / \div \mathrm{var}[X(j)/\mu,\tau]$$

where $E(X(j)/\mu,\tau)=\mu$ and $\mathrm{var}(X(j)/\mu,\tau)=1/\tau$, with the posterior distribution of $D[(\mu,\tau)/z]$, where y and z are the original and predicted set of observations. The BUGS code below, computes the posterior distribution of $D[(\mu,\tau)/x]$ and $D[(\mu,\tau)/z]$ with respect to the posterior distribution of μ and τ. The D are similar to goodness of fit type statistics, but in Bayesian guise. For more about this type of goodness of fit, see Gelman et al.[16]

The number of observations generated is 25,000, using the first 1000 to burn in the process, with a refresh of 100, and the characteristics of the marginal posterior distributions are shown in Table A.2.

BUGS CODE A.2

Model;

```
# The Newcomb speed of light study
{for(j in 1:66){

x[j]~dnorm(mu, tau.c)
d[j]<-(x[j]-mu)*(x[j]-mu)
p[j]<-(z[j]-mu)*(z[j]-mu)

# the x[] vector is the set of original observations
# the z[] vector is a set of predicted observations

mu~dnorm(0.000,.001)}
tau.c~dgamma(0.0001,.0001)

D<-sum(d[])*tau.c
D1<-sum(p[])*tau.c}
```

$$\# \ D = D[(\mu,\tau)/x]$$
$$\# \ D1 = D[(\mu,\tau)/z]$$

The effect of the two negative observations is detected by the posterior means of D and $D1$, because the average standardized distance between the original set and the population mean is greater than the corresponding standardized distance between the predicted observations and the population mean μ, as measured by $D1$.

TABLE A.2

Posterior Analysis for Speed of Light

Parameter	Mean	SD	2½	Median	97½
D	65.97	11.44	45.32	65.25	90.34
D1	54.95	10.5	36.52	54.1	77.7
μ	26.16	1.348	23.52	26.16	28.8
τ	0.008656	0.001512	0.005932	0.0088563	0.01188

If the two negative observations are deleted, the mean of the 64 observations is 27.75 and the standard deviation is 5.083. The predictive distribution is a univariate t, with 63 degrees of freedom, mean 29.253, and precision 0.038089. Another way to look at the discrepancy is to compare the sum of squared deviations between the original and predicted observations which give a value of 12,177 for the original 66, but when the two outliers are deleted, the sum of squared deviations is 3,3152.3, a reduction of 72.7%. Now the question is: is it reasonable to delete the negative observations? By generating many replications of future observations, Carlin et al. perform a more detailed analysis than provided here and the reader should refer to their presentation.

A7.4 A Poisson Population

It is assumed the sample is from a Poisson population, however, actually, it is generated from a uniform discrete population over the integers from 0 to 10. The sample of size 25 is $x = (8,3,8,2,6,1,0,2,4,10,7,9,5,4,8,4,0,9,0,3,7,10,7,5,1)$, with a sample mean of 4.92 and standard deviation 3.278. When the population is Poisson, $P(\theta)$, and an uninformative prior

$$\xi(\theta) = 1/\theta, \theta > 0$$

is appropriate, the posterior density is gamma with parameters alpha $= \sum_{i=1}^{i=25} = 123$ and beta $= n = 25$. Observations z from the predictive distribution are generated by taking a sample θ from the gamma posterior density, then selecting a z from the Poisson distribution $P(\theta)$. This was repeated 25 times to give $z = (2,5,6,2,4,3,5,3,2,3,3,6,7,5,5,3,1,5,7,3,5,3,6,4,5)$, with a sample mean of 4.48 and standard deviation 1.896.

The most obvious difference show a symmetric sample from the discrete uniform population, but on the other hand, box plots of the predicted observations reveal a slight skewness to the right. The largest difference is in the inter quartile ranges being (2,8) for the original observations and (3,5.5) for the predictive sample. Although there are some differences, to declare that the Poisson assumption is not valid, might be premature. One should generate more replicate samples from the predictive distribution to reveal additional information.

A7.5 Measuring Tumor Size

A study of agreement involving the lesion sizes of five radiologists was described in Section 1.3 of Chapter 1, and it was assumed that the observations were normally distributed. Is this assumption valid? A probability plot of 40 lesion sizes of one replication (there were two) of reader 1 would show the normal distribution a reasonable assumption. Is this the implication from the Bayesian predictive density? The original observations are:

$x = (3.5,3.8,2.2,1.5,3.8,3.5,4.2,5.4,7.6,2.8,5.0,2.3,4.4,2.5,5.2,1.7,4.5,4.5,6.0,3.3,7.0,$
$4.0,4.8,5.0,2.7,3.7,1.8,6.3,4.0,1.5,2.2,2.2,2.6,3.7,2.5,4.8,8.0,4.0,3.5,4.8).$

Descriptive statistics are: mean $= 3.920$ cm, standard deviation 1.612 cm, and inter quartile range (2.525, 4.800). The basic statistics for the predicted observations are: mean $= 4.017$ and standard deviation 1.439 cm, with an inter quartile range of (4.069, 4.930) and were based on the future observations

$z = (4.85,4.24,3.32,1.84,5.56,3.40,4.02,1.38,4.21,6.26,0.55,4.56,5.09,4.51,3.28,3.9$
$4,5.05,7.23,4.19,4.85,4.24,2.86,3.98,2.00,2.99,3.50,2.53,1.95,6.07,4.68,5.39,1.89,5.79,$
$5.86,2.85,3.62,4.95,4.46, 4.22,4.33).$

A comparison of the histograms for the original and predicted observations would show some small difference in the two distributions, but the differences would not be striking, however, the predicted observations appear to be somewhat more symmetric about the mean than those of the original observations. In addition, the corresponding sample means, standard deviations, modes, and inter quartile ranges are quite alike for the two samples. There is nothing that stands out that implies questioning the validity of the normality assumption. Of course, additional replicates from the posterior predictive density should be computed for additional information about discrepancies between the two.

A7.6 Testing the Multinomial Assumption

Shoukri[24] compares two ways of staging prostate cancer by ultrasound and pathology, and the information is portrayed in Table A.3, where 245 patients are in the study. An ultrasound image is read by a team of radiologists, who stage the tumor as advanced or localized, and, in addition, a team of pathologists grade tissue from the lesion as advanced or localized. This is a study of the accuracy of ultrasound, using pathology as a gold standard. That is, one views an accuracy problem as one of agreement between the results of a diagnostic test and the gold standard. Usually, however, test accuracy is measured by such things as sensitivity and ROC areas, but Kappa also gives a measure of test accuracy.

A multinomial model is valid if the 245 patients are selected at random from some well-defined population, if the responses of an individual are independent of the other respondents, and if the probability of an event (advanced or localized) is the same for all individuals. Another sampling

TABLE A.3

Staging of Prostate Cancer with Ultrasound

Ultrasound Stage	Pathological Stage		Total
	Advanced	Localized	
Advanced	45	50	95
Localized	60	90	150
Total	105	140	245

scheme is to sample 95 patients with an advanced stage of prostate cancer, as determined by ultrasound, and record the number of advanced and localized lesions, as determined by pathology, with a similar approach for the 150 patients selected at random from a population with localized disease as determined by ultrasound.

In some studies it is difficult to know if the multinomial population is valid, for example, with so-called chart reviews, where medical records are selected, not necessarily at random, but from some population determined by the eligibility criterion of the study. Thus, in the case of agreement studies, such as above, the best way to determine validity of the multinomial model is to know the details of how the study was designed and conducted. The crucial issue in the multinomial model is independence, that is, given the parameters of the multinomial, the results of one patient are independent of those of another. It is often the case, that the details of the study are not available. The other important aspect of a multinomial population is that the probability of a particular outcome is constant over all patients. One statistical way to check is to look for runs in the sequence, etc.

In Table A.3, one could condition on the row totals, and check if the binomial model is valid for each row.

A8 Sample Size Problems

A8.1 Introduction

When designing an agreement study, one must take into consideration the reliability of the information from the study. The more subjects in the study, the more information is available for estimating the parameters and for testing hypotheses about those parameters. The main focus will be in justifying the sample size so that one may minimize the errors in testing hypotheses.

For example, suppose two imaging modalities (e.g., CT versus MRI) for diagnosing lung cancer are to be compared on the basis of Kappa, then how many patients are sufficient to reject the null hypothesis that the probability of agreement is no more than some given value? Such examples will be introduced in the next section using the one and two-sample binomial studies of the previous section. The main objective of this section is to introduce Bayesian testing principles based on the frequentist concepts of the probability of a type I error and of the probability of rejecting the null hypothesis for a set of alternative values, or power.

The last chapter of the book emphasizes the design of experiment for agreement studies and will encompass sample size requirements for a large variety of designs.

A8.2 The One Sample Binomial

Consider a random sample from a Bernoulli population with parameters n and θ, where n is the number of patients and θ is the probability of a response. Let X be the number of responses among n patients, and suppose the null hypotheses is

$$H: \theta \leq \theta_0$$

versus the alternative

$$A: \theta > \theta_0$$

From previous related studies and the experience of the investigators, the prior information for θ is determined to be Beta (a, b), thus the posterior distribution of θ is Beta $(x+a, n-x+b)$. The null hypothesis is rejected in favor of the alternative when

$$Pr[\theta > \theta_0/x, n] > \gamma \qquad (A.32)$$

where γ is usually chosen as some "large" value as 0.90, 0.95, or 0.99. Equation A.32 determines the critical region of the test, thus the power function of the test is

$$g(\theta) = Pr_{X/\theta}\{Pr[\theta > \theta_0/x, n] > \gamma\} \qquad (A.33)$$

where the outer probability is with respect to the conditional distribution of X given θ. The power at a given value of θ is interpreted as a simulation as follows:

(a) select n and θ and set $I=0$,
(b) generate a X~Binomial (θ, n), $\qquad\qquad\qquad\qquad$ (A.34)
(c) generate a θ~Beta $(x+a, n-x+b)$,
(d) if $Pr[\theta > \theta_0/x, n] > \gamma$, let the counter $I=I+1$, otherwise let $I=I$,
(e) repeat (b)–(d) M times, where M is 'large', and
(f) select another θ and repeat (b)–(d), then the power is estimated as I/M.

We consider a case where the historical value for the probability of success is 0.20. The null and alternative hypotheses are given as

$$H: \theta \leq 0.20 \text{ and } A: \theta > 0.20 \qquad (A.35)$$

where θ is the probability of a success. The null hypothesis is rejected if the posterior probability of the alternative hypothesis is greater than the threshold value γ.

When a uniform prior is appropriate, the power curve for the following scenarios is computed using the simulation scenario (Equation A.34) with sample sizes $n=125$, 205, and 500, threshold values $\gamma=0.90$, 0.95, 0.99, $M=1000$, and null value $\theta_0=0.20$. See Table A.4 for the power of the test for various alternative values of θ.

For example, the power of the test at $\theta=0.30$ and $\gamma=0.95$, is 0.841, 0.958, and 0.999 for sample sizes $N=125$, 205, and 500, respectively. The Bayesian test behaves in a reasonable way. For the conventional type I error of 0.05, a sample size of $N=125$ would be sufficient to detect the difference 0.3 versus 0.2 with a power of 0.841. On the other hand, in order to detect the alternative 0.4 with 125 patients, the power is essentially 1. In order to estimate the sample size for scenarios other than those given by the table, one must use the simulation (Equation A.34). When employing a conventional sample size program such as PASS®, the power is 0.80 for detecting a success probability of 0.3, which is comparable to the 0.841 with the Bayesian simulation.

A8.3 One Sample Binomial with Prior Information

Suppose we consider the same problem as above, but where prior information is available from an earlier study, where among 50 trials, there are ten successes. The null and alternative hypotheses are as above, however the null is rejected whenever

$$\Pr[\theta > \phi / x, n] > \gamma \tag{A.36}$$

where θ is independent of $\phi \sim$ Beta (10, 40). This can be considered as a one-sample problem where a future study is to be compared to a historical control. As above, using the simulation rules of Equation A.34, the power function for the critical region (Equation A.36) is computed. See Table A.5 with the same sample sizes and threshold values as in Table A.4. According to the

TABLE A.4

Power Function for H Versus A, $N=125, 205, 500$

	γ		
θ	0.90	0.95	0.99
0	0,0,0	0,0,0	0,0,0
0.1	0,0,0	0,0,0	0,0,0
0.2	0.107,0.099,0.08	0.047,0.051,0.05	0.013,0.013,0.008
0.3	0.897,0.97,1	0.841,0.958,0.999	0.615,0.82,0.996
0.4	1,1,1	1,1,1	0.996,1,1
0.5	1,1,1	1,1,1	1,1,1
0.6	1,1,1	1,1,1	1,1,1
0.7	1,1,1	1,1,1	1,1,1

TABLE A.5

Power for One-Sample Binomial with Prior Information

θ	γ		
	0.90	0.95	0.99
0	0,0,0	0,0,0	0,0,0
0.1	0,0,0	0,0,0	0,0,0
0.2	0.016,0.001,0.000	0.002,0.000,0.000	0.000,0.000,0.000
0.3	0.629,0.712,0.850	0.362,0.374,0.437	0.004,0.026,0.011
0.4	0.996,0.999,1	0.973,0.998,1	0.758,0.865,0.982
0.5	1,1,1	1,1,1	0.999,1,1
0.6	1,1,1	1,1,1	1,1,1
0.7	1,1,1	1,1,1	1,1,1

table, the power of the test is 0.758, 0.865, and 0.982 for $\theta = 0.4$ and sample sizes $N = 125$, 205, and 500, respectively.

We see how important prior information is for testing hypotheses. If the hypothesis is rejected with the critical region

$$\Pr[\theta > 0.2/x, n] > \gamma \tag{A.37}$$

the power (see Table A.4) will be larger than the corresponding power (see Table A.5) determined by the critical region (Equation A.36), because of the additional variability introduced by the historical information contained in ϕ. Thus, larger sample sizes are required with the critical region (Equation A.36) to achieve the same power as with the test given by the rejection region (Equation A.37).

On the other hand, if the prior information is incorporated directly into the likelihood function, the power function is higher for all values of θ, compared to the values in Table A.5, because of the increased sample size. Of course if this is done, one is ignoring the prior variability of the historical control. This puts us into somewhat of a dilemma that is not easily resolved.

A8.4 Comparing Two Binomial Populations

The case of two binomial populations was introduced in a previous section, where Equation A.25 determines the power function for testing the null hypothesis

$$H: \theta_1 = \theta_2$$

versus the alternative

$$A: \theta_1 \neq \theta_2$$

TABLE A.6

Power of Two-Sample Binomial

θ_1	θ_2									
	0.1	0.2	0.3	0.4	0.5	0.6	0.7	0.8	0.9	1
0.1	0.004	0.032	0.135	0.360	0.621	0.842	0.958	0.992	1	1
0.2	0.031	0.011	0.028	0.106	0.281	0.536	0.744	0.913	0.997	1
0.3	0.171	0.028	0.006	0.029	0.107	0.252	0.487	0.767	0.961	1
0.4	0.368	0.098	0.025	0.013	0.028	0.075	0.244	0.542	0.847	0.999
0.5	0.619	0.289	0.100	0.022	0.007	0.017	0.108	0.291	0.640	0.981
0.6	0.827	0.527	0.237	0.086	0.035	0.005	0.027	0.116	0.357	0.882
0.7	0.950	0.775	0.464	0.254	0.113	0.037	0.013	0.049	0.171	0.587
0.8	0.996	0.928	0.768	0.491	0.316	0.132	0.028	0.010	0.040	0.205
0.9	1	0.996	0.946	0.840	0.647	0.359	0.156	0.037	0.006	0.014
1	1	1	1	1	0.984	0.873	0.567	0.200	0.017	0.000

The power function of the test at (θ_1, θ_2) is

$$g(\theta_1, \theta_2) = \Pr_{x_1, x_2 / \theta_1, \theta_2} [P > \gamma / x_1, x_2, n_1, n_1], \ (\theta_1, \theta_2) \in (0,1) \times (0,1) \qquad \text{(A.38)}$$

where P is given by Equation A.23.

Suppose $n_1 = 20 = n_2$ are the sample sizes of the two groups, and that the prior probability of the null hypotheses is 0.5. The power at each point (θ_1, θ_2) is calculated via simulation, similar to that given by Equation A.34, with $\gamma = 0.90$, and the values are given in Table A.6. When the power is calculated with Pass® for the two-sample, two-tailed binomial test with alpha $= 0.013$, sample sizes $n_1 = 20 = n_2$, and $(\theta_1, \theta_2) = (0.3, 0.9)$, the power is 0.922, which is less than the power of the Bayesian test. Adjustments in γ of the Bayesian test would give a power equal to that of the conventional test. For the θ_1 and θ_2 values considered, the maximum type I error for the Bayesian test is approximately 0.013.

This type of test for comparing two binomial populations is applicable for comparing the agreement between two raters. This introduction serves as the foundation for the chapter on designing studies for agreement.

A9 Computing

A9.1 Introduction

This section introduces the computing algorithms and software that will be used for the Bayesian analysis of problems encountered in agreement investigations. In the previous sections of the chapter, direct methods (non-iterative) of computing the characteristics of the posterior distribution were demonstrated with some standard one sample and two sample problems. An example of this is the posterior analysis of a normal population, where the posterior distribution of the mean and variance is generated from its posterior distribution by the t-distribution random number generator in Minitab. In addition to some direct methods, iterative algorithms are briefly explained.

MCMC methods (an iterative procedure) of generating samples from the posterior distribution is introduced, where the Metropolis-Hasting algorithm and Gibb sampling are explained and illustrated with many examples. WinBUGS uses MCMC methods such as the Metropolis-Hasting and Gibbs sampling techniques, and many examples of a Bayesian analysis are given. An analysis consists of graphical displays of various plots of the posterior density of the parameters, by portraying the posterior analysis with tables that list the posterior mean, standard deviation, median, and lower and upper 2½ percentiles, and of other graphics that monitor the convergence of the generated observations.

A9.2 Direct Methods

To illustrate the direct method for the Bayesian analysis of a problem, the two raters are compared (Table A.7). Shoukri[24] presents the familiar 2×2 table which classifies subjects by the ratings of both raters, $X=1$ or $X=2$, by rater 1, and $Y=1$ or $Y=2$ by rater 2. Thus each rater assigns a 1 or 2 to each of n subjects, and the subjects are selected at random from some well defined population.

Suppose the subjects are classified by each rater, then let the corresponding joint probabilities of the two raters be denoted by θ_{ij} in Table A.8, where $i, j = 1$ or 2.

The probability that a subject will be assigned a 1 by both raters is $P[X=1$ and $Y=1)=\theta_{11}$, the probability that a subject will receive a score of 1 by the first rater and 2 by the second is θ_{12}, etc. The probability a subject will receive a score of 1 by the second rater is $\theta_{\cdot 1}$. In order to perform a Bayesian analysis of agreement, a prior probability density must be assigned to the parameters, which when combined with the likelihood function

TABLE A.7

Distribution of Patients for Agreement

Rater 1	Rater 2		Row Total
	$Y=1$	$Y=2$	
$X=1$	n_{11}	n_{12}	$n_{1\cdot}$
$X=2$	n_{21}	n_{22}	$n_{2\cdot}$
Column total	n_{21}	$n_{\cdot 2}$	n

TABLE A.8

Joint Probabilities of Agreement

		Rater 2	
Rater 1	$Y=1$	$Y=2$	Row Total
$X=1$	n_{11}, θ_{11}	n_{12}, θ_{12}	$n_1, \theta_1.$
$X=2$	n_{21}, θ_{21}	n_{22}, θ_{22}	$n_2, \theta_2.$
Column total	$n_{.1}, \theta_{.1}$	$n_{.2}, \theta_{.2}$	n

TABLE A.9

Results of Two Raters

		Rater 2	
Rater 1	$Y=1$	$Y=2$	Row total
$X=1$	$n_{11}=1$	$n_{12}=9$	$n_1.=10$
$X=2$	$n_{21}=9$	$n_{22}=81$	$n_2.$
Column total	$n_{21}=10$	$n_{.2}=90$	$n=100$

$$L(\theta/n) \propto \theta_{11}^{n_{11}} \theta_{12}^{n_{12}} \theta_{21}^{n_{21}} \theta_{22}^{n_{22}}$$

via Bayes theorem, yields the posterior density of $\theta = (\theta_{11}, \theta_{12}, \theta_{21}, \theta_{22})$, and the likelihood function is based on the joint multinomial distribution of $n = (n_{11}, n_{12}, n_{21}, n_{22})$, the number of patients in each category, given θ. If a uniform prior density is appropriate, the joint density of the parameters is

$$\xi(\theta/n) = \{\Gamma(n_{11} + n_{12} + n_{21} + n_{22})/\Gamma(n_{11})\Gamma(n_{12})\Gamma(n_{21})\Gamma(n_{22})\}\theta_{11}^{n_{11}} \theta_{12}^{n_{12}} \theta_{21}^{n_{21}} \theta_{22}^{n_{22}} \quad (A.39)$$

where $0 \leq_{ij} \leq 1$ and $\sum_{i=1}^{i=2}\sum_{j=1}^{j=2}\theta_{ij} = 1$. Therefore,

$$\theta \sim \text{Dir}(n_{11} + 1, n_{12} + 1, n_{21} + 1, n_{22} + 1)$$

a Dirichlet distribution, and all inferences will be based on it.

Suppose the results for the two raters are as in Table A.9 with the objective of estimating Kappa, a measure of agreement between the two raters:

$$\text{Kappa} = [(\theta_{11} + \theta_{22}) - (\theta_1. \theta_{.1} + \theta_2. \theta_{.2})]/[1 - (\theta_1. \theta_{.1} + \theta_2. \theta_{.2})]$$

The numerator of Kappa is the difference in two terms: the first the sum of the diagonal elements, the probability of agreement, and the second, which is the probability of agreement if the raters are independent in their assignment of rating to subjects. This second term gives the probability of agreement that will occur by chance, thus Kappa is a chance corrected measure of agreement. Note that Kappa varies over $(-\infty, 1]$ and is sometimes given the interpretation in Table A.10. See Kundel and Ploansky[25] for a discussion of Table A.10.

Minitab generated 1000 samples from the joint posterior distribution of the parameters, which is Dirichlet (2,10,10,82), assuming a uniform prior distribution. Using Minitab, this is done by first generating four independent columns with gamma variates with first shape parameters (2,10,10,82) and a common second shape parameter 2. The four columns are then divided by the total of the four columns, resulting in random samples from the appropriate Dirichlet distribution. From the four columns of Dirichlet variates, the Kappa value is computed, giving 1000 samples from the posterior distribution of that parameter. The posterior analysis is shown in Table A.11.

Note that the posterior mean of the agreement parameter $(\theta_{11} + \theta_{22})$ and of the chance agreement parameter $(\theta_1 \theta_{.1} + \theta_2 \theta_{.2})$ have almost the same posterior mean and standard deviation. The table was constructed so that the readers are providing approximate independent rating. There is a slight right skewness to the posterior distribution of Kappa indicated by a median of 0.043 compared a mean of 0.057, but overall, Kappa implies only slight agreement between the two raters. This is an example of direct computation of a posterior distribution, and is repeated using the following WinBUGS code. The posterior analysis is performed with a sample size of 25,000, using the first 1000 as a burn in and a refresh of 100 observations.

TABLE A.10

Degrees of Agreement for Kappa

K value	Degree of Agreement
<0	Poor
0–0.20	Slight
0.21–0.40	Fair
0.41–0.60	Moderate
0.61–0.80	Substantial
0.80–1.00	Perfect

TABLE A.11

Posterior Distribution for Agreement

Parameter	Mean	Median	SD	95% CI
Agreement	0.806	0.808	0.039	(0.723,0.875)
Chance agreement	0.794	0.794	0.034	(0.724,0.858)
Kappa	0.057	0.043	0.032	$(-0.105, 0.297)$

BUGS CODE A.3

```
model;
{
g11 ~ dgamma(a11,b11);
g12 ~ dgamma(a12,b12);
g21 ~ dgamma(a21,b21);
g22 ~ dgamma(a22,b22);
h<-g11 + g12 + g21 + g22;

theta11 <-g11/h
theta12 <-g12/h
theta21 <-g21/h
theta22 <-g22/h
theta1. <-theta11 + theta12
theta.1 <-theta11 + theta21
theta2. <-theta21 + theta22
theta.2 <-theta12 + theta22

# the theta are from Table A.10

kappa1 <-theta11 + theta11- theta1.*theta.1-theta2.*theta.2
kappa2 <-1-theta1.*theta.1-theta2.*theta.2
kappa <-kappa1/kappa2

agree <-theta11 + theta22

cagree <-theta1.*theta.1 + theta2.*theta.2

phi[1,1] <-(theta11-theta1.*theta.1)* (theta11-theta1.*theta.1)
phi[1,2] <-(theta12-theta1.*theta.2)* (theta12-theta1.*theta.2)
phi[2,1] <-(theta21-theta2.*theta.1)* (theta21-theta2.*theta.1)
phi[2,2] <-(theta22-theta2.*theta.2)* (theta22-theta2.*theta.2)

g<-sum(phi[,])

}
list(a11 = 2, b11 = 2, a12 = 10, b12 = 2, a21 = 10, b21 = 2,

a22 = 82, b22 = 2)

list(g11 = 2, g12 = 2, g21 = 2, g22 = 2)
```

Upon execution of the above statements, the posterior analysis is almost the same as that in Table A.11, however the Minitab computations were based on a sample size of 1000, compared to 24,000 with BUGS. One would expect some difference because of large differences in simulated sample sizes.

The BUGS code above also tests for independence between the two raters by determining the posterior distribution of

$$g(\theta) = \sum_{i,j=1}^{i,j=2} (\theta_{ij} - \theta_{i.}\theta_{.j})^2$$

On can show a 95% credible interval for g is (0.000000367, 0.00446), that is, 95% of the posterior probability is between these two limits, indicating independence between the two raters. Of course, by design the value of g is zero, and the computations agree with this, to the limits of accuracy given by WinBUGS.

A9.3 Monte Carlo Markov Chain

A9.3.1 Introduction

The direct sampling approach described above will be used frequently, however, it has some limitations. For example, when considering a hierarchical model with many levels of parameters, it is more efficient to use a MCMC technique such as Metropolis-Hasting or Gibbs sampling iterative procedure in order to sample from the many posterior distributions. It is very difficult, if not impossible to use non-iterative direct methods for complex models.

A way to draw samples from a target posterior density $\xi(\theta,x)$ is to use Markov chain techniques, where each sample only depends on the last sample drawn. Starting with an approximate target density, the approximations are improved with each step of the sequential procedure. Or in other words, the sequence of samples is converging to samples drawn at random from the target distribution. A random walk from a Markov chain is simulated, where the stationary distribution of the chain is the target density, and the simulated values converge to the stationary distribution or the target density. The main concept in a Markov chain simulation is to devise a Markov process whose stationary distribution is the target density. The simulation must be long enough so that the present samples are close enough to the target. It has been shown that this is possible and that convergence can be accomplished. The general scheme for a Markov chain simulation is to create a sequence θ_t, $t = 1, 2, \ldots$ by beginning at some value θ_0 and at the t-th stage select the present value from a transition function $Q_t(\theta_t/\theta_{t-1})$, where the present value θ_t only depends on the previous one, via the transition function. The value of the starting value θ_0 is usually based on a good approximation to the target density. In order to converge to the target distribution the transition function much be selected with care. The account given here is a summary of Gelman et al.[16], who presents a very complete account of MCMC. Metropolis-Hasting is the general name given to methods of choosing appropriate transition functions, and two special cases of this are the Metropolis algorithm and the other is referred to as Gibbs sampling.

A9.3.2 The Metropolis Algorithm

Suppose the target density is $\xi(\theta,x)$ can be computed, then the Metropolis technique generates a sequence θ_t, $t = 1, 2, \ldots$ with a distribution that converges to a stationary distribution of the chain. Briefly, the steps taken to construct the sequence are:

    a. Draw the initial value θ_0 from some approximation to the target density.

    b. For $t = 1, 2,...$ generate a sample θ_* from the jumping distribution $G_t(\theta_*/\theta_{t-1})$.

    c. Calculate the ratio $s = \xi\,(\theta_*/x) \,/\, \xi(\theta_{t-1}/x)$.

    d. Let $\theta_t = \theta_*$ with probability $\min(s,1)$ or let $\theta_t = \theta_{t-1}$.

To summarize the above, if the jump given by (b) increases the posterior density, let $\theta_t = \theta_*$, on the other hand, if the jump decreases the posterior density, let $\theta_t = \theta_*$ with probability s, otherwise let $\theta_t = \theta_{t-1}$. One must show the sequence generated is a Markov chain with a unique stationary density that converges to the target distribution. For more information see Gelman et al.[16] There is a generalization of the above Metropolis algorithm to the Metropolis-Hasting method.

A9.3.3 Gibbs Sampling

Another MCMC algorithm is Gibbs sampling that is quite useful for multi dimensional problems and is an alternating conditional sampling way to generate samples from the joint posterior distribution. Gibbs sampling can be thought of as a practical way to implement the fact that the joint distribution of two random variables is determined by the two conditional distributions.

The two variable case is first considered by starting with a pair (θ_1,θ_2) of random variables. The Gibbs sampler generates a random sample from the joint distribution of θ_1 and θ_2 by sampling from the conditional distributions of θ_1 given θ_2 and from θ_2 given θ_1. The Gibbs sequence of size k

$$\theta_2^0,\theta_1^0;\theta_2^1,\theta_1^1;\theta_2^2,\theta_1^2;...;\theta_2^k,\theta_1^k \tag{A.40}$$

is generated by first choosing the initial values θ_2^0,θ_1^0 while the remaining are obtained iteratively by alternating values from the two conditional distributions. Under quite general conditions, for large enough k the final two values θ_2^k,θ_1^k are samples from their respective marginal distributions. To generate a random sample of size n from the joint posterior distribution, generate the above Gibbs sequence n times. Having generated values from the marginal distributions with large k and n, the sample mean and variance will converge to the corresponding mean and variance of the posterior distribution of (θ_1, θ_2).

Gibbs sampling is an example of a MCMC because the generated samples are drawn from the limiting distribution of a 2×2 Markov chain. See Casella and George[26] for a proof that the generated values are indeed values from the appropriate marginal distributions. Of course, Gibbs sequences can be generated from the joint distribution of three, four, and more random variables.

The Gibbs sampling scheme is illustrated with two examples: (a) a case of three random variables for the common mean of two normal populations; and (b) an example taken from WinBUGS estimating variance components and the intraclass correlation coefficient for agreement between radiologists who are measuring lesion sizes of lung cancer patients aided by MRI images.

A9.3.4 The Common Mean of Normal Populations

Gregurich and Broemeling[27] describe the various steps in Gibbs sampling to determine the posterior distribution of the parameters in independent normal populations with a common mean. Consider k independent populations normally distributed $N(\theta, \tau_i^{-1})$ for $i = 1, 2, ..., k$ where the parameters $\tau_1, \tau_2, ..., \tau_k$ are k unknown precisions and θ is the unknown common mean value. Suppose the x_{ij} are k independent samples where and $j = 1, 2, ..., n_i$, then the joint density of the sample variables is

$$p(\text{data}/\theta, \tau_1, \tau_2, ... \tau_k) \propto \prod_{i=1}^{k} \tau_i^{n_i/2} \exp -\frac{\tau_i}{2} \left[(n_i - 1)s_i^2 + n_i(\theta - \bar{x}_i)^2 \right]$$

$$\text{where } s_i^2 = \sum_{j=1}^{n_i} (x_{ij} - \bar{x}_i)^2 / (n_i - 1) \quad \text{and} \quad \bar{x}_i = \sum_{j=1}^{n_i} \frac{x_{ij}}{n_i}.$$

According to Bayes' formula, the posterior distribution of the parameters is proportional to the product of the prior distribution of the parameters and the likelihood function of the parameters given the data. The likelihood function for the parameters $\theta, \tau_1, \tau_2, ..., \tau_k$ of the k independent populations is

$$L(\theta, \tau_1, \tau_2, ..., \tau_k / \text{data}) \propto \prod_{i=1}^{k} \tau_i^{n_i/2} \exp -\frac{\tau_i}{2} \left[(n_i - 1)s_i^2 + n_i(\theta - \bar{x}_i)^2 \right].$$

A vague prior density is assumed for the parameters $\theta, \tau_1, \tau_2, ..., \tau_k$, namely

$$p(\theta, \tau_1, \tau_2, ..., \tau_k) \propto \prod_{i=1}^{k} 1/\tau_i$$

for $i = 1, 2, ..., k, \tau_i > 0$ and $\theta \in \Re$.

The joint posterior distribution for $\theta, \tau_1, \tau_2, ..., \tau_k$ is

$$g(\theta, \tau_1, \tau_2, ..., \tau_k) \propto \prod_{i=1}^{k} \tau_i^{\frac{n_i}{2}-1} \exp -\frac{\tau_i}{2} \left[(n_i - 1)s_i^2 + n_i(\theta - \bar{x}_i)^2 \right]$$

Determining the marginal posterior distribution of τ is obtained by integrating the joint density with respect to θ. This is accomplished by using the properties of the normal density. The result is a non-standard distribution that is difficult to manipulate except by numerical integration.

Similarly, the marginal posterior distribution of θ is

$$p(\theta / \text{data}) \propto \prod_{i=1}^{k} \frac{1}{\left[1 + \dfrac{n_i(\theta - \bar{x}_i)^2}{(n_i - 1)s_i^2}\right]^{\frac{n_i}{2}}}, \quad \theta \in \Re \tag{A.41}$$

This distribution is the product of k t-densities called a poly-t-distribution. It is difficult to work with analytically, and numerical integration must be used to determine the normalizing constant, its moments and other characteristics. This is a good case for using MCMC.

In situations where the integration of the joint density is extremely difficult, an algorithm known as the Gibbs sampler has proven to be a good alternative. The Gibbs sampler generates a sample from the joint density by sampling instead from the conditional densities which are often known. According to Casella and George[26], by generating a large enough sample, characteristics of the marginal density and even the density itself can be obtained. Since the conditional posterior distributions are easily obtained, the Gibbs sampling method will be used.

In general, given θ the conditional posterior distributions of the parameters τ_i are independent gamma densities

$$\tau_i / \theta \sim \text{Gamma}\left[\frac{n_i}{2}, \frac{(n_i - 1)s_i^2 + n_i(\theta - \bar{y}_i)^2}{2}\right] \tag{A.42}$$

for $i = 1, 2,..., k$.

The conditional posterior distribution of θ given the k parameters $\tau_1, \tau_2, ..., \tau_k$ is a normal density defined as

$$\theta / \tau_1, \tau_2, ..., \tau_k \sim N\left[\sum_{i=1}^{k}\left(\frac{n_i\tau_i\bar{y}_i}{n_i\tau_i}\right), \left(\sum_{i=1}^{k} n_i\tau_i\right)^{-1}\right] \tag{A.43}$$

The Gibbs sampling approach can best be explained by illustrating the procedure using two normal populations with a common mean θ. Thus, let $y_{ij}, j = 1, 2, ..., n_i$ be a random sample of size n_i from a normal population for $i = 1, 2$.

The likelihood function for $\theta, \tau_1,$ and τ_2 is

$$L(\theta, \tau_1, \tau_2 / \text{data}) \propto \tau_1^{\frac{n_1}{2}} \exp -\frac{\tau_1}{2}\left[(n_1 - 1)s_1^2 + n_1(\theta - \bar{y}_1)^2\right] * \tau_2^{\frac{n_2}{2}}$$

$$\exp -\frac{\tau_2}{2}\left[(n_2 - 1)s_2^2 + n_2(\theta - \bar{y}_2)^2\right]$$

where, $\theta \in \Re$, $\tau_1 > 0$, $\tau_2 > 0$, $s_1^2 = \sum_{j=1}^{n_1}\left(y_{1j} - \bar{y}_1\right)^2 / (n_1 - 1)$, and $s_2^2 = \sum_{j=1}^{n_2}\left(y_{2j} - \bar{y}_2\right)^2 / (n_2 - 1)$.

The prior distribution for the parameters θ, τ_1, and τ_2 is assumed to be a vague prior defined as

$$g(\theta, \tau_1, \tau_2) \propto \frac{1}{\tau_1}\frac{1}{\tau_2}, \quad \tau_i > 0$$

Then combining the above gives the posterior density of the parameters as

$$p(\theta, \tau_1, \tau_2 / \text{data}) \propto \prod_{i=1}^{2} \tau_i^{\frac{n_i-1}{2}} \exp{-\frac{\tau_i}{2}\left[(n_i - 1)s_i^2 + n_i\left(\theta - \bar{y}_i\right)^2\right]}$$

Therefore, the conditional posterior distribution of τ_1 and τ_2 given θ is

$$\tau_i / \theta \sim \text{Gamma}\left[\frac{n_i}{2}, \frac{(n_i - 1)s_i^2 + n_i\left(\theta - \bar{y}_i\right)^2}{2}\right] \tag{A.44}$$

for $i = 1, 2$ and τ_1 / θ and τ_2 / θ are independent.

The conditional posterior distribution of θ given τ_1 and τ_2 is normal. It can be shown that

$$\theta / \tau_1, \tau_2 \sim N\left[\frac{n_1 \tau_1 \bar{y}_1 + n_2 \tau_2 \bar{y}_2}{n_1 \tau_1 + n_2 \tau_2}, (n_1 \tau_1 + n_2 \tau_2)^{-1}\right] \tag{A.45}$$

Given the starting values $\tau_1^{(0)}$, $\tau_2^{(0)}$, and $\theta^{(0)}$ where

$$\tau_1^{(0)} = 1/s_1^2, \quad \tau_2^{(0)} = 1/s_2^2, \text{ and } \theta^{(0)} = \frac{n_1 \bar{y}_1 + n_2 \bar{y}_2}{n_1 + n_2},$$

draw $\theta^{(1)}$ from the normal conditional distribution (Equation A.45) of θ, given $\tau_1 = \tau_1^{(0)}$ and $\tau_2 = \tau_2^{(0)}$. Then draw $\tau_1^{(1)}$ from the conditional gamma distribution (Equation A.44), given $\theta = \theta^{(1)}$. And lastly draw $\tau_2^{(1)}$ from the conditional gamma distribution of τ_2 given $\theta = \theta^{(1)}$. Then generate

$$\theta^{(2)} \sim \theta / \tau_1 = \tau_1^{(1)}, \tau_2 = \tau_2^{(1)}$$

$$\tau_1^{(2)} \sim \tau_1 / \theta = \theta^{(2)}$$

$$\tau_2^{(2)} \sim \tau_2 / \theta = \theta^{(2)}$$

Continue this process until there are t iterations $\left(\theta^{(t)}, \tau_1^{(t)}, \tau_2^{(t)}\right)$. For large t, $\theta^{(t)}$ would be one sample from the marginal distribution of θ, $\tau_1^{(t)}$ from the marginal distribution of τ_1, and $\tau_2^{(t)}$ from the marginal distribution of τ_2.

Independently repeating the above Gibbs process m times produces m 3-tuple parameter values $\left(\theta_j^{(t)}, \tau_{1j}^{(t)}, \tau_{2j}^{(t)}\right)$, $j = 1, 2, \ldots, m$ which represents a random sample of size m from the joint posterior distribution of (θ, τ_1, τ_2) The statistical inferences are drawn from the m sample values generated by the Gibbs sampler.

The statistical inferences can be drawn from the m sample values generated by the Gibbs sampler. The Gibbs sampler will produce three columns of samples, shown in Table A.12. Each row is a sample drawn from the posterior distribution of (θ, τ_1, τ_2). The first column is the sequence of the sample m, the second column is a random sample of size m from the poly-t-distribution of θ, the third and fourth columns are also random samples of size m but from the marginal posterior distributions of τ_1 and τ_2, respectively.

To obtain the characteristics of the marginal posterior distribution of a parameter such as the mean and variance it should be noted that the Gibbs sampler generates a sample of values of a marginal distribution from the conditional distributions without the actual marginal distribution. By simulating a large enough sample the characteristics of the marginal can be calculated. If m is "large" the sample mean of the column is

$$E(\theta/\text{data}) = \sum_{j=1}^{m} \theta_j^t / m = \bar{\theta}$$

and as the mean of the posterior distribution of θ, the sample variance

$$(m-1)^{-1} \sum_{j=1}^{m} \left[\theta_j^t - \bar{\theta}\right]^2$$

is the variance of the posterior distribution of θ.

Additional characteristics such as the median, mode, and the 95% credible region of the posterior distribution of the parameter θ can be calculated from the samples generated by the Gibbs technique. Hypothesis testing can also

TABLE A.12

Random Samples from Posterior Distribution of (θ, τ_1, τ_2)

#	θ	θ_1	θ_2
1	θ_1^t	τ_{11}^t	τ_{21}^t
2	θ_2^t	τ_{12}^t	τ_{22}^t
⋮	⋮	⋮	⋮
m	θ_m^t	τ_{1m}^t	τ_{2m}^t

TABLE A.13

Results from Gibbs Sampler for θ, Box and Tiao 'The Weighted Mean Problem'
(95% Credible Region)

m	Mean	SD	SEM	Lower	Upper
250	108.42	1.04	0.07	106.03	110.65
500	108.31	0.94	0.04	106.35	110.21
750	108.31	0.90	0.03	106.64	110.15
1500	108.36	0.94	0.02	106.51	110.26

be performed. Similar characteristics of the parameters $\tau_1, \tau_2, \ldots, \tau_k$ can be calculated from the samples resulting from the Gibbs method.

The example is from Box and Tiao[10]. It is referred to as 'the weighted mean problem'. It has two sets of normally distributed independent samples with a common mean and different variances. Samples from the posterior distributions were generated from Gibbs sequences using the statistical software Minitab®. The final value of each sequence was used to approximate the marginal posterior distribution of the parameters $\theta, \tau_1, \ldots, \tau_k$. All Gibbs sequences were generated holding the value of $t = 50$. Each example has the results of the parameters using four different Gibbs sampler sizes, where the sample size $m = 250, 500, 750$, and 1500.

The "weighted mean problem" has two sets of normally distributed independent observations with a common mean and different variances. The estimated values of θ determined by the Gibbs sampling method are reported in Table A.13. The mean value of the posterior distribution of θ generated from the 250 Gibbs sequences is 108.42 with 0.07 as the standard error of the mean. The mean value of θ generated from 500 and 750 Gibbs sequences have the same value of 108.31, and the standard errors of the mean equal 0.04 and 0.03, respectively. The mean value of θ generated from 1500 Gibbs sequences is 108.36 and a standard error of the mean of 0.02. Box and Tiao determined the posterior distribution of θ using the t-distribution as an approximation to the target density. They estimated the value of θ to be 108.43. This is close to the value generated using the Gibbs sampler method. The exact posterior distribution of θ is the poly-t-distribution. The effect of m appears to be minimal indicating that 500–750 iterations of the Gibbs sequence are sufficient.

A9.3.5 MCMC Sampling with WinBUGS

The steps involved in Gibbs sampling were described in the previous example. The real power of Gibbs sampling and MCMC in general is when the posterior analysis is based on a hierarchical model that has several levels of parameters. In the example given below, the data is from a study by Kundra[28] with three radiologists, who are measuring the lesion size of 22 lung cancer patients aided by MRI images.

BUGS CODE A.4

```
# WinBUGS progam for Kundra inter reader agreement
# three readers
# 22 images
# response is lesion size in mm

Model;

{

for( i in 1 : images) {
m[i] ~ dnorm(theta, tau.b)
for(j in 1 : radiologists) {
y[i , j] ~ dnorm(m[i], tau.w)

                    }

                    }

sigma2.w <- 1 / tau.w
sigma2.b <- 1 / tau.b

tau.w ~ dgamma(0.001, 0.001)
tau.b ~ dgamma(0.001, 0.001)
theta ~ dnorm(0.0, 1.0E-10)
icc <- sigma2.b/(sigma2.b + sigma2.w)

}

data
list(images = 22, radiologists = 3,
y = structure(.Data = c( 26.00,22.00,10.00,NA, 25.00,25.00,
NA, NA,10.00,NA,17.00,19.00,
20.00,    17.00, NA,30.00,28.00,NA,
39.00,    44.00,    40.00,33.00,        53.00,    40.00,
46.00,    47.00,    27.00,26.00,        25.00,    26.00,
50.00,    51.00,    49.00,44.00,        48.00,    47.00,
77.00,    75.00,    74.00,116.00,44.00,104.00,
11.00,    8.00,     12.00, 10.00,12.00,9.00,
12.00,    12.00,    15.00,61.00,        56.00,    53.00,
21.00,    21.00,    27.00,13.00,        13.00,    15.00,
36.00, NA,NA, 36.00,37.00,36.00), .Dim = c(22, 3)))

Inits
list(theta = 34.48 ,tau.w = 1, tau.b = 1)

# NA denotes a missing observation
```

How well do the readers agree? One approach is to employ a one-way random model and estimate the intraclass correlation coefficient ρ. This coefficient is the common correlation between the three pairs of radiologists. With a burn in of 1000 observations and generating 26,000 observations, with a refresh of 100, the posterior analysis for the mean, variance components, and the intra class correlation is shown in Table A.14.

The efficiency of MCMC is now apparent, because it is not obvious how to determine the various marginal posterior distributions of the parameters, based on the one-way random model.

Let y_{ij} denote the observation in the i-th row and j-th column, where

$$y_{ij} = \theta_i + e_{ij} \tag{A.45}$$

where the $\theta_i \sim nid(\theta, \sigma_b^2)$ and are independent of the $e_{ij} \sim nid(0, \sigma_w^2)$, for $i = 1, 2, ..., r$ and $j = 1, 2, ..., c$. The intra class correlation coefficient is

$$\rho = \sigma_b^2 / (\sigma_b^2 + \sigma_w^2) \tag{A.46}$$

There are two levels of parameters, the first level comprising the θ_i and σ_w^2 (the within variance component), while the second level consists of θ and σ_b^2 (the between variance component). For such hierarchical models, MCMC algorithms are necessary and provide a convenient way to do a Bayesian analysis. Note that the prior distributions for the variance components are vague, where the between and within precision parameters are specified as gammas with both parameters set at 0.001, while the overall mean, theta, is given a normal prior with mean 0 and precision 10^{-10}. The model assumes the observations are normally distributed and that there is a common correlation between all pairs of raters.

For an experienced user of the software, the above actions will be familiar and easy to do, however for the novice it will appear quite strange. It requires much experience, and the beginner can download it at http://www.mrc-bsu.cam.ac.uk/bugs. The download will have a users manual and a very useful help menu, where many examples are provided that will be of invaluable assistance to the new user. Appendix B provides an introduction to WinBUGS.

TABLE A.14

Agreement Among Three Radiologists

Parameter	Mean	SD	2.5%	Median	97.5%
θ	33.22	4.761	23.62	33.25	42.63
σ_b^2	447.5	167.5	220.5	415.4	868.2
σ_w^2	108.4	26.88	67.74	104.5	171.6
ρ	0.7912	0.0706	0.6301	0.8000	0.9031

Exercises

1. For the Beta density with parameters α and β, show that the mean is $[\alpha/(\alpha+\beta)]$ and the variance is $[\alpha\beta/(\alpha+\beta)^2(\alpha+\beta+1)]$.

2. From Equation A.11, show the following. If the normal distribution is parameterized with μ and the precision $\tau=1/\sigma^2$, the conjugate distribution is as follows: (a) the conditional distribution of μ given τ is normal with mean $\bar{x}$ and precision $n\tau$, and (b) the marginal distribution of τ is gamma with parameters $(n-1)/2$ and $\sum_{i=1}^{i=n}(x_i-\bar{x})^2/2=(n-1)S^2/2$, where S^2 is the sample variance.

3. Verify Table A.1.

4. Verify the following statement: To generate values from the $t(n-1,\bar{x},n/S^2)$ distribution, generate values from Student's t-distribution with $n-1$ degrees of freedom and multiply each by $S/\sqrt{n}$ and then add $\bar{x}$ to each.

5. Refer to the binomial testing problem and verify Equation A.22.

6. Derive Equation A.19 and Equation A.20.

7. Refer to Section A.5.3.5 and compute the posterior probability of the alternative hypothesis for comparing two normal populations.

8. Refer to Section A.6.2, the binomial prediction problem and Equation A.29 and verify the probability mass function (Equation A.29) of Z.

9. Verify Equation A.30, the predictive density of a future observation Z from a normal population with both parameters unknown.

10. Verify Table A.2 of the speed of light example.

11. Suppose $x_1, x_2,...,x_n$ are independent and that $x_i \sim$ gamma(α_i, β), and show that $y_i = x_i/(x_1+x_2+...+x_n)$ jointly have a Dirichlet distribution with parameter $(\alpha_1, \alpha_2,...,\alpha_n)$. Describe how this can be used to generate samples from the Dirichlet distribution.

12. Verify the posterior analysis of Table A.11. Execute the corresponding BUGS program, using a 1000 observation burn in, generating 35,000 from the posterior of Kappa and a refresh of 100. Are your results the same? If not, explain why there are differences.

13. Derive the posterior distribution (Equation A.41) of the common mean of several normal populations.

14. Write a BUGS program and compute the posterior distribution of the common mean appearing in Table A.13. Compare the results with those given by Minitab in Table A.13. Use a sample of size 25,000, with 1000 as a burn in and a refresh of 100.

15. Repeat the analysis for the intraclass correlation ρ, but with a simulation sample of size 55,000, using 5000 as a burn in, and 200 as a

refresh, compare your results to those in Table A.14. Explain differences between the two results. See BUGS CODE A.4.

16. Refer to Section A7.2 of the exponential sample, assuming a normal population. Produce ten replications of the histogram of the predictive distribution similar to Figure A.2. Use 10 different predictive samples and check if the normal assumption is justified.

17. Refer to BUGS CODE A.3 of Section A.9.2, and show that the two raters are giving independent scores by finding the posterior distributions of g and p. Verify the credible intervals stated for g and p.

18. Suppose $(X_1, X_2, \ldots, X_k)$ is multinomial with parameters n and $(\theta_1, \theta_2, \ldots, \theta_k)$, where $\sum_{i=1}^{i=k} X_i = n$, $0 < \theta_i < 1$, and $\sum_{i=1}^{i=k} \theta_i = 1$. Show that $E(X_i) = n\theta_i$, $\text{var}(X_i) = n\theta_i(1 - \theta_i)$, and $\text{cov}(X_i, X_j) = -n\theta_i\theta_j$. What is the marginal distribution of θ_i?

19. Suppose $(\theta_1, \theta_2, \ldots, \theta_k)$ is Dirichlet with parameters $(\alpha_1, \alpha_2, \ldots, \alpha_k)$, where $\alpha_l > 0$, $\theta_i > 0$, and $\sum_{i=1}^{i=k} \theta_i = 1$. Find the mean and variance of θ_i and covariance between θ_i and θ_j, $i \neq j$.

20. Show the Dirichlet family is conjugate to the multinomial family.

21. Suppose $(\theta_1, \theta_2, \ldots, \theta_k)$ is Dirichlet with parameters $(\alpha_1, \alpha_2, \ldots, \alpha_k)$. Show the marginal distribution of θ_i is beta and give the parameters of the beta. What is the conditional distribution of θ_i given θ_j?

References

1. Bayes, T. 1764. An essay towards solving a problem in the doctrine of chances. *Phil. Trans. Roy. Soc. London* 53, 370.
2. Laplace, P. S. 1778. Mémoire des probabilités, Mémoires de l'Academie des sciences de Paris, 227.
3. Stigler, M. 1986. *The history of statistics. The measurement of uncertainty before 1900.* Harvard: The Belknap Press of Harvard University Press.
4. Hald, A. 1990. *A history of mathematical statistics from 1750–1930.* London: Wiley Interscience.
5. Lhoste, E. 1923. Le calcul des probabilités appliqué à l'artillerie, lois de probabilité a priori. Revue d'artillerie, Mai, 405–23.
6. Jeffreys, H. 1939. *An introduction to probability.* Oxford: Clarendon Press.
7. Savage, L. J. 1954. *The foundation of statistics.* New York: John Wiley & Sons.
8. Lindley, D. V. 1965. *Introduction to probability and statistics from a Bayesian viewpoint, volumes I and II.* Cambridge: Cambridge University Press.
9. Broemeling, L. D., and A. L. Broemeling. 2003. Studies in the history of probability and statistics XLVIII: The Bayesian contributions of Ernest Lhoste, *Biometrika* 90(3), 728.
10. Box, G. E. P., and G. C. Tiao. 1973. *Bayesian inference in statistical analysis.* Reading, MA: Addison Wesley.

11. Zellner, A. 1971. *An introduction to Bayesian inference in econometrics.* New York: JohnWiley & Sons.
12. Broemeling, L. D. 1985. *The Bayesian analysis of linear models.* New York: Marcel-Dekker.
13. Hald, A. A. 1998. *History of mathematical statistics before 1750.* London: Wiley Interscience.
14. Dale, A. I. 1991. *A history of inverse probability from Thomas Bayes to Karl Pearson.* Berlin: Springer-Verlag.
15. Leonard, T., and J. S. J. Hsu. 1999. *Bayesian methods. An analysis for statisticians and interdisciplinary researchers.* Cambridge UK: Cambridge University Press.
16. Gelman, A., J. B. Carlin, H. S. Stern, and D. B. Rubin. 1997. *Bayesian data analysis.* New York: Chapman & Hall/CRC.
17. Congdon, P. 2001. *Bayesian statistical modeling.* London: John Wiley & Sons.
18. Congdon, P. 2003. *Applied Bayesian modeling.* New York: John Wiley & Sons.
19. Congdon, P. 2005. *Bayesian models for categorical data.* New York: John Wiley & Sons.
20. Carlin, B. P., and T. A. Louis. 1996. *Bayes and empirical bayes for data analysis.* New York: Chapman & Hall.
21. Gilks, W. R., S. Richardson, and D. J. Spiegelhalter. 1996. *Markov Chain Monte Carlo in practice.* Boca Raton: Chapman & Hall/CRC.
22. Lehmann, E. L. 1959. *Testing statistical hypotheses.* New York: John Wiley & Sons.
23. Lee, P. M. 1997. *Bayesian statistics, an introduction.* 2nd ed. London: Arnold, a member of the Hodder Headline Group.
24. Shoukri, M. M. 2004. *Measures of interobserver agreement.* Boca Raton: Chapman &Hall/CRC.
25. Kundel, H. L., and M. Polansky. 2003. Measurements of observer agreement. *Radio* 228–303.
26. Casella, G., and E. I. George. 2004. Explaining the Gibbs sampler. *The American Statistician* 46, 167–74.
27. Gregurich, M. A., and Broemeling, L. D. 1997. A Bayesian analysis for estimating the common mean of independent normal populations using the Gibbs sampler. *Comm. Stat.* 26(1), 25–31.
28. Kundra, V. 2006. Personal communication. The University of Texas Cancer Center.

Appendix B: Introduction to WinBUGS

B1 Introduction

WinBUGS is the statistical package that is used throughout this book and it is important that the novice user be introduced to the fundamentals of working in the language. This is a brief introduction to the package and for the first-time user it will be necessary to gain more knowledge and experience by practicing with the numerous examples provided in the download. WinBUGS is specifically designed for Bayesian analysis and is based on Markov Chain Monte Carlo (MCMC) techniques for simulating samples from the posterior distribution of the parameters of the statistical model. See A9.3 of Appendix A for additional information about MCMC algorithms employed by WinBUGS. It is quite versatile and once the user has gained some experience, there are many rewards.

Once the package has been downloaded, the essential features of the program are described, first by explaining the layout of the BUGS document. The program itself is made up of two parts, one part for the program statements, and the other for the input for the sample data and the initial values for the simulation. Next to be described are the details of executing the program code and what information is needed for the execution. Information needed for the simulation are the sample sizes of the simulated posterior distribution and the number of such observations that will apply to the posterior distribution.

After execution of the program statements, certain characteristics of the posterior distribution of the parameters are computed including the posterior mean, median, credible intervals, and plots of posterior densities of the parameters. In addition, WinBUGS provides information about the posterior distribution of the correlation between any two parameters and information about the simulation. For example, one may view the record of simulated values of each parameter and the estimated error of estimation of the process. These and other activities involving the simulation and interpretation of the output will be explained.

Examples based on agreement studies illustrate the use of WinBUGS, and includes estimation of the Kappa coefficient, and modeling of two observers assigning depression scores to patients. Of course, this is only a brief introduction, but should be sufficient for the beginner to start the adventure of analyzing data. Because the book's examples provide the necessary code for all examples, the program can easily be executed by the user. After the book is completed by the dedicated student, they will have a good understanding of WinBUGS and the Bayesian approach to agreement.

B2 Download

I downloaded the latest version of WinBUGS at:http://www.mrc-bsu.cam.
ac.uk/bugs/winbugs/contents.shtml, which you can install in you program
files. You will be requested to download a decoder which allows the activa-
tion of the full capabilities of the package.

B3 The Essentials

The essential feature of the package is a WinBUGS file or document that contains the program statements and space for input information.

B3.1 Main Body

The main body of the software is a WinBUGS document, which contains the
program statements, the major part of the document, and a list statement
or statements, which include the data values and some initial values for the
MCMC simulation of the posterior distribution. The document is given a
title and saved as a WinBUGS file, which can be accessed as needed. For
example, with an agreement study using the logistic linear model, the program statements determine the model and identifies model parameters to be
estimated.

B3.2 List Statements

List statements allow the program to incorporate certain necessary information that is required for successful implementation. For example, experimental or study information is usually inputed with a list statement. For
example, in a typical agreement study, the cell frequencies of the two raters
are contained in the list statement. As another example, consider an agreement study involving estimation of the sample size, where data from previous related studies are used to induce the posterior distribution of the
sample size; it is those experimental values that are inputed with a list
statement.

B4 Execution

The execution of the code is explained in terms of the following statements
that make up the WinBUGS file or document.

WinBUGS Document

model;

Sample size formulas for m and n

{

th1~dnorm(0, 10)

th2~dnorm(.5,10)

th21 < - th2- th1

tau~dgamma(31,30)

sigma < -1/tau

n < - (2*sigma*2.48*2.48)/(th21*th21)

n1 < - 2*sigma*2.48*2.48

n2 < - th21*th21

m < - log(n1)- log(n2)

thsq < - th21*th21

power < - phi(sqrt((n*th21*th21/2*sigma) + 1.96))

}

list(th2 = .5, th1 = 0, tau = 1)

The objective of this program is to determine the posterior distribution of the sample size for comparing the means θ_1 and θ_2 of two normal populations, based on the familiar formula

$$n = 2\sigma^2 (z_{1-\alpha/2} + z_{1-\beta})^2 / (\theta_1 - \theta_2)^2$$

where σ^2 is the common variance, θ_1 and θ_2 are the means, α is the chance of a Type I error and β the probability of a Type II error.

The program statements are included between the first and last curly brackets { }, and the formula for the sample size is

n < - (2*sigma*2.48*2.48)/(th21*th21),

where sigma is σ^2, $\alpha = 0.05$, $\beta - 0.20$, and th21 $= \theta_2 - \theta_1$.

The prior distributions are as follows: th1~dnorm(0, 10) and th2~dnorm(.5,10) for θ_1 and θ_2, respectively, and tau~dgamma(31,30) for the inverse of σ^2. The prior distributions are based on past information, where θ_1 and θ_2 are given normal distributions with mean 0 and variance 1/10, while the common variance σ^2 is given an inverse gamma with mean 30/(31 − 1) and variance 1/(31 − 2)$^2 \cong 0.001$.

The statement for m, is for the natural log of the sample size and is given by

m < - log(n1)- log(n2),

where n1 and n2 are the logs of the numerator and denominator of n, respectively. The list statement is list(th2 = .5, th1 = 0, tau = 1) and specifies the starting values for the MCMC simulation of the three parameters.

B4.1 The Specification Tool

The tool bar of WinBUGS is labeled as follows, from left to right: file, edit, attributes, tools, info, model, inference, doodle, maps, text, windows, examples, manuals, and help, and I have highlighted the model and inferences labels. When the user clicks on one of the labels, a pop up menu appears. In order to execute the program, the user clicks on model, then clicks on specification, and the specification tool appears (Figure B.1).

The specification tool is used together with the BUGS document as follows: (a) click on the word "model" of the document; (b) click on the check model box of the specification tool; (c) click on the compile box of the specification tool; (d) click on the word "list" of the list statement of the document, and lastly; (e) click on load inits box of the tool. Now close the specification tool and go to the next step below.

B4.2 The Sample Monitor Tool

The sample tool is activated by first clicking on the inference menu of the tool bar, then click on sample, and the sample monitor tool appears as in Figure B.2. Type m in the node box and click on the set box. Type n in the node box and click on set, then type an* in the node box. Type 5000 in the beg box, which means the first 5001 observations generated for the posterior distribution of m and n will not be used when reporting the results of the simulation. The 5000 observations typed in beg are referred to as the 'burn in'.

B4.3 The Update Tool

In order to activate the update tool, click on the **model** menu of the tool bar, then click on updates, which appears in Figure B.3.

Suppose you want to generate 45,000 observations from the posterior distributions of n and m, using the statements which are listed in the document above, then type 45,000 in the updates box, and 100 for refresh. In order to

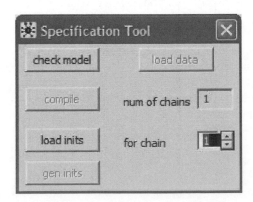

FIGURE B.1
The specification tool.

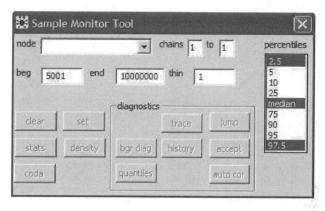

FIGURE B.2
The sample monitor tool.

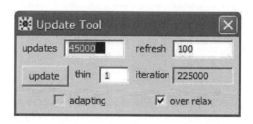

FIGURE B.3
The update tool.

execute the simulation using the program statements in the document click on update of the update tool.

B5 Output

After the 45,000 observations have been generated from the joint posterior distribution of m and n, click on the stats box of the sample monitor tool, see Figure B.2. Certain characteristics of the joint distribution are displayed. Upon clicking on the history box, the values of 40,000 observations from the joint density of m are displayed. The output for the posterior analysis will look like Table B.1.

The first entry 4.341 for m is the posterior mean of m, which is the natural log of the sample size, and the second entry is 2.079, the standard deviation of the posterior distribution of m, the next entry is the 0.01155, which is the MCMC simulation error for m, while 1.815 is the 2½ percentile of the posterior distribution of m. The median of the posterior distribution of m is 3.83, and the 97½ percentile is 9.81. The last two entries for m are 5000, and 40,0001, where the former refers to the first 5000 observations (the 'burn in') that were not used for

TABLE B.1

Posterior Analysis

Parameter	Mean	SD	MC error	2½	Median	97½
m	4.341	2.079	0.0115	1.815	3.83	9.81
n	$1.085*10^6$	$1.217*10^8$	605400	6.14	46.04	18330

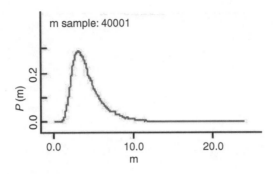

FIGURE B.4
Posterior density of m.

computing the posterior characteristics of m, and the 40,0001 which refers to the number of observations that were actually used for computing the posterior characteristics. Note the huge values for the posterior mean and standard deviation of n, this is because the denominator of n can be quite small, resulting in 'large' values of n, which influences the posterior mean of n. By clicking on the history box of the sample monitor tool, values from the posterior density of m are displayed and one can check the stability of the generated sequence.

The posterior distribution of m is that of the log of the sample size, and is the difference in the logs of the numerator and denominator of n. When the posterior mean, standard deviation, median, and percentiles of m are exponentiated, the posterior characteristics of the sample size are: mean $= \exp(4.341) = 82.2$, SD $= \exp(2.079) = 7.99$, 2½ $= \exp(1.815) = 6.14$, median $= \exp(3.83) = 44.70$, and 97½ $= \exp(9.81) = 18.214$. Thus, because of the extreme right skewness, the median 44.70 is the reasonable choice for the sample size estimate.

Another feature of the output is that the posterior densities of the nodes can be displayed by clicking on the 'density' box of the sample monitor tool. For example, Figure B.4 presents the posterior density of m, which demonstrates extreme right skewness.

B6 Examples

In order to illustrate the use of WinBUGS and to gain additional insight, two examples of agreement studies will be presented. The first has to do with estimating the Kappa coefficient for the agreement of two raters, and the second with using the linear logistic model to compare two raters.

B6.1 Kappa

BUGS DOCUMENT KAPPA

```
model; {
    g[1,1]~dgamma(12,2)
    g[1,2]~dgamma(3,2)
    g[1,3]~dgamma(20,2)
    g[2,1] ~dgamma(2,2)
    g[2,2] ~dgamma(4,2)
    g[2,3]~dgamma(4,2)
    g[3,1]~dgamma(1,2)
    g[3,2]~dgamma(9,2)
    g[3,3]~dgamma(83,2)
    h <- sum(g[,])
    for( i in 1 : 3 ) { for( j in 1 :3 ){ theta[i,j] < - g[i,j]/h }}
    theta1. <- sum(theta[1,])
    theta.1 <- sum(theta[,1])
    theta2. <- sum(theta[2,])
    theta.2 <- sum(theta[,2])
    theta3. <- sum(theta[3,])
    theta.3 <- sum(theta[,3])
    kappa <- (agree- cagree)/(1- cagree)
    agree <- theta[1,1] + theta[2,2] + theta[3,3]
    cagree <- theta1.*theta.1 + theta2.*theta.2 + theta3.*theta.3
    list( g = structur e(.Data = c(2,2,2,2,2,2,2,2,2), .Dim = c(3,3)))
```

The BUGS document Kappa is for an agreement study taken from Chapter 2 where two raters are involved and score 129 patients with three depression values and with cell frequencies: 11,2,19,1,3,3,0,8,82 that appear in nine cells. The main focus is on three parameters or nodes: kappa, agree, and cagree, where a Bayesian approach is taken, with a uniform prior placed on the cell probabilities, and the study information from the table is placed directly into the g[i,j] parameters above. The Dirichlet posterior for the cell probabilities theta[i,j] is generated by a transformation

```
for( i in 1 : 3 ) {for( j in 1 :3 ){ theta[i,j] < - g[i,j]/h
```

from the gamma variables g[i,j] to the Dirichlet variables, theta[i,j].

Kappa is coded as

kappa <- (agree- cagree)/(1- cagree)

and depends on two nodes agree and cagree, where the probability of agreement is coded as

agree <- theta[1,1] + theta[2,2] + theta[3,3]

and is the sum of the three diagonal cell probabilities, and

cagree <- theta1.*theta.1 + theta2.*theta.2 + theta3.*theta.3

is the probability of agreement, assuming independence between the two raters.

To begin the analysis, click on the model menu of the tool bar and pull down the specification tool: (a) click on the word 'model' of the BUGS document; (b) click on the check model box of the specification tool, see Figure B.1; (c) activate the word 'list' of the first list statement; and (d) click on the compile box of the specification tool; and lastly (e) click on the word 'list' of the second list statement of the document. You are now ready to execute the analysis.

In order to continue the program, pull down from the inference menu, the sample monitor tool (see Figure B.2) below and type kappa in the node box, followed by clicking the set box, then repeat the operation for the nodes agree, and cagree. For the final operation, put an * in the node box, and type 5000 in the beg box for the burn in.

Pull down the update tool (see Figure B.3) from the model menu and type 45,000 in the updates box, put 100 in the refresh box, and click on the update box. The simulation now begins with 45,000 observations generated from the posterior distribution of kappa, agree, and cagree. After the simulation is completed, click on the stats box of the sample monitor tool, and the output from Table B.2 will appear.

Note that the posterior mean (SD) of kappa is .3581(0.072), for agree it is 0.7174(0.0381), and is 0.5594(0.0370) for cagree. In order to see the posterior densities of the three nodes, click on the density box of the sample monitor tool and by clicking on the trace box of the sample monitor tool, one will see the 'last' 200 observations generated from the posterior distribution of kappa.

TABLE B.2

Posterior Analysis for Kappa

Parameter	Mean	SD	MC error	2½	Median	97½
agree	0.7174	0.03815	0.000198	0.6399	0.7184	0.7889
cagree	0.5594	0.0370	0.000192	0.4889	0.55587	0.6333
Kappa	0.3581	0.072	0.000382	0.2189	0.358	0.4994

B6.2 Logistic Linear Model

The example of two raters assigning depression scores (1, 2, or 3) to 129 patients is expanded by a comparison of two psychiatrists with the aid of logistic linear model, which allows one to model the effects of the raters on the logit scale so that the pattern in the way the two assign scores may be studied. Consider the WinBUGS document:

BUGS DOCUMENT LOGISTIC LINEAR

Model;
# von eye and mun example page 33{
for (i in 1:N){y[i]~dbin(p[i],129)}
for(i in 1:N) {logit(p[i])<- b[1]*X1[i] + b[2]*X2[i] + b[3]*X3[i] + b[4]*X4[i] + b[5]*X5[i] }
# b[2] is the effect of the contrast (1-3) for rater A
# b[3] is the effect of the contrast (2-3) for rater A
# b[4] is the effect of the contrast (1-3) for rater B
# b[5] is the effect of the contrast (2-3) for rater B
# prior distribution for the coefficients
for(i in 1:5){ b[i]~dnorm(0.0,.000001)}
d24 <- b[2]-b[4]
d35 <- b[3]-b[5]
}
list(N = 9,y = c(11,2,19,1,3,3,0,8,82),
X1 = c(1,1,1,1,1,1,1,1,1),X2 = c(1,1,1,0,0,0,– 1,– 1,– 1),
X3 = c(0,0,0,1,1,1,– 1,– 1,– 1), X4 = c(1,0,– 1,1,0,– 1,1,0,– 1),
X5 = c(0,1,– 1,0,1,– 1,0,1,– 1))
# X1 is the vector for the constant
# X2 is the contrast between the score 1 and 3 for rater A
# X3 is the contrast between the score 2 and 3 for rater A
# X4 is the contrast between the score 1 and 3 for rater B
# X5 is the contrast between the score 2 and 3 for rater B
list(b = c(0,0,0,0,0))

This example is taken from Chapter 5, where the logistic linear model was introduced as

$$\log\mathrm{it}(\phi_i) = \sum_{j=1}^{j=p} b[j]X_j[i]$$

where the cell probabilities are denoted by ϕ_i, $i = 1, 2, \ldots, 9$, and the $b[j]$ measure the effects of the $p = 5$ factors. The cell frequencies are the values for the dependent variable y and are given in the first list statement of the document as

$$y = c(11,2,19,1,3,3,0,8,82).$$

Model parameters are the scalars $b[j]$, $j = 1, 2, 3, 4, 5$, which are given non informative normal $(0,0.000001)$ prior distributions coded as

for(i in 1:5){ b[i]~dnorm(0.0,.000001)}.

The column vectors of the design matrix are denoted by the $X_j[]$ and are recorded in the first list statement of the document. For example, the second vector is listed as

$$X2 = c(1,1,1,0,0,0,-1,-1,-1),$$

which is the 1 minus 3 contrast (the score 1 minus the score 3) assigned by the first rater, and the corresponding contrast of the second rater is denoted by vector

$$X4 = c(1,0,-1,1,0,-1,1,0,-1),$$

which is also listed in the list statement.

Major emphasis will be placed on the posterior distribution of the rater effects $b[j]$, $j = 2, 3, 4, 5$, and the differences

d24 <- b[2]-b[4]

and

d35 <- b[3]-b[5],

where the first difference is the comparison of the effects of the contrast 1 minus 3 between the two raters. Thus for the analysis, the posterior distributions of the $b[j]$ coefficients and the two differences d24 and d35 will be determined via MCMC simulations.

To begin the analysis, click on the model menu of the tool bar and pull down the specification tool (see Figure B.1): (a) click on the word 'model' of BUGS document; (b) click on the check model box of the specification tool; (c) activate the word 'list' of the first list statement; and (d) click on the compile box of the specification tool; and lastly (e) click on the word 'list' of the second list statement of the document. You are now ready to execute the analysis.

In order to continue the analysis, click on the model menu of the tool bar and pull up the sample monitoring tool (see Figure B.2) and type 'b' in the node box and click on the set box. Repeat this operation for the b24 and b35 nodes, and type * in the node box, followed by typing 5000 in the beg box for 'burn in'.

You are ready to execute the analysis by clicking on the model menu of the tool bar and bringing up the update tool (Figure B.3), and typing 45000 in the updates box, 100 in the refresh box, and finally, clicking on the update box to initiate the posterior analysis.

TABLE B.3

Posterior Analysis Logistic Regression

Parameter	Mean	SD	MC error	2½	Median	97½
$b[1]$	−3.219	0.1907	0.0022	−3.607	−3.213	−2.861
$b[2]$	0.07458	0.1898	0.0024	−0.296	0.0726	0.4514
$b[3]$	−1.662	0.2722	0.0038	−2.232	−1.65	−1.162
$b[4]$	−0.9545	0.2309	0.0025	−1.428	−0.9487	−0.5233
$b[5]$	−0.8656	0.2259	0.0025	−1.319	−0.8613	−0.4368
d24	1.029	0.2937	0.0034	0.4694	1.024	1.62
d35	−0.796	0.3467	0.0042	−1.494	−0.7908	−0.1324

The posterior analysis appears as Table B.3. Note that the posterior mean of $b[2]$ is 0.0745, which is the effect of the first rater of the 1 minus three contrast, while the same effect of rater 2 has a mean of −0.9545, but the major focus of the analysis is on d34, which has a posterior mean of 1.029 and a 95% credible interval of (0.4694, 1.62), implying a difference in the effects of the two raters. By clicking on the density box of the sample monitoring tool (see Figure B.2), plots of the posterior densities of all nodes will be generated.

B7 Summary

Appendix B introduces the reader to WinBUGS and the novice should be able to begin Chapter 1 and learn the main topic, namely, how a Bayesian analyzes agreement studies. To gain additional experience, refer to the manual and to the numerous examples that come with the downloaded version of the package. Practice, practice, and more practice is the key to understanding the importance of analyzing actual data with a Bayesian approach. There are many references about WinBUGS, including Broemeling[1], Woodworth[2], and the WinBUGS website[3], which in turn refer to many books and other resources about the package.

References

1. Broemeling, L. D. 1996. *Bayesian biostatistics and diagnostic medicine*. Boca Raton: Chapman & Hall/CRC, Taylor & Francis Group.
2. Woodworth, G. G. 2004. *Biostatistics, a Bayesian introduction*. Hoboken, NJ: John Wiley & Sons.
3. The BUGS Project Resource at http://www.mrc-bsu.cam.ac.uk/bugs/winbugs/contents.shtml. March 2008.

Index

A

Agreements
  Bayesian approach
    multivariate techniques, 220–22
  BUGS CODE 6.1, 198
  BUGS CODE 6.2, 201–202
  chance corrected measures of, 41–42
  computing correlation between
    pairs of, 212–213
  and correlated observations
    correlation model common,
      generalization of, 119–123
    Oden Pooled Kappa and
      Schouten weighted Kappa,
      118–119
    paired observations, 112–118
  data sources of, 22–23
  fixed and mixed effects of, 216–220
  forgery cases and, 21–22
  G coefficient and, 124–125
  graphical techniques, 15
  homogeneity and, 125–130
  information sources of, 22–23
  intraclass correlation coefficient,
    204–209
  Lennox Lewis–Evander Holyfield
    bout, 7
    Bayesian methods and latent
      variable method, 9
    score cards for, 8
  logistic regression and, 130
    Bayesian posterior analysis for,
      132
    BUGS CODE 4.5, 133–135
    BUGS CODE 4.6, 136–137
    posterior analysis for, 138
    Shoukri example, 131
  matching problems and, 103
  in medicines, 2–3
  model building strategy with,
    182–183
  modeling patterns, nominal
    responses, 147, 165–167

    base model for, 158
    BUGS CODE 5.1, 150–153
    BUGS CODE 5.2, 159
    BUGS CODE 5.3, 164–165
    chi-square and likelihood ratio
      goodness of fit, 155
    data model, 157
    equal weight vector, 156
    observed and expected
      observations, 154, 162
    posterior analysis for logistic
      base model of depression,
      151–154
    Von Eye and Mun model, 148–149
    weight vector, 160–161
  notation and statistics, 3
  Oscar de la Hoya–Felix Trinidad
    bout, 9
  patterns of, 177–182
  phase II clinical trial for cancer
    research and, 10
  with potential covariates, 209–212
  with quantitative scores, 193–194
    regression and correlation,
      194–199
    variance approach, analysis,
      199–204, 215–216
  rater
    histological grades, 173–177
    with ordinal scores, 167–173
  regression and correlation models, 1
  review committee and, 11
  software packages and, 23–24
  statistical methods and ANOVA
    techniques, 4–5
  studies, sample size, 235–237
    Bayesian issues of, 266–267
    confidence interval, effect of, 251
    G coefficient, 245–246
    intraclass correlation, 260–266
    kappa, 246–248
    power analysis, 238–239
    regression and correlation
      methods, 255–260

Printed and bound by CPI Group (UK) Ltd, Croydon, CR0 4YY

24/10/2024

01778278-0014